International Conference on Mathematical Sciences and Statistics 2013

Adem Kilicman • Wah June Leong
Zainidin Eshkuvatov
Editors

International Conference on Mathematical Sciences and Statistics 2013

Selected Papers

Springer

Editors
Adem Kilicman
University Putra Malaysia
Serdang, Selangor
Malaysia

Wah June Leong
University Putra Malaysia
Serdang, Selangor
Malaysia

Zainidin Eshkuvatov
University Putra Malaysia
Serdang, Selangor
Malaysia

ISBN 978-981-10-1221-1 ISBN 978-981-4585-33-0 (eBook)
DOI 10.1007/978-981-4585-33-0
Springer Singapore Heidelberg New York Dordrecht London

Printed on acid-free paper

Springer is part of Springer Science+Business Media (www.springer.com)

Preface

The 2013 International Conference on Mathematical Sciences and Statistics (ICMSS2013) was held in Kuala Lumpur, Malaysia on 5–7th February, 2013. This conference is aimed to bring together scholars, leading researchers, and experts from diverse backgrounds and applications areas in mathematical sciences and statistics. Special emphasis is placed on promoting the interaction between the theoretical, experimental, and applied communities, so that a high-level exchange in new and emerging areas within mathematics, statistics, and its related applications in sciences is achieved. We are pleased that many friends and colleagues, postgraduate students as well as many practitioners and other scientists, were attracted by the goals of this recent event and offered to contribute to its success.

This volume contains 29 articles chosen from 172 contributions presented at ICMSS 2013. These were selected with a careful eye to their scientific quality and relevance to the specific topics of the conference, or more general methods and tools for statistical analysis. The articles explore a variety of topics, including mathematical analyses, probability, statistics, algebra, geometry, mathematical physics, wave propagation, stochastic processes, ordinary and partial differential equations, boundary value problems, linear operators, cybernetics, and number and functional theory.

We are grateful to the Department of Mathematics, Faculty of Science, Universiti Putra Malaysia (UPM), and all those who have supported the conference, and in particular the members of the organizing committee, for their dedication and activity. We would like to thank the authors for their valuable contributions, the referees for their priceless efforts of reviewing the manuscripts, and Professor Dr. Fudziah Ismail, Head of Department of Mathematics, UPM for encouraging us and making this volume realized.

November, 2013

Adem Kilicman
Wah June Leong
Zainidin Eshkuvatov

Contents

On Seven-Dimensional Filiform Leibniz Algebras

Abdulafeez O. Abdulkareem, I. S. Rakhimov and S. K. Said Husain

Abstract This paper gives complete classification of a subclass of seven-dimensional complex filiform Leibniz algebras denoted by TLb_n in fixed dimension n. The classification is carried out by choosing a basis which is adapted. Through this basis, an appropriate table of multiplication of the algebra is constructed. This leads to establishment of necessary and sufficient condition for any two algebras to be isomorphic. According to this condition, we break this class into 30 subsets and as a result, 10 of these subsets are single orbits and the remaining are union of parametric family of orbits. The single orbits are outlined from the parametric ones and their respective representatives are given. In parametric orbits case, the invariants that characterize the parameter are given. The filiform Lie algebras in this dimension are pointed out.

1 Introduction

The concept of Leibniz algebra was introduced by Loday [8] in the study of Leibniz (co)homology as a non-commutative analogue of Lie algebras (co)homology. A Leibniz algebra over a field K is a vector space over K equipped with a K-bilinear map $[\cdot,\cdot]$: $L \times L \longrightarrow L$ satisfying the Leibniz identity $[x,[y,z]] = [[x,y],z] - [[x,z],y]$, for all $x, y, z \in L$. Clearly, a Lie algebra is a Leibniz algebra, and conversely, a Leibniz algebra L over the field K, $Char K \neq 2$ with property $[x,y] = -[y,x]$, for all $x, y \in L$ is a Lie algebra. Any Leibniz algebra L gives rise to a Lie algebra L_{Lie}, which is obtained as the quotient of L by relation $[x,x] = 0$. Let I be the ideal of L generated by all squares. Then I is the minimal ideal with respect to the property that $\mathfrak{g} := L/I$ is a Lie algebra. The quotient map $\pi : L \longrightarrow \mathfrak{g}$ is a homomorphism of Leibniz algebras.

A. O. Abdulkareem (✉)
Institute for Mathematical Research, Universiti Putra Malaysia,
Serdang 43400, Malaysia
e-mail: abkaaz4success@yahoo.com

I. S. Rakhimov · S. K. Said Husain
Department of Mathematics, Faculty of Science, Universiti Putra Malaysia,
Serdang 43400, Malaysia
e-mail: risamiddin@gmail.com

S. K. Said Husain
e-mail: kartini@upm.edu.my

A. Kilicman et al. (eds.), *International Conference on Mathematical Sciences and Statistics 2013*, DOI 10.1007/978-981-4585-33-0_1,

The term filiform Lie algebra was introduced by M. Vergne while studying the variety of nilpotent Lie algebras laws through which she gave the first exact classification of nilpotent complex Lie algebras in dimension $m \leq 6$ (see also[6]). Filiform Leibniz algebras of n-dimension have been divided into three disjoint classes [2], [7] which were sourced from naturally graded non-Lie filiform Leibniz algebra and naturally graded filiform Lie algebra.

The first and second classes which arise from naturally graded non-Lie filiform Leibniz algebra have been classified up to dimension $n \leq 10$.

The present paper is however concerned with subclass of filiform Leibniz algebras. This subclass arises from naturally graded filiform Lie algebras. In dimension n it has been denoted by TLb_n and treated before in [1], [3], and [4]. In case of TLb_n, the ideal I generated by squares coincides with the center of the Leibniz algebra. Therefore, derived algebras are one-dimensional Leibniz central extensions of the Lie algebra L_{Lie}. The aim of this paper is to give complete lists of isomorphism classes of one-dimensional Leibniz central extensions of filiform Lie algebras of dimension six. As for the lower dimensions they either have been treated before [3] or have been included in the list of nilpotent Leibniz algebras [9], [10]. Another motivation to study TLb_n comes out from the fact that it contains the class of all n-dimensional filiform Lie algebras. Classifying TLb_n, we once again examine thoroughly and recover the lists of low-dimensional filiform Lie algebras given in [6].

Organization of this paper is as follows: Sect. 2 is devoted to providing necessary results which are needed in our classification. In the third section, the complete classification is given with the proof of some propositions that shall be given therein. We describe the orbits of the base change for single orbits, and for parametric family of orbits cases, we give the respective invariant functions to distinguish the orbits.

2 Preliminaries

In this section we give some elementary facts on Leibniz algebras that will be used throughout the study.

Definition 1 An algebra L over a field K is called a Leibniz algebra, if its bilinear operation $\lambda(\cdot, \cdot)$ satisfies the following Leibniz identity:

$$\lambda(x, \lambda(y, z)) = \lambda(\lambda(x, y), z) - \lambda(\lambda(x, z), y), \quad \text{for any } x, y, z \in L.$$

A Leibniz algebra on n-dimensional vector space V over a field K can be regarded as a pair $L = (V, \lambda)$, where λ is a Leibniz algebra law on V, the underlying vector space of L. We denote by LB_n, the set of Leibniz algebra laws on V. It is a subspace of the linear space of all bilinear mappings $V \times V \longrightarrow V$.

Definition 2 Two laws λ_1 and λ_2 from LB_n are said to be isomorphic, if there is $g \in GL_n(K)$ such that

$$\lambda_2(x, y) = (g * \lambda_1)(x, y) = g^{-1}(\lambda_1(g(x), g(y)))$$

for all $x, y \in V$. Thus, we get an action of $GL_n(K)$ on LB_n.

Let $O(\lambda)$ be the set of the laws isomorphic to λ, it is called the orbit of λ. Let us fix a basis $\{e_1, e_2, e_3, \ldots, e_n\}$ of V. The structure constants of $\lambda \in LB_n$ are $\gamma_{ij}^k \in K$ given by

$$\lambda(e_i, e_j) = \sum_{k=1}^{n} \gamma_{ij}^k e_k, \ \ i, j = 1, 2, \ldots, n.$$

Once a basis is fixed, we can identify the law λ with its structure constants. These constants γ_{ij}^k satisfy:

$$\sum_{l=1}^{n} (\gamma_{jk}^l \gamma_{il}^m - \gamma_{ij}^l \gamma_{lk}^m + \gamma_{ik}^l \gamma_{lj}^m) = 0, \ \ i, j, k, m = 1, 2, \ldots, n.$$

Then LB_n appears as a subvariety of the algebraic variety, $Alg_n(K)$, of all algebraic structures on V. Let $\lambda \in LB_n$ and G_λ be the subgroups of $GL_n(K)$ defined by $G_\lambda = \{f \in GL_n(K) \mid f * \lambda = \lambda\}$. The orbit $O(\lambda)$ of λ with respect to the action of $GL_n(K)$ is isomorphic to the homogeneous space $GL_n(K)/G_\lambda$.

Definition 3 A function $f\colon LB_n \longrightarrow K$ is said to be invariant (or orbit) function with respect to an action of a subgroup G of $GL_n(K)$ on LB_n if $f(g * \lambda) = f(\lambda)$ for $g \in G$.

In the foregoing, all algebras are assumed to be over the fields of complex numbers $\mathbb{C}$ and the bracket notation is used to describe the law: $[x, y] = \lambda(x, y)$ for $x, y \in L$.

Let L be a Leibniz algebra. Define

$$L^1 = L, \ L^{k+1} = [L^k, L], \ k \geq 1. \tag{1}$$

Clearly,

$$L^1 \supseteq L^2 \supseteq \cdots$$

Definition 4 A Leibniz algebra L is said to be a nilpotent, if there exists $s \in \mathbb{N}$, such that

$$L^1 \supset L^2 \supset \ldots \supset L^s = \{0\}. \tag{2}$$

Definition 5 A Leibniz algebra L is said to be a filiform, if $dimL^i = n - i$, where $n = dimL$ and $2 \leq i \leq n$.

According to [1], [9] the class of $(n+1)$-dimensional filiform Leibniz algebras arising from naturally graded filiform Lie algebras, denoted by TLb_{n+1}, admits a so-called adapted basis $\{e_0, e_1, \ldots, e_n\}$ such that the table of multiplication with respect to this basis is as given in [1].

We note from [5] that the algebras from TLb_{n+1} are one-dimensional Leibniz central extension of the Lie algebra μ_n. In this situation, we consider another cocycle which is Leibniz 2-cocycle, ψ satisfying the relation,

$$\psi \circ \psi(x, y, z) = \psi(x, \psi(y, z)) + \psi(\psi(x, z), y) - \psi(\psi(x, y), z) = 0. \tag{3}$$

One-dimensional Leibniz central extension of linear deformation of the Lie algebra μ_n has been denoted by Ced(μ_n) in [5].

Definition 6 Let $\{e_0, e_1, \ldots, e_n\}$ be an adapted basis of $L \in TLb_{n+1}$. Then a nonsingular linear transformation $f: L \to L$ is said to be adapted if the basis $\{f(e_0), f(e_1), \ldots, f(e_n)\}$ is adapted.

The subgroup of GL_{n+1} consisting of all adapted transformations are denoted by G_{ad}. The following proposition proved in [3] specifies the elements of G_{ad}.

Proposition 1 *Let $\{e_0, e_1, \ldots, e_n\}$ be an adapted basis of $L \in TLb_{n+1}$ and f be an adapted transformation. Then, f can be represented as follows:*

$$f(e_0) = e_0' = \sum_{i=0}^{n} A_i e_i,$$

$$f(e_1) = e_1' = \sum_{i=1}^{n} B_i e_i,$$

$$f(e_i) = e_i' = [f(e_{i-1}), f(e_0)], \qquad 2 \le i \le n$$

$A_0, A_i, B_j, (i, j = 1, \ldots, n)$ are complex numbers and $A_0 B_1(A_0 + A_1 b) \neq 0$.

In G_{ad}, we specify the following transformations called elementary:

$$\tau(a,b,c) = \begin{cases} \tau(e_0) = ae_0 + be_1, \\ \tau(e_1) = ce_1, \quad ac \neq 0, \\ \tau(e_{i+1}) = [\tau(e_i), \tau(e_0)], \quad 1 \le i \le n-1, \end{cases}$$

$$\sigma(a,k) = \begin{cases} \sigma(e_0) = e_0 + ae_k, \quad 2 \le k \le n, \\ \sigma(e_1) = e_1, \\ \sigma(e_{i+1}) = [\sigma(e_i), \sigma(e_0)], \quad 1 \le i \le n-1, \end{cases}$$

$$\phi(c,k) = \begin{cases} \phi(e_0) = e_0, \\ \phi(e_1) = e_1 + ce_k, \quad 2 \le k \le n, \\ \phi(e_{i+1}) = [\phi(e_i), \phi(e_0)], \quad 1 \le i \le n-1, \end{cases}$$

where $a, b, c \in \mathbb{C}$.

Proposition 2 *Let L be an algebra from TLb_{n+1}, then any adapted transformation f of L can be represented as the composition:*

$$\begin{aligned} f = \phi(B_n, n) \circ \phi(B_{n-1}, n-1) \circ \ldots \circ \phi(B_2, 2) \circ \sigma(A_n, n) \\ \circ \sigma(A_{n-1}, n-1) \circ \ldots \circ \sigma(A_2, 2) \circ \tau(A_0, A_1, B_1) \end{aligned} \tag{4}$$

Proof The proof can be obtained by direct verification. □

Proposition 3 *The transformations*

$$\begin{aligned} g = \phi(B_n, n) \circ \phi(B_{n-1}, n-1) \circ \phi(B_{n-2}, n-2) \circ \ldots \circ \sigma(A_n, n) \\ \circ \sigma(A_{n-1}, n-1) \circ \ldots \circ \sigma(A_2, 2) \end{aligned} \tag{5}$$

do not change the structure constants of algebras from TLb_{n+1}.

For the proof of Proposition 3, we refer the reader to [4]. The proof of the following lemma can be found easily by a simple computation.

Lemma 1 *Let* $\{e_0, e_1, \ldots, e_n\} \longrightarrow \{e'_0, e'_1, \ldots, e'_n\}$ *be an adapted base change,* $b_{00}, b_{01}, b_{11}, \ldots$ *and* $b'_{00}, b'_{01}, b'_{11}, \ldots$ *be the respective structure constants. Then, for* b'_{00}, b'_{01} *and* b'_{11} *one has*

$$\begin{aligned} b'_{00} &= \frac{A_0^2 b_{00} + A_0 A_1 b_{01} + A_1^2 b_{11}}{A_0^{n-2} B_1 (A_0 + A_1\, b)}, \\ b'_{01} &= \frac{A_0 b_{01} + 2A_1 b_{11}}{A_0^{n-2}(A_0 + A_1\, b)}, \\ b'_{11} &= \frac{B_1 b_{11}}{A_0^{n-2}(A_0 + A_1\, b)}. \end{aligned}$$

3 Isomorphism Classes in *TLb*$_7$

3.1 Isomorphism Criterion for TLb_7

Due to the observation in [5] as pointed out in section (2), an algebra L from TLb_7 is represented as a one-dimensional Leibniz central extension ($C(L) =< e_6 >$) of the filiform Lie algebra with law $\mu_n + \psi$, where ψ is the appropriate cocycle on the adapted basis $\{e_0, e_1, \ldots, e_6\}$, the class TLb_7 can be represented as follows:

$$TLb_7 = \begin{cases} [e_i, e_0] = e_{i+1}, & 1 \le i \le 5, \\ [e_0, e_i] = -e_{i+1}, & 2 \le i \le 5, \\ [e_0, e_0] = b_{00} e_6, & \\ [e_0, e_1] = -e_2 + b_{01} e_6, & \\ [e_1, e_1] = b_{11} e_6, & \\ [e_1, e_2] = -[e_2, e_1] = a_{14}\, e_4 + a_{15}\, e_5 + b_{12} e_6, & \\ [e_1, e_3] = -[e_3, e_1] = a_{14}\, e_5 + b_{13} e_6, & \\ [e_1, e_4] = -[e_4, e_1] = -a_{25}\, e_5 + b_{14} e_6, & \\ [e_2, e_3] = -[e_3, e_2] = a_{25}\, e_5 + b_{23} e_6, & \\ [e_1, e_5] = -[e_5, e_1] = b_{15} e_6, & \\ [e_2, e_4] = -[e_4, e_2] = b_{24} e_6. & \end{cases}$$

Observe that the structure constants a_{ij} and b_{st} in the table above are not free. The interrelations between them are defined as follows:

Lemma 2 *The structure constants of algebras from* TLb_7 *satisfy the following constraints:*

1. $b_{13} = a_{15}$,
2. $b_{14} = a_{14} - b_{23}$ and
3. $b_{15} = b_{24} = a_{25} = 0$.

Proof The first and second relations occur if we apply the Leibniz identity to the triples $\{e_0, e_1, e_2\}$ and $\{e_0, e_1, e_3\}$ of the basis vectors respectively, and the last relation is a result of application of the Leibniz identity to $\{e_0, e_2, e_3\}$, $\{e_0, e_1, e_4\}$, and $\{e_1, e_2, e_3\}$. □

Further unifying the table of multiplication above, we rewrite it via new notations $c_{00}, c_{01}, c_{11}, c_{12}, c_{13}, c_{14}, c_{23}$ for the structure constants:

$$TLb_7 = \begin{cases} [e_i, e_0] = e_{i+1}, & 1 \le i \le 5, \\ [e_0, e_i] = -e_{i+1}, & 2 \le i \le 5, \\ [e_0, e_0] = c_{00} e_6, \\ [e_0, e_1] = -e_2 + c_{01} e_6, \\ [e_1, e_1] = c_{11} e_6, \\ [e_1, e_2] = -[e_2, e_1] = c_{12}\, e_4 + c_{13}\, e_5 + c_{14} e_6, \\ [e_1, e_3] = -[e_3, e_1] = c_{12}\, e_5 + c_{13} e_6, \\ [e_1, e_4] = -[e_4, e_1] = (c_{12} - c_{23})\, e_6, \\ [e_2, e_3] = -[e_3, e_2] = c_{23}\, e_6. \end{cases}$$

An algebra from TLb_7 with the structure constants $c_{00}, c_{01}, c_{11}, c_{12}, c_{13}, c_{14}, c_{23}$ is codified by $L(C)$, where $C = (c_{00}, c_{01}, c_{11}, c_{12}, c_{13}, c_{14}, c_{23})$. The next theorem represents the action of the adapted base change on the parameters $c_{00}, c_{01}, c_{11}, c_{12}, c_{13}, c_{14}, c_{23}$ of an algebra from TLb_7.

Theorem 1 (*Isomorphism criterion for* TLb_7)
Two filiform Leibniz algebras, $L(C')$ *and* $L(C)$, *where* $C' = (c'_{00}, c'_{01}, c'_{11}, c'_{12}, c'_{13}, c'_{14}, c'_{23})$ *and* $C = (c_{00}, c_{01}, c_{11}, c_{12}, c_{13}, c_{14}, c_{23})$, *from* TLb_7 *are isomorphic if and only if there exist* $A_0, A_1, B_1, B_2, B_3 \in \mathbb{C}$ *such that* $A_0 B_1 \neq 0$ *and the following hold:*

$$c'_{00} = \frac{A_0^2 c_{00} + A_0 A_1 c_{01} + A_1^2 c_{11}}{A_0^5 B_1}, \tag{6}$$

$$c'_{01} = \frac{A_0 c_{01} + 2\, A_1 c_{11}}{A_0^5}, \tag{7}$$

$$c'_{11} = \frac{B_1 c_{11}}{A_0^5}, \tag{8}$$

$$c'_{12} = \frac{B_1 c_{12}}{A_0^2}, \tag{9}$$

$$c'_{13} = \frac{B_1\, c_{13} + B_2\, c_{12}}{A_0^3}, \tag{10}$$

$$c'_{14} = \frac{1}{A_0^5 B_1} \big\{ A_0 B_1^2 c_{14} + B_2 (A_0 B_1 c_{13} + A_1 B_1 c_{12} c_{23} + A_0 B_2 c_{23} - A_1 B_1 c_{12}^2) - B_3 (2 A_0 B_1 c_{23} - A_0 B_1 c_{12}) \big\} \tag{11}$$

$$c'_{23} = \frac{B_1 c_{23}}{A_0^2} \tag{12}$$

Proof "If" part. The equations in (6), (7), and (8) occur due to **Lemma 1**. Note that according to Propositions (2) and (3), the adapted transformations in TLb_7 can be taken in the form of the base change

$$\begin{cases} e'_0 & = f(e_0) = A_0\, e_0 + A_1\, e_1, \\ e'_1 & = f(e_1) = B_1\, e_1 + B_2\, e_2 + B_3\, e_3, \\ e'_{i+1} & = f(e_{i+1}) = [f(e_i), f(e_0)], \quad 1 \leq i \leq 6, \end{cases} \tag{13}$$

where $A_0\, B_1 \neq 0$. Consider

$$\begin{aligned} [e'_1, e'_2] &= A_0\, B_1^2(c_{12}e_4 + c_{13}e_5 + c_{14}e_6) + A_0 B_1 B_2(c_{12}e_5 + c_{13}e_6) + B_1(A_2 B_1 c_{12} \\ &\quad - A_1 B_2 c_{12} + A_0 B_3)(c_{12} - c_{23})e_6 + A_0\, B_2^2 c_{23}e_6 - A_0 B_1 B_3 c_{23} e_6 \\ &= c'_{12}e'_4 + c'_{13}e'_5 + c'_{14}e'_6 \end{aligned} \tag{14}$$

but,

$$\begin{aligned} c'_{12}e'_4 &= c'_{12}\left(A_0^3 B_1\, e_4 + \left(A_0^3 B_2 - 2\, A_0^2 A_1 B_1 c_{12}\right) e_5 + \left(A_0^3 B_3 - 2\, A_1 A_0^2 B_1 c_{13}\right.\right. \\ &\quad \left.\left. + A_0 A_1^2 B_1 c_{12}^2 - A_0 A_1^2 B_1 c_{12} c_{23} - 3\, A_1 A_0^2 B_2 c_{12} + A_0^2 A_1 B_2 c_{23}\right) e_6\right) \\ c'_{13}e'_5 &= c'_{13}\left(A_0^4\, B_1\, e_5 + \left(A_0^4 B_2 - 3\, A_1 A_0^3 B_1 c_{12} + A_1 A_0^3 B_1 c_{2,3}\right) e_6\right) \\ c'_{14}e'_6 &= c'_{14}\left(A_0^5\, B_1\, e_6\right). \end{aligned}$$

On comparing the coefficients of e_4, e_5 and e_6 with the corresponding coefficients in (14) after rearrangement, we get the equalities (9), (10) and (11), respectively. The last equality follows from

$$\begin{aligned} [f(e_2), f(e_3)] &= c'_{23} f(e_6) \\ A_0^3 B_1^2 c_{23} e_6 &= c'_{23} A_0^5 B_1 e_6 \\ Hence,\ c'_{23} &= \frac{B_1\, c_{23}}{A_0^2}. \end{aligned} \tag{15}$$

"Only if" part. Let the equalities (6)–(12) hold. Then the base change above is adapted and $L(C)$ is transformed to $L(C')$.

Indeed,

$$\begin{aligned} [e'_0, e'_0] &= [A_0 e_0 + A_1 e_1,\ A_0 e_0 + A_1 e_1] \\ &= A_0^2[e_0, e_0] + A_0 A_1[e_0, e_1] + A_0 A_1[e_1, e_0] + A_1^2[e_1, e_1] \\ &= \left(A_0^2 c_{00} + A_0 A_1 c_{01} + A_1^2 c_{11}\right) e_6 \\ &= c'_{00} A_0^5 B_1 e_6 \\ &= c'_{00} e'_6. \end{aligned}$$

Other brackets are obtained similarly. □

4 The Isomorphism Classes in TLb_7

In this subsection, we give the list of isomorphism classes in TLb_7. For simplification purpose, we introduce the function $\chi(X) = \left(4x_{00}x_{11} - x_{01}^2\right)$ where $X = (x_{00}, x_{01}, x_{11}, x_{12}, x_{13}, x_{14}, x_{23})$. By denoting $\Delta = \chi(C)$ and $\Delta' = \chi(C')$, we present TLb_7 as a union of the following subsets:

$$
\begin{aligned}
U_7^1 &= \{L(C) \in TLb_7 : c_{23} \neq 0,\ c_{11} \neq 0,\ c_{12} \neq 0\};\\
U_7^2 &= \{L(C) \in TLb_7 : c_{23} \neq 0,\ c_{11} \neq 0,\ c_{12} = 0, \Delta \neq 0\};\\
U_7^3 &= \{L(C) \in TLb_7 : c_{23} \neq 0,\ c_{11} \neq 0, c_{12} = \Delta = 0\};\\
U_7^4 &= \{L(C) \in TLb_7 : c_{23} \neq 0,\ c_{11} = 0, c_{01} \neq 0, c_{12} \neq 0\};\\
U_7^5 &= \{L(C) \in TLb_7 : c_{23} \neq 0, c_{11} = 0, c_{01} \neq 0, c_{12} = 0, c_{13} \neq 0\};\\
U_7^6 &= \{L(C) \in TLb_7 : c_{23} \neq 0,\ c_{11} = 0, c_{01} \neq 0, c_{12} = c_{13} = 0\};\\
U_7^7 &= \{L(C) \in TLb_7 : c_{23} \neq 0,\ c_{11} = c_{01} = 0, c_{00} \neq 0, c_{12} \neq 0\};\\
U_7^8 &= \{L(C) \in TLb_7 : c_{23} \neq 0, c_{11} = c_{01} = 0, c_{00} \neq 0, c_{12} = 0, c_{13} \neq 0\};\\
U_7^9 &= \{L(C) \in TLb_7 : c_{23} \neq 0, c_{11} = c_{01} = 0, c_{00} \neq 0, c_{12} = 0, c_{13} = 0\};\\
U_7^{10} &= \{L(C) \in TLb_7 : c_{23} \neq 0, c_{11} = c_{01} = c_{00} = 0, c_{12} \neq 0, \};\\
U_7^{11} &= \{L(C) \in TLb_7 : c_{23} \neq 0, c_{11} = c_{01} = c_{00} = c_{12} = 0, c_{13} \neq 0\};\\
U_7^{12} &= \{L(C) \in TLb_7 : c_{23} \neq 0, c_{11} = c_{01} = c_{00} = c_{12} = c_{13} = 0\};\\
U_7^{13} &= \{L(C) \in TLb_7 : c_{23} = 0, c_{11} \neq 0, c_{12} \neq 0\};\\
U_7^{14} &= \{L(C) \in TLb_7 : c_{23} = 0, c_{11} \neq 0, c_{12} = 0, \Delta \neq 0\};\\
U_7^{15} &= \{L(C) \in TLb_7 : c_{23} = 0, c_{11} \neq 0, c_{12} = \Delta = 0, c_{13} \neq 0\};\\
U_7^{16} &= \{L(C) \in TLb_7 : c_{23} = 0, c_{11} \neq 0, c_{12} = \Delta = c_{13} = 0, c_{14} \neq 0\};\\
U_7^{17} &= \{L(C) \in TLb_7 : c_{23} = 0,\ c_{11} \neq 0,\ c_{12} = \Delta = c_{13} = c_{14} = 0\};\\
U_7^{18} &= \{L(C) \in TLb_7 : c_{23} = c_{11} = 0, c_{01} \neq 0, c_{12} \neq 0\};\\
U_7^{19} &= \{L(C) \in TLb_7 : c_{23} = c_{11} = 0, c_{01} \neq 0, c_{12} = 0, c_{13} \neq 0\};\\
U_7^{20} &= \{L(C) \in TLb_7 : c_{23} = c_{11} = 0, c_{01} \neq 0, c_{12} = c_{13} = 0, c_{14} \neq 0\};\\
U_7^{21} &= \{L(C) \in TLb_7 : c_{23} = c_{11} = 0, c_{01} \neq 0, c_{12} = c_{13} = c_{14} = 0\};\\
U_7^{22} &= \{L(C) \in TLb_7 : c_{23} = c_{11} = c_{01} = 0,\ c_{00} \neq 0,\ c_{12} \neq 0\};\\
U_7^{23} &= \{L(C) \in TLb_7 : c_{23} = c_{11} = c_{01} = 0,\ c_{00} \neq 0,\ c_{12} = 0, c_{13} \neq 0\};\\
U_7^{24} &= \{L(C) \in TLb_7 : c_{23} = c_{11} = c_{01} = 0,\ c_{00} \neq 0,\ c_{12} = 0, c_{13} = 0, c_{14} \neq 0\};\\
U_7^{25} &= \{L(C) \in TLb_7 : c_{23} = c_{11} = c_{01} = 0, c_{00} \neq 0, c_{12} = c_{13} = c_{14} = 0\};\\
U_7^{26} &= \{L(C) \in TLb_7 : c_{23} = c_{11} = c_{01} = c_{00} = 0,\ c_{12} \neq 0\}.\\
U_7^{27} &= \{L(C) \in TLb_7 : c_{23} = c_{11} = c_{01} = c_{00} = c_{12} = 0, c_{13} \neq 0, c_{14} \neq 0\}.\\
U_7^{28} &= \{L(C) \in TLb_7 : c_{23} = c_{11} = c_{01} = c_{00} = c_{12} = c_{13} \neq 0, c_{14} = 0\}.\\
U_7^{29} &= \{L(C) \in TLb_7 : c_{23} = c_{11} = c_{01} = c_{00} = c_{12} = c_{13} = 0, c_{14} \neq 0\}.\\
U_7^{30} &= \{L(C) \in TLb_7 : c_{23} = c_{11} = c_{01} = c_{00} = c_{12} = c_{13} = c_{14} = 0\}.
\end{aligned}
$$

The adapted group action on TLb_7 is induced on each of these disjoint subsets. The next propositions give isomorphism criterion for one subset from each type, that is, from infinitely many orbits case and from single orbits case, since other cases can be proved similarly.

Proposition 4

1. a. *Two algebras $L(C')$ and $L(C)$ from U_7^1 are isomorphic, if and only if*

$$\left(\frac{c'_{23}}{c'_{11}}\right)^{\frac{8}{3}} \Delta' = \left(\frac{c_{23}}{c_{11}}\right)^{\frac{8}{3}} \Delta$$

and

$$\frac{c'_{12}}{c'_{23}} = \frac{c_{12}}{c_{23}}$$

b. *For any $\lambda_1, \lambda_2, \in \mathbb{C}$, there exists $L(C) \in U_7^1$:*

$$\frac{1}{4}\left(\frac{c_{23}}{c_{11}}\right)^{\frac{8}{3}} \Delta = \lambda_1, \quad \frac{c_{12}}{c_{23}} = \lambda_2.$$

Then orbits in U_7^1 can be parameterized as $L(\lambda_1, 0, 1, \lambda_2, 0, 0, 1)$, $\lambda_1, \lambda_2 \in \mathbb{C}$.

2. *The subset U_7^6 is a single orbit with the representative* $L(0, 1, 0, 0, 0, 0, 1)$.

Proof

1. a. The "If" part of the proposition is due to Theorem 1 if one substitutes the expressions for $c'_{00}, c'_{01}, c'_{11}, c'_{12}, c'_{13}, c'_{23}$ into

$$\frac{1}{4}\left(\frac{c'_{23}}{c'_{11}}\right)^{\frac{8}{3}} \Delta' \text{ and } \frac{c'_{12}}{c'_{23}}, \text{ respectively.}$$

" Only if " part. Let the equalities $\left(\frac{c'_{23}}{c'_{11}}\right)^{\frac{8}{3}} \Delta' = \left(\frac{c_{23}}{c_{11}}\right)^{\frac{8}{3}} \Delta$ and $\frac{c'_{12}}{c'_{23}} = \frac{c_{12}}{c_{23}}$ hold. Consider the base change (13) in the proof of Theorem 1 with

$$A_1 = \frac{-A_0 c_{01}}{2c_{11}}, \quad B_1 = \frac{A_0^2}{c_{23}},$$

and

$$B_3 = \frac{1}{2A_0B_1c_{23} - A_0B_1c_{12}} \{A_0B_1^2c_{14} + B_2(A_0B_1c_{13} + A_1B_1c_{12}c_{23} + A_0B_2c_{23} - A_1B_1c_{12}^2)\}$$

This base change transforms $L(C)$ into

$$L\left(\left(\frac{c_{23}}{c_{11}}\right)^{\frac{8}{3}} \Delta, 0, 1, \frac{c_{12}}{c_{23}}, 0, 0, 1\right).$$

Table 1 Single orbits

Subsets	Representative of orbits	Subsets	Representative of orbits
U_7^6	$L(0, 1, 0, 0, 0, 0, 1)$	U_7^9	$L(1, 0, 0, 0, 0, 0, 1)$
U_7^{11}	$L(0, 0, 0, 0, 1, 0, 1)$	U_7^{21}	$L(0, 1, 0, 0, 0, 0, 0)$
U_7^{12}	$L(0, 0, 0, 0, 0, 0, 1)$	U_7^{22}	$L(1, 0, 0, 1, 0, 0, 0)$
U_7^{15}	$L(0, 0, 1, 0, 1, 0, 0)$	U_7^{23}	$L(1, 0, 0, 0, 1, 0, 0)$
U_7^{16}	$L(0, 0, 1, 0, 0, 1, 0)$	U_7^{24}	$L(1, 0, 0, 0, 0, 1, 0)$
U_7^{17}	$L(0, 0, 1, 0, 0, 0, 0)$	U_7^{25}	$L(1, 0, 0, 0, 0, 0, 0)$
U_7^{18}	$L(0, 1, 0, 1, 0, 0, 0)$	U_7^{26}	$L(0, 0, 0, 1, 0, 0, 0)$
U_7^{19}	$L(0, 1, 0, 0, 1, 0, 0)$	U_7^{27}	$L(0, 0, 0, 0, 1, 0, 0)$
U_7^{20}	$L(0, 1, 0, 0, 0, 1, 0)$	U_7^{28}	$L(0, 0, 0, 0, 1, 1, 0)$
U_7^{30}	$L(0, 0, 0, 0, 0, 0, 0)$	U_7^{29}	$L(0, 0, 0, 0, 0, 1, 0)$

An analogous base change with "prime" transforms $L(C')$ into

$$L\left(\left(\frac{c'_{23}}{c'_{11}}\right)^{\frac{8}{3}}\Delta', 0, 1, \frac{c'_{12}}{c'_{23}}, 0, 0, 1\right).$$

Since,

$$\left(\frac{c'_{23}}{c'_{11}}\right)^{\frac{8}{3}}\Delta' = \left(\frac{c_{23}}{c_{11}}\right)^{\frac{8}{3}}\Delta$$

and

$$\frac{c'_{12}}{c'_{23}} = \frac{c_{12}}{c_{23}},$$

the algebras are isomorphic.

b. Obvious.

2. To prove it, it is enough to put appropriate values of A_0, A_1, B_1 in the base change (13).

   By an easy calculations, it is easy to see that $A_0 = c_{01}^{\frac{1}{4}}, A_1 = -A_0\frac{c_{00}}{c_{01}}, B_1 = \frac{c_{01}^{\frac{1}{2}}}{c_{23}}$ and $B_2, B_3 \in \mathbb{C}$. □

5 Conclusion

This section summarizes the result of this study by listing its milestones. The isomorphism classes in TLb_7 are then presented in two tables as shown in Table 1 and Table 2 below.

1. In TLb_7, we outline 30 isomorphism classes (10 parametric family of orbits and 20 single orbits) of seven-dimensional complex filiform Leibniz algebras and these classes exhaust all possible cases.
2. The seven-dimensional filiform Lie algebras are covered by U_7^{10}, U_7^{11}, U_7^{12}, U_7^{26}, U_7^{27}, U_7^{28}, U_7^{29}, and U_7^{30}.
3. The list of resulting filiform Lie algebras in this classification agrees with the list given in [6].

Table 2 Parametric family of orbits

Subsets	Representative of orbits	Invariant functions
$U_7^1 = \bigcup\limits_{\lambda_1,\lambda_2,} O_7^1(\lambda_1,\lambda_2,)$	$L(\lambda_1,0,1,\lambda_2,0,0,1)$	$f_1(X) = \left(\frac{x_{2,3}}{x_{11}}\right)^{\frac{8}{3}} \chi(x),\ f_2(X) = \frac{x_{12}}{x_{23}}$
$U_7^2 = \bigcup\limits_{\lambda_1,\lambda_2} O_7^2(\lambda_1,\lambda_2)$	$L(1,0,1,0,\lambda_1,0,\lambda_2)$	$f_1(X) = \frac{x_{13}}{x_{11}}[\chi(x)]^{\frac{1}{4}},\ f_2(X) = \frac{x_{23}}{x_{11}}[\chi(x)]^{\frac{3}{8}}$
$U_7^3 = \bigcup\limits_{\lambda} O_7^3(\lambda)$	$L(0,0,1,0,\lambda,0,1)$	$f(X) = \frac{x_{13}}{\left(x_{11}x_{23}^2\right)^{\frac{1}{3}}}$
$U_7^4 = \bigcup\limits_{\lambda} O_7^4(\lambda)$	$L(0,1,0,\lambda,0,0,1)$	$f(X) = \frac{x_{12}}{x_{23}}$
$U_7^5 = \bigcup\limits_{\lambda} O_7^5(\lambda)$	$L(0,\lambda,0,0,1,0,1)$	$f(X) = x_{01}\left(\frac{x_{23}}{x_{13}}\right)^4$
$U_7^7 = \bigcup\limits_{\lambda} O_7^7(\lambda)$	$L(1,0,0,1,0,0,\lambda)$	$f(X) = \frac{x_{23}}{x_{12}}$
$U_7^8 = \bigcup\limits_{\lambda} O_7^8(\lambda)$	$L(\lambda,0,0,0,1,0,1)$	$f(X) = \frac{x_{00}x_{23}^6}{x_{13}^5}$
$U_7^{10} = \bigcup\limits_{\lambda} O_7^{10}(\lambda)$	$L(0,0,0,1,0,0,\lambda)$	$f(X) = \frac{x_{23}}{x_{12}}$
$U_7^{13} = \bigcup\limits_{\lambda} O_7^{13}(\lambda)$	$L(\lambda_1,0,1,1,0,0,0)$	$f(X) = \left(\frac{x_{12}}{x_{11}}\right)^{\frac{8}{3}} \chi(x)$
$U_7^{14} = \bigcup\limits_{\lambda} O_7^{14}(\lambda)$	$L(1,0,1,0,\lambda,0,0,)$	$f(X) = \left(\frac{x_{13}}{x_{11}}\right)[\chi(x)]^{\frac{1}{4}}$

Acknowledgement This research was supported by the Universiti Putra Malaysia (UPM) through the research grant 05-02-12-2188RU.

References

1. Omirov, B.A., Rakhimov, I.S.: On Lie-like complex filiform Leibniz algebras. Bull. Aust. Math. Soc. **79**, 391–404 (2009)
2. Rakhimov, I.S., Bekbaev, U.D.: On classification of finite dimensional complex filiform Leibniz algebras. Commun. Algebra. **38**(12), 4705–4738 (2010)
3. Rakhimov, I.S., Hassan, M.A.: On low- and their invariants. Bull. Malays. Math. Sci. Soc. **34**(3), 475–485 (2011)
4. Rakhimov, I.S., Hassan, M.A.: On one dimensional Leibniz central extensions of a naturally graded filiform Lie algebra. Bull. Aust. Math. Soc. **84**, 205–224 (2011)
5. Rakhimov, I.S., Hassan, M.A.: On isomorphism criteria for Leibniz central extensions of a linear deformation of μ_n. Int. J. Algebra Comput. **21**(5), 715–729 (2011)
6. Gómez, J.R., Jimenez-Merchan, A., Khakimdjanov, Y.: Low-dimensional filiform Lie algebras. J. Pure Appl. Algebra. **130**, 133–158 (1998)
7. Gómez, J.R., Omirov, B.A.: On classification of complex filiform Leibniz algebras. http://front.math.ucdavis.edu/, arXiv:0612735v1 [math.RA] (2006)
8. Loday, J.-L., Pirashvili, T.: Universal enveloping algebras of Leibniz algebras and (co)homology. Math. Ann. **296**, 139–158 (1993)
9. Albeverio, S., Omirov, B.A., Rakhimov, I.S.: Varieties of nilpotent complex Leibniz algebras of dimension less than five. Commun. Algebra. **33**, 1575–1585 (2005)
10. Albeverio, S., Omirov, B.A., Rakhimov, I.S.: Classification of four-dimensional nilpotent complex Leibniz algebras. Extracta. Math. **21**(3), 197–210 (2006)

On Exponential Stability of Stochastic Control Systems

Fakhreddin Abedi and Wah June Leong

Abstract In this paper, we study the exponential input-to-state stability in probability of a wider class of composite stochastic control system. Our aim is to establish sufficient conditions for exponential input-to-state stability in probability of this composite system. We also give a numerical example illustrating our results.

1 Introduction

In this paper, we study the exponential input-to-state stability in probability in the r-th mean (rEISSP) of a composite stochastic control system (CSCS). We employ the stochastic version of control Lyapunov function and extend the exponential input-to-state stability in probability (EISSP) results proved by Spiliotis and Tsinias [11] for stochastic control system (SCS) to the class of CSCS driven by two independent Wiener processes. We establish sufficient conditions for rEISSP of CSCS.

Recently, Spiliotis and Tsinias [11] used the notion of control Lyapunov function and established the rEISSP of stochastic differential system (SDS). The asymptotic and exponential stability of stochastic systems have been derived by Mao [8], Liu and Raffool [7], Lan and Dang [6], Abedi et al. [1], Khasminiskii [4] and Kushner [5]. Michel [9] established asymptotic and exponential stability in probability for some classes of continuous and discrete parameters in stochastic composite system. The sufficient conditions for exponential stability in probability of stochastic system and composite stochastic system have been developed by Boulanger [3] and Rusinek [10].

In this paper, we first introduce a class of CSCS and also recall some basic definitions and results concerning rEISSP property. Finally, we state and prove the main results of the paper.

F. Abedi (✉) · W. J. Leong
Department of Mathematics, Faculty of Science,
Universiti Putra Malaysia, Serdang 43400, Malaysia
e-mail: f_abedi1352@yahoo.com

W. J. Leong
e-mail: wjleong@science.upm.edu.my

A. Kilicman et al. (eds.), *International Conference on Mathematical Sciences and Statistics 2013*, DOI 10.1007/978-981-4585-33-0_2,

2 Exponential Stability

Let (Ω, F, P) be a complete probability space and $(w_t)_{t\geq 0}$ denote a standard R^m-valued Wiener process defined on this space.

We consider the SDS

$$dx = f(x)dt + h(x)dw_t, \tag{1}$$

where the following conditions hold true:

(i) x is given in R^n,
(ii) $f : R^n \to R^n$ and $h : R^n \to R^{n\times m}$ are Lipschitz functionals in R^n with $f(0) = 0, h(0) = 0$, and there exists a constant $C \geq 0$ such that the following growth condition holds true:

$$|f(x) - f(y)| + |h(x) - h(y)| \leq C|x - y|, \quad \forall x, y \in R^n. \tag{2}$$

We recall the following definition of exponential stability in mean square given by Khasminskii [4] as follows.

Definition 1 The equilibrium state $x_t = 0$ of the SDS (1) is exponentially stable in probability in mean square if, and only if, there exist constants $c_1, c_2 > 0$ such that

$$E(|x_t(t, t_0, x)|^2) \leq c_1|x|^2 e^{-c_2(t-t_0)}, \quad \forall x \in R^n, \quad t \geq t_0. \tag{3}$$

Let $(\beta_t)_{t\geq 0}$ denote a standard R^q-valued Wiener process defined on (Ω, F, P). Now, we consider the SCS

$$d\upsilon = F(\upsilon, u)dt + G(\upsilon)d\beta_t, \tag{4}$$

where
(1) u is a R^p valued measurable control law,
(2) $F : R^r \times R^p \to R^r$ and $G : R^r \to R^r \times R^q$ are Lipschitz functionals mapping with $F(0, 0) = G(0) = 0$, and there exists a constant $C \geq 0$ such that for any $\upsilon \in R^r$ and $u \in R^p$ the following growth condition holds true:

$$|F(\upsilon, u)| + |G(\upsilon)| \leq C(1 + |\upsilon| + |u|).$$

For simplicity, let $V(t) = V(t, t_0, \upsilon, u)$ and we recall the following definition of rESP derived by Spiliotis and Tsinias [11] as follows.

Definition 2 The equilibrium state $\upsilon_t = 0$ of the SCS (4) is rESP for some $r > 0$ if, and only if, there exist constants $c_1, c_2 > 0$ such that

$$E(|V(t)|^r) \leq c_1|\upsilon|^r e^{-c_2(t-t_0)}, \quad \forall \upsilon \in R^r, \quad t \geq t_0. \tag{5}$$

Definition 3 The SCS (4) is said to satisfy the exponential Lyapunov condition if there exists a Lyapunov function $\Phi : R^r \to R^+$ of class $C^2(R^r\backslash\{0\})$ and positive constants $a_i, 1 \leq i \leq 5$, such that

$$a_1|\upsilon|^r \leq \Phi(\upsilon) \leq a_2|\upsilon|^r, \tag{6}$$

$$|\nabla\Phi(\upsilon)| \leq a_3|\upsilon|^{r-1}, \quad |\nabla^2\Phi(\upsilon)| \leq a_4|\upsilon|^{r-2}, \tag{7}$$

$$\mathbf{Y}\Phi(\upsilon) = \sum_{i=1}^{n} F(\upsilon,u)\frac{\partial\Phi(\upsilon)}{\partial\upsilon_i} + \frac{1}{2}\sum_{i,j=1}^{n} G(\upsilon)G(\upsilon)^T\frac{\partial^2\Phi(\upsilon)}{\partial\upsilon_i\partial\upsilon_j} \leq -a_5|\upsilon|^r, \tag{8}$$

where $\mathbf{Y}$ is the infinitesimal generator for the stochastic process solution of the SCS (4).
Note that Definition 2 obtained in Spiliotis and Tsinias [11] is an extension of Definition 1 established in Khasminskii [4].

Definition 4 The equilibrium state $\upsilon_t = 0$ of the SCS (4) is rEISSP for some $r > 0$, if there exist a positive definite function $\gamma : R^+ \to R^+$ and constants $c_1, c_2 > 0$ such that (5) holds true and

$$|u(t)| \leq \gamma(|V(t)|), \quad \forall\upsilon \in R^r, \quad t \geq t_0. \tag{9}$$

We shall now turn the attention to a general composite stochastic system.

Let $\{\beta_t, t \in R^+\}$ be a standard R^q-valued Wiener process defined on the space (Ω, F, P) independent of the Wiener process $\{w_t, t \in R^+\}$. Consider the pair of stochastic processes solution $(x_t, \upsilon_t) \in R^n \times R^r$ of the CSCS

$$\begin{cases} dx = (f(x) + g(x,\upsilon)D\upsilon)dt + h(x)dw_t, \\ d\upsilon = F(\upsilon,u)dt + G(\upsilon)d\beta_t, \end{cases} \tag{10}$$

where
(1) $x \in R^n$, $\upsilon \in R^r$, and D is a matrix function with value in $M_{r\times r}(R)$,
(2) f and h are functionals in $C^2(R^n, R^n)$ and $C^2(R^n, R^{n\times m})$, respectively, such that $f(0) = 0$ and $h(0) = 0$,
(3) $g : R^n \times R^r \to R^{n\times r}$ is Lipschitz functional mapping such that there exists a nondecreasing scalar function $\alpha(|\upsilon|) \geq 0$ bounded for all υ such that

$$|g(x,\upsilon)| \leq \alpha(|\upsilon|)|x|, \quad \forall(x,\upsilon) \in R^n \times R^r,$$

(4) u is a R^p-valued measurable control law,
(5) $F : R^r \times R^p \to R^r$ and $G : R^r \to R^{r\times q}$ are Lipschitz functionals mapping vanishing at the origin and there exists a constant $C \geq 0$ such that the following growth condition holds true:

$$|F(\upsilon,u)| + |G(\upsilon)| \leq C(1 + |\upsilon| + |u|), \quad \forall(\upsilon,u) \in R^r \times R^p. \tag{11}$$

Suppose that there exist functionals $F_1 : R^r \to R^r$ and $F_2 : R^r \to R^{r\times p}$ such that

$$F(\upsilon,u) = F_1(\upsilon) + F_2(\upsilon)u,$$

and

$$d\upsilon = (F_1(\upsilon) + F_2(\upsilon)u)dt + G(\upsilon)d\beta_t, \tag{12}$$

for any $(\upsilon, u) \in R^r \times R^p$. Then, the CSCS (10) is rewritten as

$$\begin{cases} dx = (f(x) + g(x,\upsilon)D\upsilon)dt + h(x)dw_t, \\ d\upsilon = (F_1(\upsilon) + F_2(\upsilon)u)dt + G(\upsilon)d\beta_t. \end{cases} \tag{13}$$

Considering the EISSP results in [11], we can easily establish the following elementary result.

Lemma 1 *The CSCS (13) satisfies the rEISSP property if and only if* $0 \in R^n$ *is rESP for the system*

$$\begin{cases} dx = (f(x) + g(x,\upsilon)D\upsilon)dt + h(x)dw_t, \\ d\upsilon = (F_1(\upsilon) + F_2(\upsilon)\gamma(|\upsilon|)\theta)dt + G(\upsilon)d\beta_t, \end{cases} \tag{14}$$

$$\theta \in I = \{\theta \in R^p : |\theta| \leq 1\}.$$

In the next section, we will derive a state feedback law that renders the satisfaction of rEISSP property for CSCS (13).

3 Main Results

In the following theorem, we suppose that the function g is bounded on $R^n \times R^p$, U is the set of admissible control, and sufficient conditions are established for rEISSP property of CSCS (13). Theorem 1 is the stochastic extension of Proposition 4.1 and Theorem 4.1 established in Spiliotis and Tsinias [11] and Boulanger [3], respectively, to the CSCS (13). In addition, we can consider the exponential stability in mean square results of Boulanger [3] as a special case of our rEISSP results (Theorem 1) where $r = 2$.

Theorem 1 *Suppose that the function* $\gamma : R^+ \to R^+$ *is positive definite with bounded first derivative* $\gamma^{(1)}$, *and for all* $|u| \leq \gamma(|\upsilon|)$, *there exists a* $C^2(R^r \backslash \{0\})$ *function* Φ *satisfying the exponential Lyapunov condition. In addition, suppose that*

$$D\upsilon = F_2(\upsilon)^T \nabla \Phi_2(\upsilon). \tag{15}$$

Then, the control law

$$k(x,\upsilon) = k_1(\upsilon) - g(x,\upsilon)^T \nabla \Phi_1(x), \tag{16}$$

where Φ_1 *is a smooth Lyapunov function corresponding to the SDS (1) and* $k_1 : R^r \to R^p$ *is a control law corresponding to the closed-loop system deduced from (12), which guarantees that the CSCS (13) satisfies the rEISSP property.*

Proof In the proof of this theorem, we shall employ Lemma 1 and show that the rEISSP is satisfied for the CSCS (13), if and only if the origin is rESP for the CSCS (14). Clearly, by our hypothesis, the function $\Phi(V(t))$ satisfies the exponential Lyapunov condition with respect to the CSCS (14). We now show that the origin satisfies rESP for the CSCS (14). Since the exponential Lyapunov condition is fulfilled for the CSCS (14), there exists a Lyapunov function $\Phi(V(t))$ and positive constants a_i, $1 \leq i \leq 5$, such that (6)–(8) hold true. Consider the composite Lyapunov function

$$\Phi(V(t)) = \Phi_1(x) + \Phi_2(\upsilon), \quad \forall (x, \upsilon) \in R^n \times R^r, \tag{17}$$

where Φ_1 and Φ_2 is the Lyapunov function corresponding to the SDS (1) and SCS (12), respectively. Substituting $\theta(x, \upsilon) = \frac{k(x,\upsilon)}{\gamma(|\upsilon|)}$ into the closed-loop system deduced from CSCS (14), we obtain

$$\begin{aligned} dx &= (f(x) + g(x, \upsilon)D\upsilon)dt + h(x)dw_t, \\ d\upsilon &= (F_1(\upsilon) + F_2(\upsilon)k_1(\upsilon) - F_2(\upsilon)g(x, \upsilon)^T \nabla \Phi_1(x) + G(\upsilon)d\beta_t. \end{aligned} \tag{18}$$

Denoting $\mathbf{D}_\upsilon$ as the infinitesimal generator of the stochastic process solution of the closed-loop system (18) yields

$$\begin{aligned} \mathbf{D}_\upsilon \Phi(V(t)) = \mathbf{D}\Phi_1(x) &+ \nabla \Phi_1(x)^T g(x, \upsilon)D\upsilon + \mathbf{Y}_2 \Phi_2(\upsilon) \\ &- \nabla \Phi_2(\upsilon)^T F_2(\upsilon)g(x, \upsilon)^T \nabla \Phi_1(x), \end{aligned} \tag{19}$$

Substituting (15) into (19) we get

$$\mathbf{D}_\upsilon \Phi(V(t)) = \mathbf{D}\Phi_1(x) + \mathbf{Y}_2 \Phi_2(\upsilon). \tag{20}$$

Then, from (8) and (20) we have

$$\begin{aligned} \frac{d}{dt} E\Phi(V(t)) &= E(\mathbf{D}_\upsilon \Phi(V(t))) \\ &= E(\mathbf{D}\Phi_1(x) + \mathbf{Y}_2 \Phi_2(\upsilon)) \leq -a_5 E(|x|^r) - a_5' E(|\upsilon|^r). \end{aligned} \tag{21}$$

The desired condition (5) is a direct consequence of inequality (6) and (21). Thus, CSCS (14) satisfies the rESP property at the origin. It turns out by Lemma 1 that CSCS (13) satisfies the rEISSP property.

Remark 1 (1) Theorem 1 is the stochastic extension of Proposition 4.1 and Theorem 4.1 established in Spiliotis and Tsinias [11] and Boulanger [3], respectively, to the CSCS (13). In addition, we can consider the exponential stability in mean square results of Boulanger [3] as a special case of our rEISSP results (Theorem 1) where $r = 2$.

(2) The exponential stability results obtained in [3], [10], [11] do not permit us to make a conclusion about rEISSP while the results of this paper are still valid.

Finally, we give a numerical example illustrating our results.

Example 1 Consider the multi-input composite system

$$\begin{cases} dx = (-2x + 2\upsilon)dt + x dw_t, \\ d\upsilon = (-2\upsilon + \upsilon u)dt + 2\upsilon d\beta_t, \end{cases} \tag{22}$$

where $\{w_t, t \in R^+\}$ and $\{\beta_t, t \in R^+\}$ are two independent standard real-valued Wiener processes, $x, \upsilon \in R$, u is a real-valued measurable control law, $g(x, \upsilon) = 2$, and $D = I$ is the identity matrix in $M_{r\times r}(R)$. Consider the Lyapunov function

$$\Phi(V(t)) = \Phi_1(x) + \Phi_2(\upsilon) = \frac{1}{2}(x^2 + \upsilon^2).$$

A simple calculation shows that the Lyapunov function $\Phi(V(t))$ satisfies the exponential Lyapunov condition and

$$k_1(\upsilon) = \xi(\mathbf{D}\Phi_2(\upsilon), (\mathbf{D}_z\Phi_2(\upsilon))^2)(\mathbf{D}_z\Phi_2(\upsilon)) = -\frac{1}{1+\upsilon^2}, \tag{23}$$

is a control law corresponding to the closed-loop system deduced from

$$d\upsilon = (-2\upsilon + \upsilon u)dt + 2\upsilon d\beta_t, \tag{24}$$

where

$$\mathbf{D}_z\Phi_2(\upsilon) = \sum_{i=1}^{n} K_z^i(\upsilon)\frac{\partial\Phi_2(\upsilon)}{\partial\upsilon_i} = \upsilon^2, \tag{25}$$

$$\mathbf{D}\Phi_2(\upsilon) = \sum_{i=1}^{n} H^i(\upsilon)\frac{\partial\Phi_2(\upsilon)}{\partial\upsilon_i} + \frac{1}{2}\sum_{i,j=1}^{n}\sum_{k=1}^{m} M_k^i M_k^j \frac{\partial^2\Phi_2(\upsilon)}{\partial\upsilon_i\partial\upsilon_j} = 0, \tag{26}$$

$$\xi(\mathbf{D}\Phi_2, \mathbf{D}_z\Phi_2) = \begin{cases} -\frac{\mathbf{D}\Phi_2 + \sqrt{\mathbf{D}\Phi_2^2 + \mathbf{D}_z\Phi_2^2}}{\mathbf{D}_z\Phi_2(1+\sqrt{1+\mathbf{D}_z\Phi_2})} & \mathbf{D}_z\Phi_2 > 0, \\ 0 & \mathbf{D}_z\Phi_2 = 0, \end{cases} \tag{27}$$

and $H(\upsilon) = -2\upsilon, K(\upsilon) = \upsilon, M(\upsilon) = 2\upsilon$.

(For more information about the control law $k_1(\upsilon)$, we refer to Theorem 3.2 obtained by Abedi et al. [2]). Then, by Theorem 1, the control law

$$\begin{aligned} k(x, \upsilon) &= k_1(\upsilon) - g(x, \upsilon)^T \nabla\Phi_1(x) \\ &= -\frac{1}{1+\upsilon^2} - 2x, \end{aligned} \tag{28}$$

renders the composite system (22) satisfying the rEISSP property.

4 Conclusions

In this paper, we have studied the exponential input-to-state stability in probability of a class of composite stochastic system. We have employed the stochastic version of control Lyapunov function and extended the EISSP results proved by Spiliotis and Tsinias [11] for SCS to the class of composite stochastic system driven by two independent Wiener processes. We have also established sufficient conditions for the rEISSP of composite system.

References

1. Abedi, F., Abu Hassan, M., Arifin, M.N.: Lyapunov function for nonuniform in time global asymptotic stability in probability with application to feedback stabilization. Acta. Appl. Math. **116**(1), 107–117 (2011)
2. Abedi, F., Abu Hassan, M., Arifin, M.N.: Control Lyapunov function for feedback stabilization of affine in the control stochastic time-varying systems. Int. J. Math. Anal. **5**(4), 175–188 (2011)
3. Boulanger, C.: Stabilization of a class of nonlinear composite stochastic systems. Stochastic Anal. App. **18**(5), 723–735 (2000)
4. Khasminskii, Z.R.: Stochastic stability of differential equations, 2nd edition, Springer, New York (2012).
5. Kushner, J.H.: On the stability of processes defined by stochastic difference-differential equation. J. Diff. Equ. **4**, 424–443 (1968)
6. Lan, T.T.T., Dang, N.H.: Exponential stability of nontrivial solutions of stochastic differential equations Series A. Math. Sci. **21**, 97–106 (2011)
7. Liu, R., Raffoul, Y.: Boundedness and exponential stability of highly nonlinear stochastic differential equations. Electron. J. Differ. Equ. **143**, 1–10 (2009)
8. Mao, X.: Exponential Stability of Stochastic Differential Equations. Marcel Dekker Inc., New York (1994)
9. Michel, N.A.: Stability analysis of stochastic composite systems. IEEE Trans. Autom. Control. **20**, 246–250 (1975)
10. Rusinek, R.L.: String stability of singularly perturbed stochastic systems. Annales Mathematicae Silesianae. **16**, 43–55 (2003)
11. Spiliotis, J., Tsinias, J.: Notions of exponential robust stochastic stability, ISS and their Lyapunov characterizations. Int. J. Robust Nonlinear Control. **13**, 173–187 (2003)

On the Solution of Singular Ordinary Differential Equations Using a Composite Chebyshev Finite Difference Method

A. Kazemi Nasab, Z. Pashazadeh Atabakan, A. Kilicman and Zainidin K. Eshkuvatov

Abstract In this paper, a numerical algorithm based upon a hybrid of Chebyshev polynomials and block-pulse functions is proposed for solving both linear and nonlinear singular boundary value problems. Composite Chebyshev finite difference method is indeed an extension of the well-known Chebyshev finite difference method. We take advantage of the useful properties of Chebyshev polynomials and finite difference method to reduce the computation of the problem to a set of algebraic equations simplifying the problem. Several examples are included to illustrate the applicability and accuracy of the introduced method. Convergence analysis is presented.

1 Introduction

Many phenomena in physics, chemistry and mechanics such as Thomas–Fermi differential equation, Ginzburg–Landau equation, Lane–Emden equation can be modeled by singular boundary value problems of second order. The existence of singularities makes the approximate solution lose its accuracy in the neighbourhood of the singular points. Even for local methods, such as the finite difference or finite element methods, spurious oscillations appearing near the singularity may distort the solution in the whole domain. This phenomenon is even more critical for global solution methods, such as the Chebyshev method, whose accuracy depends on the regularity of the solution. For a solution with a low regularity, the "infinite accuracy" commonly associated to spectral methods is lost and such methods show little advantages over local approximation methods. Therefore, a suitable treatment of the singularities is necessary for preserving, as far as possible, the high accuracy of spectral methods [7].

We make use of the useful properties of Chebyshev polynomials and hybrid functions to solve ordinary differential equations in which, at least, one of the coefficient functions or solution function is not analytic, although spectral methods are not good to handle such problems [3], [4].

This paper is organized as follows. We first present some necessary definitions and mathematical preliminaries of block-pulse functions and hybrid functions of

---

A. K. Nasab (✉) · Z. P. Atabakan · A. Kilicman · Z. K. Eshkuvatov
Department of Mathematics, Faculty of Science, Universiti Putra Malaysia,
43400 Serdang, Selangor, Malaysia
e-mail: a.kazeminasab@gmail.com

A. Kilicman et al. (eds.), *International Conference on Mathematical Sciences and Statistics 2013*, DOI 10.1007/978-981-4585-33-0_3,

block-pulse and Chebyshev polynomials. We then describe the composite Chebyshev finite difference method (ChFD). Convergence analysis of the proposed method is advanced. We apply the presented technique to solve numerically linear and nonlinear singular boundary value problems. Several numerical examples are also included to confirm the efficiency and accuracy of the proposed method. Finally, a concise conclusion completes the paper.

2 Preliminaries and Notations

This section includes some notations and preliminary facts that will be used further in this work.

2.1 Block-Pulse Functions and Their Properties

Block-pulse functions (BPFs), $b_k(t)$, $k = 1, 2, \ldots, K$ are an orthogonal set of K piecewise constant functions which are defined on the interval $[0, T]$ as follows

$$b_k(t) = \begin{cases} 1, & \left(\frac{k-1}{K}\right) T \leq t < \left(\frac{k}{K}\right) T \\ 0, & otherwise. \end{cases} \tag{1}$$

where $t_0 = 0$ and $t_K = T$. Block-pulse functions have some nice characteristics. They are disjoint, orthogonal and complete.

For more information about block-pulse functions, interested reader is referred to [5], [6], [10], [13], [16]–[18], [20].

2.2 Hybrid Functions of Block-Pulse and Chebyshev Polynomials

Hybrid functions $b_{km}(t)$, $k = 1, 2, \ldots, K, m = 0, 1, \ldots M$, are defined on the interval $[0, T)$ as

$$b_{km}(t) = \begin{cases} T_m\left(\frac{2K}{T}t - 2k + 1\right), & t \in \left[\left(\frac{k-1}{K}\right)T, \frac{k}{K}T\right) \\ 0, & otherwise. \end{cases} \tag{2}$$

where k and m are the order of block-pulse functions and Chebyshev polynomials of the first kind, respectively. The set of hybrid functions of block-pulse and Chebyshev polynomials is a complete orthogonal set in the Hilbert space $L^2_{w_k}[0, T)$. In view of the following formula:

$$\int_0^T b_{kl}(t) b_{pq}(t) w_k(t) dt = \frac{\pi T}{4N} c_l \delta_{kp} \delta_{lq} \tag{3}$$

in which $w_k(t), k = 1, 2, \ldots, K$, are the corresponding weight functions on the kth subinterval $\left[\left(\frac{k-1}{K}\right)T, \left(\frac{k}{K}\right)T\right)$ and defined as

$$w_k(t) = \frac{1}{\sqrt{1 - \left(\frac{2K}{T}t - 2k + 1\right)^2}} \tag{4}$$

we can conclude that

$$u(t) = \sum_{n=1}^{N} \sum_{m=0}^{\infty} \hat{u}_{nm} b_{nm}(t), \tag{5}$$

where

$$\hat{u}_{nm} = \frac{(u, b_{nm})w_n}{(b_{nm}, bnm)w_n} = \frac{4N}{\pi c_m t_f} \int_{\left(\frac{n-1}{N}\right)t_f}^{\left(\frac{n}{N}\right)t_f)} u(t) b_{nm}(t) w_n(t) dt, \tag{6}$$

in which$(.,.)w_n$ denotes the weighted inner product.

3 Composite Chebyshev Finite Difference Method

In this section, we present the composite Chebyshev finite difference (ChFD) method. Consider $t_{km}, k = 1, 2, \ldots, K, m = 0, 1, \ldots, M$, as the corresponding Chebyshev–Gauss–Lobatto collocation points at the kth subinterval $\left[\frac{k-1}{K}, \frac{k}{K}\right]$ such that,

$$t_{km} = \frac{T}{2K}(t_m + 2k - 1). \tag{7}$$

where t_m are the extrema of the Mth-order Chebyshev polynomial $T_M(t)$ and defined as

$$t_m = \cos\left(\frac{m\pi}{M}\right), \quad m = 0, 1, 2, \ldots, M. \tag{8}$$

Following Clenshaw [9], a function $f(t)$ can be written in terms of hybrid basis functions as follows

$$(P_M)f(t) = \sum_{k=1}^{K} \sum_{m=0}^{M}{}'' f_{km} b_{km}(t), \tag{9}$$

where $b_{k,m}(t), k = 1, 2, \ldots, K, m = 0, 1, \ldots, M$, are defined in Eq. (2) and f_{km}, $n = 1, 2, \ldots, K, m = 0, 1, \ldots, M$, are the expansion coefficients of the function $f(t)$ at the subinterval $\left[\frac{k-1}{K}, \frac{k}{K}\right]$ which are defined as following:

$$f_{km} = \frac{2}{M} \sum_{p=0}^{M}{}'' f(t_{kp}) b_{km}(t_{kp}) = \frac{2}{M} \sum_{p=0}^{M}{}'' f(t_{kp}) \cos\left(\frac{mp\pi}{M}\right). \tag{10}$$

The first two derivatives of the function $f(t)$ at the points $t_{km}, k=1,2,\ldots,K$, $m=0,1,\ldots,M$, can be obtained as

$$f^{(i)}(t_{km}) = \sum_{j=0}^{M} d^{(i)}_{k,m,j} f(t_{kj}), \quad i=1,2 \tag{11}$$

where

$$d^{(1)}_{k,m,j} = \frac{8K\theta_j}{TM} \sum_{n=1}^{M} \sum_{\substack{l=0 \\ (n+l) odd}}^{n-1} \frac{n\theta_n}{c_l} \cos\left(\frac{nj\pi}{M}\right) \cos\left(\frac{lm\pi}{M}\right), \tag{12}$$

and

$$d^{(2)}_{m,j} = \frac{8K^2\theta_j}{T^2M} \sum_{n=2}^{M} \sum_{\substack{l=0 \\ (n+l) even}}^{n-2} \frac{n(n^2-l^2)\theta_n}{c_l} \cos\left(\frac{nj\pi}{M}\right) \cos\left(\frac{lm\pi}{M}\right). \tag{13}$$

4 Convergence Analysis

A detailed proof of the following results can be found in [18], [19].

Lemma 1. *If the hybrid expansion of a continuous function $g(t)$ converges uniformly, then it converges to the function $g(t)$.*

Theorem 1. *A function $g(t) \in L^2_{w_k}[0,T)$, with bounded second derivative, say $|g''(t)| \leq B$, can be expanded as an infinite sum of hybrid functions and the series converges uniformly to $g(t)$, that is,*

$$g(t) = \sum_{k=1}^{\infty} \sum_{m=0}^{\infty} \hat{c}_{km} b_{km}(t). \tag{14}$$

Theorem 2. *Suppose $g(t) \in L^2_{w_k}[0,T)$ with bounded second derivtive, say $|g''(t)| \leq B$, then its hybrid expansion converges uniformly to $g(t)$, that is,*

$$\sum_{k=1}^{K} {\sum_{m=0}^{\infty}}' c_{km} b_{km}(t) = g(t), \tag{15}$$

where the summation symbol with prime denotes a sum with the first term halved.

Theorem 3. *(Accuracy estimation) Suppose, $h(t) \in L^2_{w_k}[0,T)$ with bounded second derivative, say $|h''(t)| \leq B$, then we have the following accuracy estimation:*

$$\sigma_{K,M} \leq \left(S + \sum_{k=K+1}^{\infty} \sum_{m=M+1}^{\infty} \frac{C^2}{(m^2-1)^2} \times \frac{\pi T c_m}{4K} \right)^{1/2}, \tag{16}$$

where

$$\sigma_{K,M} = \left(\int_0^T \left[h(t) - \sum_{k=1}^{K} {\sum_{m=0}^{M}}'' h_{km} b_{km}(t) \right]^2 w_k(t) dt \right)^{1/2},$$

$$C = \frac{R\pi^3}{12} + \frac{BT^2}{k^2 c_m},$$

$$R = max \left\{ \frac{d^2}{d\beta^2} (h(\frac{cos(\beta) + 2k - 1}{2K} T) \, cos(m\beta)), \; 0 \leq \beta \leq \pi \right\}, \; and$$

$$S = \frac{1}{4} h_{10}^2 \frac{\pi T}{2K} + \frac{1}{4} h_{KM}^2 \frac{\pi T}{4K}$$

5 Discretization of Problem

In this section, the composite Chebyshev finite difference method (ChFD) is applied for solving linear and nonlinear singular ordinary differential equations. Consider the following general form:

$$y''(t) + p(t)y'(t) + q(t)y(t) = f(t) \quad t \in (a, b) \tag{17}$$

with the initial conditions

$$y(a) = \alpha, \; y'(a) = \beta, \tag{18}$$

or boundary conditions

$$y(a) = \alpha, \; y(b) = \beta, \tag{19}$$

where at least one of the functions $p(t)$, $q(t)$ and $f(t)$ has a singular point and a, b, α and β are constants. We suppose the interval $[0, 1]$ is divided into K subintervals $I_k = \left[\frac{k-1}{K}, \frac{k}{K}\right]$, $k = 1, 2, \ldots, K$. We also consider the Chebyshev–Gauss–Lobatto collocation points $\frac{k-1}{K} = t_{k0} < t_{k1} < \ldots < t_{k,M-1} < t_{k,M} = \frac{k}{K}$ on the kth subinterval I_k, $k = 1, 2, \ldots, K$, where t_{ks} is defined as following:

$$t_{ks} = \frac{1}{2K}(t_s + 2k - 1), \quad s = 1, 2, \ldots, M. \tag{20}$$

In order to obtain the solution $y(t)$ in Eq. (17), we first approximate $y(t)$ in terms of hybrid functions as follows

$$y(t) = \sum_{k=1}^{K} {\sum_{m=0}^{M}}'' c_{km} b_{km}(t), \tag{21}$$

where $c_{km}, k = 1, 2, \ldots, K, m = 0, 1, \ldots, M$, are defined in (15). We now collocate Eq. (17) at the Chebyshev–Gauss–Lobatto points $t_{ks}, k = 1, 2, \ldots, K, s = 1, \ldots, M$, as

$$y''(t_{ks}) + p(t_{ks})y'(t_{ks}) + q(t_{ks})y(t_{ks}) = f(t_{ks}). \tag{22}$$

Table 1 Comparison of the approximate solution of Example 1 against the exact solution

t	*Our results*	*Ref.* [14]	*Exact solution*
0.1	0.09836316430834659726	0.098363164	0.09836316430834659674
0.2	0.18709785675772780738	0.187097856	0.18709785675772780664
0.3	0.25751810740024193303	0.257518107	0.25751810740024193250
0.4	0.30273069145626279161	0.302730691	0.30273069145626279169
0.5	0.31830988618379067124	0.318309886	0.31830988618379067153
0.6	0.30273069145626279039	0.302730691	0.30273069145626279168
0.7	0.25751810740024193091	0.257518107	0.25751810740024193248
0.8	0.18709785675772780496	0.187097856	0.18709785675772780663
0.9	0.09836316430834659594	0.098363164	0.09836316430834659674
1.0	0.00000000000000000328	0.000000000	0.00000000000000000000

In addition, from given conditions, we get other two equations. Besides, it is necessary that the approximate solution and it's first derivatve be continuous at the interface of subintervals. Therefore, we have a system of $K(M+1)$ algebric equations, which can be solved for the unknowns $y(t_{ks})$. Consequently, we obtain the solution $y(t)$ to the given Eqs. (17)–(19).

6 Numerical Examples

In this section, the proposed method in this paper is used to solve some numerical examples to demonstrate the validity and applicability of the method and a comparison is made with existing results. The obtained results confirm that the new method is simple and accurate. All computations are performed using Maple 14.

Example 1. Consider the singular boundary value problem [14], [21]

$$y''(t) + \pi^3 \frac{(y(t))^2}{\sin(\pi t)} = 0, \qquad 0 < t < 1, \tag{23}$$

$$y(0) = y(1) = 0, \tag{24}$$

which has the exact solution $y(t) = \frac{1}{\pi}\sin(\pi t)$. We solve this problem using the proposed method with $M = 12$ and $N = 4$. In Table 1, we make a comparison between our obtained results with those reported in [14] and exact solution. It can be seen that the result obtained by the current method is in full agreement with the exact solution up to 18 digits. The absolute error in solution is shown in Fig. 1.

Example 2. Consider the inhomogeneous Bessel equation [1], [2], [15]

$$y''(t) - \frac{\alpha}{t}y'(t) = \beta x^{\beta-2}\left[\alpha + \beta - 1 + \beta t^{\beta}\right] y(t), \qquad 0 < t < 1,$$

$$y(0) = 1 \quad y(1) = e,$$

which has the exact solution $y(t) = e^{t^{\beta}}$. This equation has a singular point at $t = 0$.

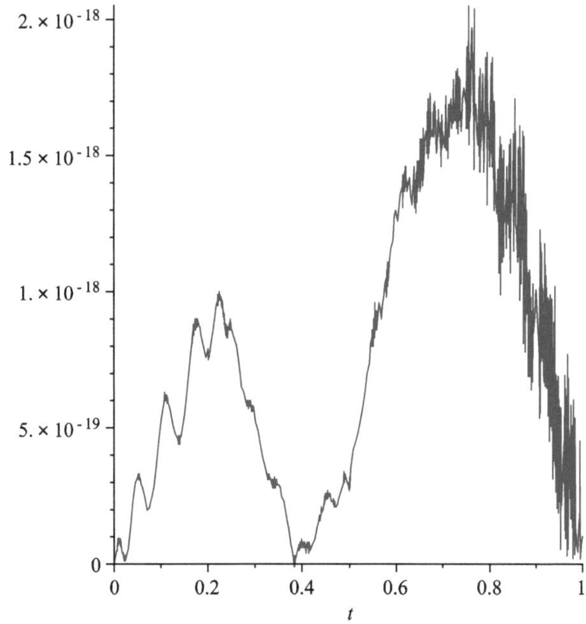

Fig. 1 The absolute error between approximate and exact solution for Example 1

Table 2 Comparison of obtained absolute errors using present method and those in [1] for Example 2

t	Present method	Ref. [1]
0.0	0	0
0.1	2.01100×10^{-16}	3.0047×10^{-5}
0.2	2.22600×10^{-16}	1.7524×10^{-4}
0.3	2.32200×10^{-16}	9.3554×10^{-6}
0.4	2.37700×10^{-16}	4.3842×10^{-5}
0.5	2.48700×10^{-16}	9.3002×10^{-5}
0.6	2.62700×10^{-16}	1.9170×10^{-4}
0.7	2.93900×10^{-16}	3.1501×10^{-4}
0.8	9.22000×10^{-17}	3.9364×10^{-4}
0.9	2.80450×10^{-15}	3.1680×10^{-4}
1.0	9.93790×10^{-15}	0

This example was solved in [1], [15] for $\alpha = 1$ and $\beta = 3$ using ADM and a three-point finite difference method, respectively. The absolute error obtained in [15] was $\|E\| = 7.1 \times 10^{-5}$ using $N = 128$. In Table 2 we compare the numerical results for the absolute errors obtained using our method and those reported in [1] using the 7-terms approximation to evaluate the decomposition series solutions. In order to confirm the accuracy and efficiency of the proposed method, we plot the graphs of absolute errors for different values of α and β in Figs. 2 and 3.

Example 3. As the last problem we consider the nonlinear, homogeneous Emden–Fowler type equation [8], [22],

$$y''(t) + \frac{5}{t}y'(t) + 8c(e^{y(t)} + 2e^{y(t)/2}) = 0, \qquad 0 \le t \le 1,$$

$$y(0) = 0 \qquad y'(0) = 0.$$

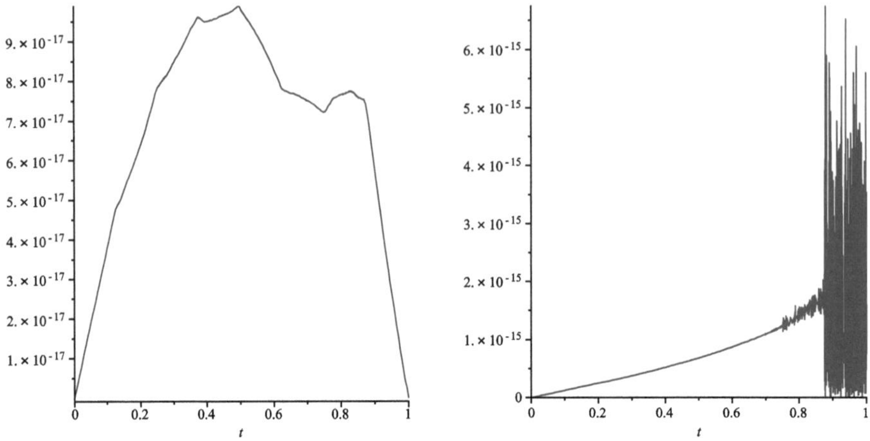

Fig. 2 The absolute error between approximate and exact solution for $\alpha = 0$ and $\beta = 1$, (*Right*) $\alpha = 0$ and $\beta = 3$ for Example 2

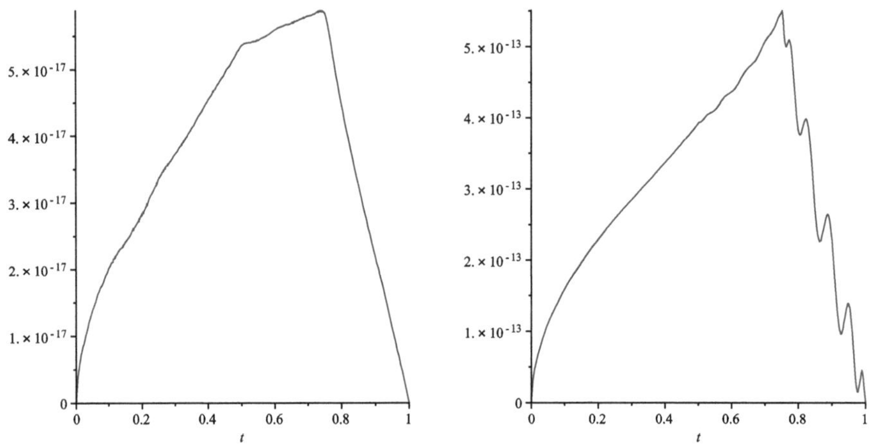

Fig. 3 The absolute error between approximate and exact solution for $\alpha = 0.5$ and $\beta = 1$, (*Right*) $\alpha = 0.5$ and $\beta = 3$ for Example 2

Wazwaz [22] obtained the closed-form solution of this problem using Adomian decomposition method (ADM):

$$y(t) = -2ln(1 + ct^2).$$

We set $M = 15$, $N = 5$ and $c = 1$ for solving this problem. In Table 3, we make a comparison between approximate solution and exact solution. In Fig. 4 the absolute errors are shown. As can be seen, the approximate and exact solutions are almost same.

Table 3 Comparison of the approximate solution of Example 3 against the exact solution

t	*Our result*	*Exact solution*
0.1	−0.0199006617063361657	−0.0199006617063361657
0.2	−0.0784414263065625925	−0.0784414263065625925
0.3	−0.1723553924821046646	−0.1723553924821046647
0.4	−0.2968400102365465560	−0.2968400102365465560
0.5	−0.4462871026284195115	−0.4462871026284195115
0.6	−0.6149693994959212861	−0.6149693994959212809
0.7	−0.7975522399147355580	−0.7975522399147355459
0.8	−0.9893924836722141377	−0.9893924836722141093
0.9	−1.1866536905554687700	−1.1866536905554687576
1.0	−1.3862943611198906426	−1.3862943611198906188

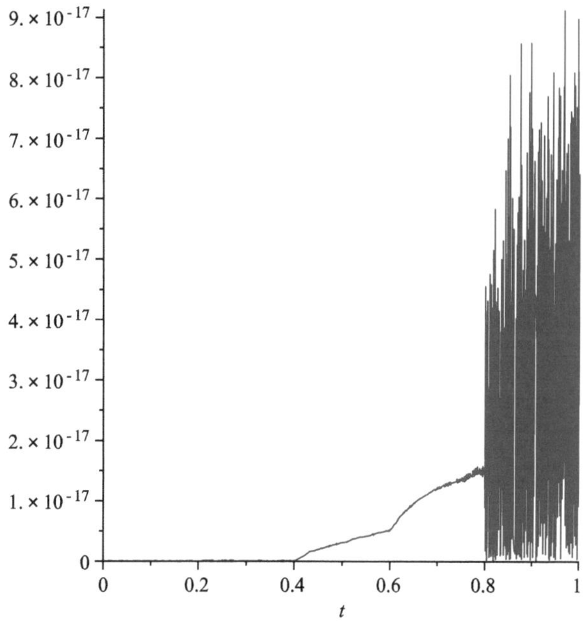

Fig. 4 Absolute error between approximate and exact solution for Example 3

7 Conclusion

In this paper, we presented the composite Chebyshev finite difference method for solving singular ordinary differential equations. This method is indeed an extension of the ChFD scheme with $K = 1$. The useful properties of Chebyshev polynomials and block-pulse functions make it a computationally efficient method to approximate the solution of singular ODE's. We converted the given problem to a system of algebric equations using collocation points. The main advantage of the present method is the ability to represent smooth and especially piecewise smooth functions properly. It was also shown that the accuracy can be improved either by increasing the number of subintervals or by increasing the number of collocation points in subintervals. Several examples have been provided to demonstrate the powerfulness of the proposed method. A comparison was made among the present method, some other well-known approaches and exact solution which confirms that the introduced method are more accurate and efficient.

References

1. Abdelhalim, E.: A new analytical and numerical treatment for singular two-point boundary value problems via the Adomian decomposition method. J. Comput. Appl. Math. **235**, 1914–1924 (2011)
2. Aziz, T., Kumar, M.: A fourth-order finite-difference method based on non-uniform mesh for a class of singular two-point boundary value problems. J. Comput. Appl. Math. **136**, 136–342 (2001)
3. Babolian, E., Hosseini, M.M.: A modified spectral method for numerical solution of ordinary differential equations with non-analytic solution. Appl. Math. Comput. **132**, 341–351 (2002)
4. Babolian, E., Bromilow, M., England, R., Saravi, M.: A modification of psedo-spectral method for solving a linear ODEs with singularity. Appl. Math. Comput. **188**, 1260–1266 (2007)
5. Babolian, E., Masouri, Z.: Direct method to solve Volterra integral equation of the first kind using operational matrix with block-pulse functions. J. Comput. Appl. Math. **220**, 51–57 (2008)
6. Beauchamp, K.G.: Applications of Walsh and Related Functions with an Introduction to Sequency Theory. Academic Press, London (1984)
7. Botella, O., Peyret, R.: Computing singular solutions of the Navier Stokes equations with the Chebyshev-collocation method. Int. J. Numer. Meth. Fluids. **36**, 125–163 (2001)
8. Chowdhury, M.S.H., Hashim, I.: Solutions of Emden-Fowler equations by homotopy-perturbation method. Nonlinear. Anal. Real. World Appl. **10**, 104–115 (2009)
9. Clenshaw, C.W., Curtis, A.R.: A method for numerical integration on an automatic computer. Numer. Math. **2**, 197–205 (1960)
10. Deb, A., Sarkar, G., Sen, S.K.: Block pulse functions, the most fundamental of all piecewise constant basis functions. Int. J. Syst. Sci. **25**(2), 351–363 (1994)
11. Elbarbary, E.M.E., El-Kady, M.: Chebyshev finite difference approximation for the boundary value problems. Appl. Math. Comput. **139**(3), 513–523 (2003)
12. Elbarbary, E.M.E.: Chebyshev finite difference method for the solution of boundary-layer equations. Appl. Math. Comput. **160**(2), 487–98 (2005)
13. Jiang, Z.H., Schaufelberger, W.: Block Pulse Functions and Their Applications in Control Systems. Springer-Verlag, Berlin (1992)
14. Kazemi Nasab, A., Kilicman, A., Babolian, E., Pashazadeh Atabakan, Z.: Wavelet analysis method for solving linear and nonlinear singular boundary value problems. Appl. Math. Model. **37**, 5876–5886 (2013).
15. Kumar, M.: A three-point finite difference method for a class of singular two-point boundary value problems. J. Comput. Appl. Math. **145**, 89–97 (2002)
16. Maleknejad, K., Mahmoudi, Y.: Numerical solution of linear Fredholm integral equation by using hybrid Taylor and block-pulse functions. Appl. Math. Comput. **149**, 799–806 (2004)
17. Maleknejad, K., Mahdiani, K.: Solving nonlinear mixed Volterra–Fredholm integral equations with two dimensional block-pulse functions using direct method. Commun. Nonlinear. Sci. Numer. Simulat. **16**, 3512–3519 (2011)
18. Marzban, H.R., Hoseini, S.M.: Solution of linear optimal control problems with time delay using a composite Chebyshev finite difference method. Optim. Control. Appl. Meth. **34**, 253–274 (2013)
19. Pashazadeh Atabakan, Z., Kazemi Nasab, A., Kilicman, A.: On solution of Fredholm integrodifferential equations using composite Chebyshev finite difference method. Abst. Appl. Anal. **2013**(2013), Article ID 694043, 11 pages, http://dx.doi.org/10.1155/2013/694043
20. Rao, G.P.: Piecewise Constant Orthogonal Functions and Their Application to Systems and Control. Springer-Verlag, New York (1983)
21. Wazwaz, A.M.: A reliable algorithm for obtaining positive solutions for nonlinear boundary value problems. Comput. Math. Appl. **41**, 1237–1244 (2001)
22. Wazwaz, A.M.: Adomian decomposition method for a reliable treatment of the Emden-Fowler equation. Appl. Math. Comput. **161**, 543–560 (2005)

On 0-Controllability and Pursuit Problems for Linear Discrete Systems Under Total Constraints on Controls

Atamurat Kuchkarov, Gafurjan Ibragimov and Akmal Sotvoldiev

Abstract We consider linear discrete control and pursuit game problems. Control vectors are subjected to total constraints, which are discrete analogues of the integral constraint. By definition, (i) the control system is 0-controllable on the whole if there is a control such that the state of the system $z(t) = 0$ at some step t, (ii) pursuit can be completed if there exists a strategy of the pursuer such that for any strategy of the evader the state of the system $y(t) = 0$. We obtained sufficient condition for equivalence of 0-controllability and completion of the game from any initial position of the space.

1 Introduction

In the case of continuous-time dynamical systems, control and pursuit problems were investigated in many works, e.g., [3], [4], [7], [8], [10], [12], [13], [14], [19].

Many papers are devoted to linear discrete control systems. Sirotin [18] studied some conditions of 0-contrallability and asymptotical controllability of discrete time linear systems. Control is bounded in Holder norm. In particular, he obtained necessary and sufficient conditions of 0-contrallability and asymptotical controllability of the system.

Sazanova [17] studied a problem of steering the state of a linear discrete system from the given point $x(k_0) = x_0$ at a given time $k = k_0$ to another given point $x(N) = x_1$ at a given time $k = N$ by a control which minimizes a given functional. She constructed an optimal program control.

A. Kuchkarov (✉) · A. Sotvoldiev
Institute of Mathematics, 29, Dorman yuli str., Tashkent, Uzbekistan 100125
e-mail: kuchkarov1@yandex.ru

G. Ibragimov
Institute for Mathematical Research, Universiti Putra Malaysia,
43400 Serdang, Selangor, Malaysia
e-mail: gafur@science.upm.edu.my

A. Sotvoldiev
e-mail: akmal.sotvoldiyev@mail.ru

A. Kilicman et al. (eds.), *International Conference on Mathematical Sciences and Statistics 2013,* DOI 10.1007/978-981-4585-33-0_4,

However, discrete linear pursuit games were studied in a few works e.g. [1], [2], [5], [6], [9], [15], [16], where the control vectors of the players are subjected to total constraints.

Satimov et al. [16] examined linear discrete pursuit games of many objects with total constraints on controls of players. They obtained sufficient conditions for completion of pursuit.

The paper of Satimov and Ibragimov [15] was also devoted to discrete time pursuit games of several players described by linear systems. They obtained sufficient conditions of completion of the game, which is necessary in the case of one pursuer. In particular, they obtained necessary and sufficient conditions of 0-controllability when coefficient-matrix of the control is an identity matrix.

In [6], the authors studied discrete-time linear pursuit game when eigenvalues of the main matrix in modulus are less than 1. A sufficient condition of completion of game was obtained.

Azamov and Kuchkarov [2] obtained necessary and sufficient conditions of solvability of both the 0-controllability problem and the problem of approaching the origin for a linear discrete system when control parameter of the pursuer is subjected to geometric constraint and that of the evader is subjected to total constraint.

In the monograph of Azamov [1], fundamentals of theory of discrete time pursuit and evasion dynamic games are given systematically. The main notions of the theory of discrete games were explained by interesting examples.

2 Statement of the Problem

In this paper, we consider discrete pursuit game described by the equation

$$z(t+1) = Az(t) + Bu(t) + Cv(t) \tag{1}$$

and control system described by the equation

$$z(t+1) = Az(t) + Bu(t) \tag{2}$$

where $z \in R^n, u \in R^m, v \in R^l$, t is the step number that belongs to nonnegative integers, A, B, C are constant matrices of corresponding dimensions, u is the control parameter of the pursuer, and v is that of the evader.

Definition 1 The sequences $u(\cdot) : N \to R^m$ and $v(\cdot) : N \to R^l$ subjected to the following total constraints

$$||u(\cdot)||_{l_p} = \left(\sum_{t=0}^{\infty} |u(t)|^p \right)^{1/p} \leq \rho, \quad \rho > 0, \tag{3}$$

$$||v(\cdot)||_{l_p} = \left(\sum_{t=0}^{\infty} |v(t)|^p \right)^{1/p} \leq \sigma, \quad \sigma \geq 0, \tag{4}$$

are referred to as the controls of the pursuer and the evader, respectively, where N is the set of nonnegative integers, and $p > 1$.

The conditions (3) and (4) are the discrete analogues of integral constraints, and they are called total constraints. If $\sigma = 0$, then clearly, the evader is absent, and Eq. (1) turns into Eq. (2). Therefore, the dynamical game becomes simply a control system with total constraints on control.

Definition 2 The function $U(t, z, v), U : R \times R^n \times R^l \to R^m$, is called a strategy of the pursuer. For a given initial point z_0, strategy U, and control of the evader $v(\cdot)$ the triple $(z_0, U, v(\cdot))$ generates a unique trajectory $z(\cdot)$ defined by the formula

$$z(t+1) = Az(t) + BU(t, z(t), v(t)) + Cv(t), z(0) = z_0.$$

The sequence $u(t) = U(t, z(t), v(t))$ is referred to as the realization of the strategy U. The strategy U is called an admissible strategy of the pursuer if all its realizations with fixed z_0 satisfy the condition (3).

Definition 3 Pursuit is said to be completed in the game (1) from the initial position z_0 if there exists a strategy of the pursuer U such that for any control of the evader v, the trajectory $z(\cdot)$ generated by z_0, U, and v satisfies the condition $z(t) = 0$ at some t.

The aim of the pursuer is to realize the equality $z(t) = 0$ at some t.

Definition 4 The system (2) is called 0-controllable if for any point z_0 there exists a control $u(\cdot) : N \to R^m$ such that the trajectory $z(\cdot)$ of the system (2) with the initial position $z(0) = z_0$ satisfies the equality $z(t) = 0$ at some t.

Problem Find conditions under which 0-controllability and pursuit problems for linear discrete systems under total constraints on controls are solvable.

3 Main Result

By a nonsingular real linear transformation, the matrix A can be transformed into the Jordan form. Therefore, we may assume that the matrix A is in real Jordan form. Then the system (2) can be written as follows:

$$x(t+1) = A_1x(t) + B_1u, \ x \in R^q, \tag{5}$$

$$y(t+1) = A_2y(t) + B_2u, \ y \in R^{n-q}, \tag{6}$$

where all eigenvalues of the real matrix A_1 are nonzero, and those of the real matrix A_2 are equal to zero. Dimensions of these matrices are $q \times q$ and $(n-q) \times (n-q)$, respectively, and dimensions B_1 and B_2 are $q \times m$ and $(n-q) \times m$, respectively.

In the future, we need the following statement, which is equivalent to Theorem (ii) in the paper [12].

Theorem 1 *The system (2) is 0-controllable if and only if*

1. all eigenvalues of the matrix A in absolute value do not exceed 1,
2. $\operatorname{rank}\|B_1, A_1 B_1, ..., A_1^{q-1} B_1\| = q$.

Let $S_r^l = \{x \in R^l \mid |x| \le r\}$ denote the unit ball of the space R^l centered at the origin.

Remark In the proof of Theorem 1, the control $u(t)$, which transfers the state of the system (1) from the initial position z_0 to $z = 0$, is constructed as follows: on each step $t = jq, j \in N$, the control is chosen in the form of synthesis, which uses only $z(jq)$, and between jq and $(j+1)q, u(t)$ is a program control. Therefore, if the conditions (i) and (ii) of Theorem 1 are satisfied, then for any $z_* \in R^n$ and $\rho_* > 0$, there exists a number $t_* \in N$ and a function $U_1(t, z_*, \rho_*), t \in N$, such that

$$\sum_{t=0}^{\infty} |U_1(t, z_*, \rho_*)|^p \le \rho_*^p,$$

and for the trajectory $z(\cdot)$ of the system (1) with $u(t) = U_1(t, z_*, \rho_*), t \in N$, and initial state $z(0) = z_*$, the equality $z(t_*) = 0$ holds.

Theorem 2 *If there is a number* $\mu > 1$ *such that*

$$\mu\sigma C S_1^q \subset \rho \operatorname{int} B S_1^p, \tag{7}$$

then the following statements are equivalent:

a. the system (2) is 0-controllable,
b. pursuit can be completed in the game (1) from any initial point of the space R^n.

Proof Let us suppose that pursuit can be completed in the game (1) from any initial point of the space R^n. Then there exists a strategy of the pursuer such that for any strategy of the evader, in particular, if $v(k) = 0$ for all $k \ge 0$, pursuit can be completed in the game (1) from any initial point of the space R^n. This means the system (2) is 0-controllable. Hence, (b) implies (a).

We now show that statement (a) implies (b). Let the hypothesis of the theorem hold and system (2) be 0-controllable. We shall have established the theorem if we prove that there exists an admissible strategy $U = U(t, z, v)$ of the pursuer such that $z(t) = 0$ at some $t \ge 0$ for the trajectory $z(\cdot)$ generated by the strategy $U = U(t, z, v)$, the control $v(\cdot)$, and initial position z_0.

We construct the strategy of the pursuer as follows. For each vector $v \in R^l$, we define the vector $U_1(v) \in R^m$ by the equation $BU_1 = -Cv$ requiring that

$$|U_1(v)| \le \frac{\rho}{\mu\sigma}|v|. \tag{8}$$

According to the inclusion (7), such choice of $U_1(v)$ is possible. Indeed, from (7) we obtain that $CR^l \subset BR^m$. Again, from (7) we obtain that for any $v \in R^l$ there exists $u_v \in S_1^m$ such that

$$\mu\sigma C\frac{v}{|v|} = \rho B u_v.$$

This implies that

$$B\left(\frac{\rho\,|v|}{\mu\sigma}u_v\right) = Cv.$$

Then

$$U_1(v) = \frac{\rho\,|v|}{\mu\sigma}u_v$$

is the desired vector.

Since $\mu > 1$, by Theorem 1 and Remark, for any $z_0 \in R^n$ there exists a control $\bar{u}(z_0,\cdot)$ such that

$$\sum_{t=0}^{\infty} |\bar{u}(z_0,t)|^p \leq \rho^p\left(1-\frac{1}{\mu}\right)^p \tag{9}$$

and the solution of the initial value problem

$$z(t+1) = Az(t) + B\bar{u}(z_0,t), z(0) = z_0,$$

turns to zero at some step $t = t_*$. We now define for the initial position z_0, the strategy U as follows

$$U(t,z,v) = \begin{cases} \bar{u}(z_0,t) + U_1(v), & z \neq 0, \\ 0, & z = 0. \end{cases} \tag{10}$$

Then combining (4), (8), and (9), we have

$$\begin{aligned} ||U(t,z,v)||l_p \leq & ||\bar{u}(z_0,t)||l_p + ||U_1(v(t))||l_p \\ \leq & \rho\left(1-\frac{1}{\mu}\right) + \frac{\rho}{\mu\sigma}||v(\cdot)||l_p \leq \rho. \end{aligned}$$

Hence, the strategy is admissible. Moreover, for the trajectory $z(\cdot)$ of the system (1) generated by the triple $(z_0, U, v(\cdot))$, we have

$$\begin{aligned} z(t+1) &= Az(t) + BU(t,v(t)) + Cv(t) \\ &= Az(t) + B(\bar{u}(z_0,t) + U_1) + Cv(t) \\ &= Az(t) + B\bar{u}(z_0,t) + (BU_1 + Cv(t)). \end{aligned}$$

Then by the choice of U_1 we obtain that

$$z(t+1) = Az(t) + B\bar{u}(z_0,t), z(0) = z_0. \tag{11}$$

As mentioned above that for the solution $z(\cdot)$ of the system (11), $z(t_*) = 0$ at some step t_*. Thus, pursuit can be completed in the game (2) from any initial position. This completes the proof of the theorem.

4 Conclusion

In this paper, we have studied the relation between 0-controllability of linear discrete systems and completion of the game, which is described by a linear discrete system, from any initial position of the space. We have obtained a condition under which 0-controllability implies completion of the game from any initial position of the space and vice versa.

Acknowledgments This research was partially supported by the Research Grant (RUGS) of the Universiti Putra Malaysia, No. 05-02-12-1868RU.

References

1. Azamov, A.A.: Fundamentals of Theory of Discrete Games. Niso Poligraf, Tashkent (2011)
2. Azamov, A.A., Kuchkarov, A.Sh.: On controllability and pursuit problems in linear discrete systems. J. Comput. Sys. Sc. Int. **49**(3), 360–365 (2010)
3. Demidovich, B.P.: Lectures on the Mathematical Theory of Stability. Nauka, Moscow (1967)
4. Filippov, A.F.: Differential Equations with Discontinuous Right-Hand Sides. Nauka, Moscow (1985)
5. Ibragimov, G.I.: Problems of linear discrete games of pursuit. Math. Notes **77**(5), 653–662 (2005)
6. Ibragimov, G.I., Kuchkarov, A.Sh.: Discrete pursuit game with total constraints. In: Proceedings of the International Symposium on New Development of Geometric Function Theory and its Applications, Universiti Kebangsaan Malaysia, Malaysia, pp. 370–374, 10–13 Nov 2008
7. Ibragimov, G.I., Satimov, N.Yu.: A multi player pursuit differential game on closed convex set with integral constraints. Abstr. Appl. Anal. **2012**, Article ID 460171, 12 p. (2012). doi:10.1155/2012/460171
8. Krasovskii, N.N.: Control of a Dynamical System. Nauka, Moscow (1985)
9. Kuchkarov, A.Sh., Ibragimov, G.I., Sotvoldiev, A.: Linear discrete pursuit game problem with total constraints. Abstr. Appl. Anal. **2013**, Article ID 840925, 5 p. (2013). (http://dx.doi.org/10.1155/2013/840925).
10. Nikol'skii, M.S.: The First Direct Method of L.S. Pontryagin in Differential Games. MSU Press, Moscow (1984)
11. Pontryagin, L.S.: Selected Scientific Papers, vol. 2. Nauka, Moscow (1988)
12. Pontryagin, L.S., Boltyanskii, V.G., Gamkrelidze, R.B., Mishchenko, E.F.: The Mathematical Theory of Optimal Processes. Nauka, Moscow (1976)
13. Pshenichnii, B.N., Ostapenko, V.V.: Differential Games. Naukova Dumka, Kiev (1992)
14. Satimov, N.Yu.: Methods for Solving a Pursuit Problem in the Theory of Differential Games. NUUz Press, Tashkent (2003)
15. Satimov, N.Yu., Ibragimov, G.I.: On a pursuit problem for the discrete games with many participants. Izv. Vyssh. Uchebn. Zaved. Mat. (Russian Math.) **12**(511), 46–57 (2004)
16. Satimov, N.Yu., Rikhsiev, B.B., Khamdamov, A.A.: On a pursuit problem for n person linear differential and discrete games with integral constraints. Math. USSR Sb. **46**(4), 456–469 (1983)
17. Sazanova, L.A.: Optimal Control of Linear Discrete Systems. Proceedings of the Steklov Institute of Mathematics (Supplementary issues), **2**, S141–S157 (2000)
18. Sirotin, A.N.: On null-controllable and asymptotically null-controllable finite-dimensional linear systems with controls bounded in the Hölder norms of control. Automat. Rem. Contr. **60**(11 Part 1), 1729–1738 (1999)
19. Subbotin, A.I., Chentsov, A.G.: Optimization of Guarantee in Control Problems. Nauka, Moscow (1981)

Construction of Strategies of Pursuers in a Differential Game of Many Players with State and Integral Constraints

Gafurjan Ibragimov, Asqar Rahmanov and Idham Arif Alias

Abstract The approach of a group of controlled objects, the pursuers, to another one, the evaders, is considered. The motions of all the objects are described by simple differential equations. The control functions of players are subjected to integral constraints. The amount of control resources such as fuel, energy etc. are described by such constraints. Given a non-empty convex subset of $\mathbb{R}^n$ all objects move in this set. If the position of each evader y_j, $j \in \{1, 2, ..., k\}$, coincides with the position of a pursuer x_i, $i \in \{1, ..., m\}$, at some time t_j, i.e. $x_i(t_j) = y_j(t_j)$, then we say that pursuit can be completed. The total resource of the pursuers is assumed to be greater than that of the evaders. We show that pursuit can be completed in this differential game.

1 Introduction

Many works were devoted to games with integral constraints on controls, e.g. see the reference.

A linear pursuit differential game of many pursuers and one evader with integral constraints on controls of players were studied by Satimov et al. [1] in $\mathbb{R}^n$. Objects move according to the following equations

$$\dot{z}_i = C_i z_i + u_i - v, \quad z_i(t_0) = z_i^0, \quad i = 1, ..., m,$$

G. Ibragimov (✉) · I. A. Alias
Institute for Mathematical Research, Universiti Putra Malaysia,
43400 Serdang, Selangor, Malaysia
e-mail: gafur@science.upm.edu.my

G. Ibragimov · A. Rahmanov · I. A. Alias
Department of Mathematics, Faculty of Science, Universiti Putra Malaysia,
43400 Serdang, Selangor, Malaysia

A. Rahmanov
Department of Informatics, Tashkent University of Information Technologies (TUIT),
ul. Amir Temur, d. 108, Tashkent, Uzbekistan
e-mail: atrahmanov@inbox.uz

I. A. Alias
e-mail: idham@science.upm.edu.my

A. Kilicman et al. (eds.), *International Conference on Mathematical Sciences and Statistics 2013,* DOI 10.1007/978-981-4585-33-0_5,

where u_i and v are the control parameters of the ith pursuer and the evader respectively. The case where eigenvalues of the matrices C_i are real is considered. If the total resource of controls of the pursuers is greater than that of the evader, then under certain conditions it was proved that pursuit can be completed.

In Ibragimov's paper [2] a pursuit differential game of m pursuers and k evaders with integral constraints described by the following systems of differential equations

$$\dot{z}_{ij} = C_{ij}z_{ij} + u_i - v_j, \quad z_{ij}(t_0) = z_{ij}^0,$$

$$i = 1, ..., m, \quad j = 1, ..., k,$$

$$\int_0^\infty |u_i(s)|^2 ds \le {\rho_i}^2, i = 1, ..., m, \int_0^\infty |v_j(s)|^2 ds \le {\sigma_j}^2, j = 1, ..., k,$$

where u_i is the control parameter of the ith pursuer, and v_j is that of the jth evader, was examined. Here, the eigenvalues of matrices C_{ij} are not necessarily real, and moreover, there can be any number of evaders. Under the assumption that the total resource of controls of the pursuers is greater than that of the evaders, that is

$$\rho_1^2 + \rho_2^2 + ... + \rho_m^2 > \sigma_1^2 + \sigma_2^2 + ... + \sigma_k^2,$$

and real parts of all eigenvalues of the matrices C_{ij} are nonpositive, it was proved that pursuit can be completed from any initial position.

The differential game of one pursuer and one evader with integral constraints studied by Ibragimov [3] occurs on a closed convex subset S of $\mathbb{R}^n$ and dynamics of the players are described by the following equations

$$\dot{x} = \alpha(t)u, x(0) = x_0, \int_0^\infty |u(s)|^2 ds \le \rho^2,$$

$$\dot{y} = \alpha(t)v, y(0) = y_0, \int_0^\infty |v(s)|^2 ds \le \sigma^2.$$

In this paper, evasion and pursuit problems are investigated. In the latter case, a formula for optimal pursuit time was found and optimal strategies of the players were constructed.

Leong and Ibragimov's paper [4] was devoted to the differential game of m pursuers and one evader with simple motions on a closed convex subset of the Hilbert space l_2. Constraints are imposed on the control functions of the players integral. It is assumed that the total resource of the pursuers is greater than that of the evader. Sufficient conditions for completion game were obtained and the strategies of pursuers were constructed.

Ibragimov and Satimov [5] studied a pursuit differential game of many pursuers and many evaders on a non-empty convex subset N of $\mathbb{R}^n, n \geq 2$. In process of the game, all players must not leave the set N. Control functions of the players are subjected to integral constraints. If the total resource of the pursuers is greater than that of the evaders, then it was shown that pursuit can be completed.

In the present paper, we consider a differential game of m pursuers and one evader with coordinate-wise integral constraints. The game occurs in a rectangle in the plane. We have obtained sufficient conditions of completion of the game. We also construct strategies for the pursuers.

2 Statement of the Problem

We consider a differential game described by the following equations

$$\begin{aligned} &\dot{x}_i = u_i, \quad x_i(0) = x_{i0}, \quad i = 1, ..., m, \\ &\dot{y} = v, \quad y(0) = y_0, \end{aligned} \tag{1}$$

where $x_i, u_i, y, v \in \mathbb{R}^2, u_i$ is control parameter of the pursuer $x_i, i = 1, \dots m$, and v is that of the evader y.

Definition 1 A Borel measurable function $u_i(t) = (u_{i1}(t), u_{i2}(t)),\ t \geq 0$, is called admissible control of the pursuer x_i if

$$\int_0^{\infty} |u_{ij}(s)|^2 ds \leq \rho_{ij}^2, \tag{2}$$

where $\rho_{ij}, i = 1, ..., m; j = 1, 2$, are given positive numbers.

Definition 2 A Borel measurable function $v(t) = (v_1(t), v_2(t)),\ t \geq 0$, is called an admissible control of the evader if

$$\int_0^{\infty} |v_j(s)|^2 ds \leq \sigma_j^2, \tag{3}$$

where $\sigma_j, j = 1, 2$, are given positive numbers.

Definition 3 A Borel measurable function $U_i(x_i, y, v)$=$(U_{i1}(x_i, y, v), U_{i2}(x_i, y, v))$, $U_i : \mathbb{R}^6 \to \mathbb{R}^2$, is called a strategy of the pursuer x_i if for any control of the evader $v(t), t \geq 0$, the initial value problem

$$\begin{aligned} &\dot{x}_i = U_i(x_i, y, v(t)), \quad x_i(0) = x_{i0}, \\ &\dot{y} = v(t), \quad y(0) = y_0, \end{aligned}$$

has a unique solution $(x_i(t),\ y(t))$ and the inequalities

$$\int_0^\infty |U_{i1}(x_i(s), y(s), v(s))|^2 ds \le \rho_{i1}^2, \int_0^\infty |U_{i2}(x_i(s), y(s), v(s))|^2 ds \le \rho_{i2}^2$$

hold.

Definition 4 We say that pursuit can be completed from the initial position $\{x_{10}, ..., x_{m0}, y_0\}$ for the time T in the game (1)–(3), if there exist strategies $U_i, i = 1, ..., m$, of the pursuers such that for any control $v = v(\cdot)$ of the evaders the equality $x_i(t) = y(t)$ holds for some $i \in \{1, ..., k\}$ at some time $t \in [0, T]$.

Given a rectangle N with the width a and height b, according to the rule of the game all players must not leave the rectangle N. This information describes a differential game of many players with integral constraints on control functions of players.

Problem Find a sufficient condition of completing pursuit in the game (1)–(3).

3 Main Result

We now formulate the main result of the paper.

Theorem *If the inequality*

$$\rho_{1j}^2 + \rho_{2j}^2 + ... + \rho_{mj}^2 > \sigma_j^2 \tag{4}$$

holds for some $j \in \{1,2\}$, then pursuit can be completed for a finite time T in the game (1)–(3) from any initial position.

Proof We prove the theorem when the inequality (4) holds at $j = 1$. The same reasoning applies to the case $j = 2$. Thus,

$$\rho_{11}^2 + \rho_{21}^2 + ... + \rho_{m1}^2 > \sigma_1^2.$$

Let

$$\sigma_{i1} = \frac{\sigma_1}{\rho_1}\rho_{i1}, \quad i = 1, 2, ..., m;\ \rho_1 = (\rho_{11}^2 + \rho_{21}^2 + ... + \rho_{m1}^2)^{1/2}.$$

It is clear that $\sigma_{i1} < \rho_{i1}$.

The strategy of each pursuer is constructed in two stages. In the first stage, the pursuer moves with constant speed along a horizontal line to become on one vertical line with the evader. In the second stage of his strategy the pursuer moves using P-strategy.

Let us construct a strategy for the pursuer $x_i, i = 1, ..., m$. This pursuer moves according to the strategy to be constructed on an interval $[\theta_i, \theta_{i+1}]$; $\theta_1 = 0$. We define θ_{i+1} inductively below. Set

$$u_{i1}(t) = \mathrm{sgn}(y_1(\theta_i) - x_{i1}(\theta_i))\frac{a}{t_{i1}}, \quad u_{i2}(t) = 0, \tag{5}$$

$$\theta_i < t \le \theta_i + \tau_{i1}, \quad i = 1, ..., m,$$

where $\tau_i, 0 \le \tau_i \le t_{i1}$, is a time for which $x_{i1}(\theta_i + \tau_{i1}) = y_1(\theta_i + \tau_{i1})$. Clearly, such a time τ_{i1} exists. We now construct the second part of the strategy of the pursuer x_i. Set

$$u_{i1}(t) = v_1(t), \quad u_{i2}(t) = \text{sgn}(y_2(\theta_i + \tau_{i1}) - x_{i2}(\theta_i + \tau_{i1}))\frac{b}{t_{i2}},$$
$$\theta_i + \tau_i < t \le \theta_{i+1}, \tag{6}$$

where

$$t_{i2} = \frac{b^2}{\rho_{i2}^2}, \quad \theta_{i+1} = \theta_i + \tau_{i1} + t_{i2}.$$

On the time interval $[\theta_i, \theta_{i+1}]$ all the other pursuers $x_j, j \in \{1, ..., i-1, i+1, ..., m\}$ do not move. In other words,

$$u_j(t) \equiv 0, \quad j \in \{1, ..., i-1, i+1, ..., m\}, \quad t \in [\theta_i, \theta_{i+1}].$$

Note that

$$x_{i1}(t) = y_1(t), \quad \theta_i + \tau_{i1} \le t \le \theta_{i+1},$$

whenever

$$\int_{\theta_i+\tau_{i1}}^{\theta_{i+1}} v_1^2(t)dt \le \sigma_{i1}^2 \tag{7}$$

since under this condition we have

$$\int_{\theta_i}^{\theta_{i+1}} u_{i1}^2(t)dt = \int_{\theta_i}^{\theta_i+\tau_{i1}} u_{i1}^2(t)dt + \int_{\theta_i+\tau_{i1}}^{\theta_{i+1}} u_{i1}^2(t)dt$$
$$= \frac{a^2}{t_{i1}} + \int_{\theta_i+\tau_{i1}}^{\theta_{i+1}} v_1^2(t)dt \le \rho_{i1}^2 - \sigma_{i1}^2 + \sigma_{i1}^2 = \rho_{i1}^2.$$

Therefore, if (7) holds, then the pursuer x_i is able to apply the strategy (6) that ensures the equality

$$x_{i2}(\theta_i + \tau_{i1} + \tau_{i2}) = y_2(\theta_i + \tau_{i1} + \tau_{i2})$$

for some $0 \le \tau_{i2} \le t_{i2}$. If (7) fails to hold for a control of the evader $v(t) = (v_1(t), v_2(t))$, then pursuit may not be completed by the pursuer x_i. However, (7) holds true at least for one value of $i \in \{1, ..., m\}$. Indeed, if

$$\int_{\theta_i+\tau_{i1}}^{\theta_{i+1}} v_1^2(t)dt > \sigma_{i1}^2$$

for all $i \in \{1, ..., m\}$, then we have

$$\int_0^{\theta_{m+1}} v_1^2(t)dt = \sum_{i=1}^{m} \int_{\theta_i}^{\theta_{i+1}} v_1^2(t)dt$$
$$\geq \sum_{i=1}^{m} \int_{\theta_i+\tau_{i1}}^{\theta_{i+1}} v_1^2(t)dt > \sum_{i=1}^{m} \sigma_{i1}^2 = \sigma_1^2,$$

which is a contradiction. Thus, the inequality (7) holds at least for one value of i and then pursuit is completed by the pursuer x_i. The proof is complete.

4 Conclusion

We have obtained a sufficient condition (4) of completion the pursuit in a differential game of m pursuers and one evader with coordinate-wise integral constraints. We have constructed strategies of the pursuers and showed that pursuit can be completed from any initial position in N. Moreover, we have constructed strategies for the pursuers. This game problem can be developed by considering the cases where $N$is any subset of $\mathbb{R}^n$.

Acknowledgments This research was partially supported by the Research Grant (RUGS) of the Universiti Putra Malaysia, No. 05-02-12-1868RU.

References

1. Satimov, N.Yu., Rikhsiev, B.B., Khamdamov, A.A.: On a pursuit problem for n-person linear differential and discrete games with integral constraints. Math. USSR Sb. **46**(4), 456–469 (1983)
2. Ibragimov, G.I.: Differential game of many persons with integral constraints on controls of players. Izv. Vuz. Mat. **1**(500), 48–52 (2004)
3. Ibragimov, G.I.: A game problem on a closed convex set. Siberian Adv. Math. **12**(3), 16–31 (2002)
4. Leong, W.J., Ibragimov, G.I.: A multiperson pursuit problem on a closed convex set in hilbert space. Far East J. Appl. Math. **33**(2), 205–214 (2008)
5. Ibragimov, G.I., Satimov, N.Yu.: A multiplayer pursuit differential game on closed convex set with integral constraints. Abstr. Appl. Anal. **2012**, 460171 (2012). doi:10.1155/2012/460171
6. Azimov, A.Ya.: Linear differential pursuit game with integral constraints on the control. Diff. Equat. **11**, 1283–1289 (1975)
7. Azamov, A.A., Samatov, B.: π-strategy. An elementary introduction to the theory of differential games. NUU Press, Tashkent (2000)
8. Chikrii, A.A., Belousov, A.A.: On linear differential games with integral constraints. Tr. Inst. Mat. Mekh. UrO RAN **15**(4), 290–301 (2009)
9. Gusiatnikov, P.B., Mohon'ko, E.Z.: On l_∞-escape in a linear many-person differential game with integral constraints. J. Appl. Math. Mech., **44**(4), 436–440 (1980)

10. Ibragimov, G.I.: On a Multiperson pursuit problem with integral constraints on the controls of the players. Math. Notes. **70**(2), 181–191 (2001)
11. Ibragimov, G.I.: An optimal pursuit problem in systems with distributed parameters. Prikl. Mat. Mekh. **66**(5), 753–759 (2002)
12. Ibragimov, G.I., Azamov, A.A., Khakestari, M.: Solution of a linear pursuit-evasion game with integral constraints. ANZIAM J. **52**(E), E59–E75 (2011)
13. Ibragimov, G.I., Salimi, M., Amini, M.: Evasion from many pursuers in simple motion differential game with integral constraints. Eur. J. Oper. Res. **218**(2), 505–511 (2012). doi:10.1016/j.ejor.2011.11.02614.
14. Krasovskii, N.N.: The Theory of Motion Control. Nauka, Moscow (1968)
15. Mesencev, A.V.: A sufficient condition for linear evasion games with integral constraints. Dokl. Akad. Nauk SSR, **218**(5), 1021–1023 (1974)
16. Nikolskii, M.S.: The direct method in linear differential games with integral constraints. Controlled systems, IM, IK, SO AN SSSR, **2**, 49–59, (1969)
17. Pshenichnii, B.N., Onopchuk, Yu.N.: Linear differential games with integral constraints. Izv. Akad. Nauk SSSR Tekhnicheskaya Kibern. 1: 13–22 (1968)
18. Satimov, N.Yu., Rikhsiev, B.B.: Methods of Solving of Evasion Problems in Mathematical Control Theory. Fan, Tashkent (2000)
19. Ushakov, V.N.: Extremal strategies in differential games with integral constraints. PMM-J. Appl. Math. Mec. **36**(1):15–23 (1972)

A Triangular Stochastic Facility Layout Problem in a Cellular Manufacturing System

Shahram Ariafar, Zahra Firoozi and Napsiah Ismail

Abstract Volatility of manufacturing environments decreases the performance of the system by degrading the efficiency and effectiveness of the layout. In this paper, in order to simulate the effects of variability of demand of products on the layout of facilities in a cellular manufacturing system (CMS), a triangular stochastic facility layout model in a CMS has been developed. To validate the model, two solution approaches have been applied; a developed enumeration method, and also Lingo 12.0 optimization software. Solving the model for a case shows that variation in demand of products may lead to a change in the layout of the facilities.

1 Introduction

Today's competitive and highly volatile market is redefining the way companies do business. The rapid introduction of new products, changes in technology, customization of the products and short product life cycle constantly results in instability in the demand of products [1]. Companies are being called upon to deliver better products faster and at less cost for customers who are more demanding in a market. To be competitive, manufacturing enterprises should have the capability to respond to uncertainties of manufacturing environments to be able to remain sustainable in competitiveness. In such situation, traditional production systems such as job shops and flow lines cannot fulfill the requirement of manufacturing systems in terms of flexibility and efficiency in production of mid-volume and mid-variety products. Group technology (GT), although not a new concept, is a response to cope with the recent challenging requirements of manufacturing environments [2].

Group Technology is a manufacturing philosophy that attempts to identify, and economically exploit similarities in product design and processing. Cellular manufacturing system (CMS) is a successful application of group technology [3]. In a

S. Ariafar (✉)
Department of Industrial Engineering, Faculty of Engineering,
Shahid Bahonar University of Kerman, Kerman, Iran
e-mail: aria@uk.ac.ir

Z. Firoozi · N. Ismail
Department of Mechanical and Manufacturing Engineering, Universiti Putra Malaysia,
43400 Serdang (UPM), Selangor, Malaysia

A. Kilicman et al. (eds.), *International Conference on Mathematical Sciences and Statistics 2013*, DOI 10.1007/978-981-4585-33-0_6,

CMS, parts are grouped based on their similarities in their characteristics and operation requirements as part families, and are assigned to a group of different machines to be produced [4]. The main aim of implementation of a CMS is to achieve a higher level of manufacturing integration by dividing the whole production environment into several subsystems [5]. The reported benefits of GT are reduction in setup time, throughput time, work in process and material handling cost, improvement in scheduling and control of the system, and increase in the workers satisfaction and productivity [6].

In the design of a CMS, two important stages are involved [7]. The first stage is how to group parts to part families and assign different function machines to produce one or more of these part families, which is named as cell formation in the literature. The second stage is the facility layout problem. In the design of a CMS, layout problem aims to arrange machines within machine cells (inter-cell), and machine cells in the shop-floor (intra-cell). Facility layout has a direct impact on the performance of every manufacturing system. Moreover, it is a long term decision, and costly proposition and any change in the arrangement of facilities provides a lot of expenses that prohibit regular rearrangement of facilities within the system [8]. However, in contrast to the importance of the layout problem in cellular manufacturing systems, it has rarely absorbed the attention of researchers in comparison to the cell formation problem [9], [10].

In the layout design of manufacturing systems, there are a lot of uncertain parameters that any change in each of them can decline the performance of the system, and lead to the inefficiency of the layout. Hence, to improve the efficacy of the layout, several methods have been developed. Dynamic facility layout problem (DFLP), stochastic facility layout problem (SFLP) and use of fuzzy sets in the development of layout models are the most popular methods that have been applied in the literature to take into account the variability and uncertainty of manufacturing environments.

DFLP considers a multi-time period for the design of manufacturing facilities. It tries to find the best arrangement of facilities based on the material flow and the cost of the rearrangement of facilities in each time period. Rosenblatt [11] was the first who introduced the DFLPs, and extended the planning horizon from a single-period to a multi-period horizon. He developed a heuristic and an optimal procedure to solve the problem [11]. Literature in the DFLP is abundant and extensive (e.g. see [12], [13], [14]). However, in DFLP, traditionally, the arrangement of facilities is based upon the product flow which should be estimated by forecasting.

Using the fuzzy theory is another approach to involve the uncertainty of the system. The main aim of applying fuzzy sets is to lead to a crisp solution from uncertain and vague parameters. Aiello and Enea (2001) proposed a fuzzy based facility layout problem to take into account the uncertainty of demand. Their model considers a limitation on the capacity of facilities [15]. In another study, Enea et al. [16] took into account the uncertainty of the products by considering a fuzzy-based facility layout model for an unequal area problem. They solved the model by a genetic algorithm solution methodology [16].

Another approach that can consider the uncertainty of manufacturing environments is the SFLP. In this method, to involve the uncertainty of the system, the probability distribution function of the uncertain parameters is taken into account. Palekar et al. [17] developed a stochastic facility layout model in a dynamic environment. They considered a probability associate with the product mix/demand and a rearrangement cost for the layout relocation and minimized the sum of expected material handling cost and rearrangement cost. They developed an exact method for small size cases, and a heuristic algorithm for larger test problems [17].

Uncertainties and changes in the demand of products also have been considered in previous researches in cellular manufacturing systems. Wang et al. (2001) developed a facility layout model in a CMS that considered demand varies over the life cycle of products. They developed a simulated annealing based algorithm to solve the model [18]. Tavakkoli-Moghaddam et al. [19] developed a stochastic facility layout model in a CMS. They considered that demand has a Normal distribution function, and formulated a mathematical model that minimized inter-cell and intra-cell material handling cost of the system. They linearized their model and solved it by Lingo optimization software [19].

Suer et al. [20] proposed a methodology that designs dedicated, shared and reminder cells for a probabilistic demand in a cellular manufacturing system. They compared their approach with the classical cellular manufacturing system to show the efficiency of their method [20]. Ariafar et al. in earlier studies, developed inter-cell and intra-cell stochastic facility layout models in cellular manufacturing systems with Uniform and Normal distribution functions that consider non-predefined locations for machine cells [21], [22].

Due to the complexity and uncertainty of manufacturing environments, formulating a model for layout problems in a manufacturing system that takes into account the uncertainty of demand, is quite prominent. Hence, the main aim of this paper is to develop a mathematical model for layout design in a cellular manufacturing system that takes into account the uncertainty in demand by considering a triangular distribution function for the demand of products. Since the model is nonlinear, two solution approaches have been applied to solve the problem. One of the approaches is the use of Lingo 12.0 optimization software to solve the problem. Another approach is to develop an enumeration method to solve the model.

2 Model Formulation

In this section, a mathematical model for facility layout design in a cellular manufacturing system is proposed that considers the demand has a triangular stochastic distribution.

2.1 Assumptions

In the development of the model, a few assumptions have to be made as follows:

- The cell formation stage is first completed, i.e. the type of facilities for each regular cell and reminder cell has been already specified and known.
- Parts are moved in batches between facilities, and the unit distance cost between each of the machines is identified.

- The layout of facilities in each cell is considered as U-shaped, i.e. the material handling system flows through a U-shape layout and the width and length of layout are known beforehand.
- The size of equipments is considered to be equal and there is not any restriction for the space and shape of the shop-floor.
- The distance between every two facilities is measured from the center of one facility to the center of the other.
- The loading point for each machine cell is at the entrance of the cell and the unloading point is at the exiting point of the cell.
- The loading and unloading points of facilities are at the center of each facility.
- The demand of products is assumed to have the triangular distribution function.

2.2 *Notations*

2.2.1 Indices

P	Index for products
R	Index for processes
i, k	Indices for locations
j, l	Indices for machines
m, n	Indices for machine cells

2.2.2 Parameters

n_m	Number of machines and also number of available locations in the system
n_c	Number of machine cells in the manufacturing system
n_P	Number of products
R_P	Number of processes in the Pth product
U_p	Upper bound of demand for the Pth product in the planning horizon
L_p	Lower bound of demand for the Pth product in the planning horizon
M_p	Demand median for the Pth product in the planning horizon
D_p	The amount of demand for the Pth product in the planning horizon
β	The cumulative probability of occurrence for the demand scenarios
C_{jl}	Unit distance cost between machine j and machine l
MF_{jl}	Material flow between machine j and machine l
$Dist_{ik}$	Distance between location i and location k
$NMC(C_m)$	The number of machines in the machine cell C_m

2.2.3 Variables

$$F_{PRj} = \begin{cases} 1 & \text{If product P is produced in its Rth process on machine j} \\ 0 & \text{Otherwise} \end{cases}$$

$$X_{ij} = \begin{cases} 1 & \text{If machine i is assigned to machine location j} \\ 0 & \text{Otherwise} \end{cases}$$

$$Y_{im} = \begin{cases} 1 & \text{If machine i belongs to machine cell } \mathrm{C_m} \\ 0 & \text{Otherwise} \end{cases}$$

2.2.4 Mathematical model

Assume that the demand for each of the products in the planning horizon has a triangular distribution function. The amount of demand for each product can be calculated as follows:

$$If \quad \beta \leq \frac{Mp - L_P}{U_P - L_P} \quad \Rightarrow \quad D_P = \frac{\beta(U_P - L_P)(Mp - Lp)}{2} + L_P \tag{1}$$

$$If \quad \beta > \frac{Mp - L_P}{U_P - L_P} \quad \Rightarrow \quad D_P = U_P - \frac{\beta(U_P - L_P)(U_P - Mp)}{2} \tag{2}$$

Then, material flow for the planning horizon is

$$MF_{jl} = \sum_{P=1}^{np} \sum_{R=1}^{R_P} D_P * F_{PRj} * F_{P(R+1)l}. \tag{3}$$

The mathematical model for the layout problem in cellular manufacturing systems is formulated as follows:

$$Min \sum_{i=1}^{n_m} \sum_{j=1}^{n_m} \sum_{k=1}^{n_m} \sum_{l=1}^{n_m} \sum_{m=1}^{n_c} \sum_{n=1}^{n_c} MF_{jl} C_{jl} Dist_{ik} X_{ij} X_{kl} Y_{im} Y_{kn} \tag{4}$$

$$\sum_{i=1}^{n_m} X_{ij} = 1 \qquad j = 1, \ldots, n_m \tag{5}$$

$$\sum_{j=1}^{n_m} X_{ij} = 1 \qquad i = 1, \ldots, n_m \tag{6}$$

$$\sum_{m=1}^{n_c} Y_{im} = 1 \qquad i = 1, \ldots, n_m \tag{7}$$

$$\sum_{i=1}^{n_m} Y_{im} = NMC(C_m) \qquad m = 1, \ldots, n_c \tag{8}$$

Table 1 Machine grouping data

Cell	Number of facilities	Facilities
1	3	1, 2, 3
2	2	4, 5
3	2	6, 7

Table 2 Products' demand and process sequences

Product	Triangular distribution			Production sequence
	Lower bound	Median	Upper bound	
1	50	80	90	1-3-2
2	20	25	40	1-2-4
3	25	30	40	3-4-5
4	50	60	70	5-4-7-6
5	20	30	50	5-6-7
6	10	30	50	7-3-6-5

The objective function (4) minimizes the total material handling cost on the shop-floor. Constraints (5) and (6) ensure that each machine is only assigned to one machine location, and each machine location is assigned to one machine. Then, constraints (7) and (8) make sure that each machine is only assigned to one machine cell, and to each cell assigns the same number of machines belongs to that machine cell.

3 Illustration

In this section, the proposed mathematical model is solved for a case by two solution approaches; Lingo 12.0 optimization software, and an enumeration method on a T2600 2.16 GHz with 2 GB RAM. The case includes of six products ($n_P = 6$) which are produced on seven machines ($n_m = 7$). As, it can be seen from Table 1, the system includes three machine cells ($n_c = 3$); three machines (1, 2, 3) for the first machine cell and two machines (4, 5) and (6, 7) for the second and third machine cells. The demand and production sequence for each of the products are given in Table 2.

4 Results and Discussions

In this section, the case has been solved by two approaches for different demand scenarios. Table 3 presents the obtained results in terms of the inter-cell layout, intra-cell layout and material handling cost.

The results in Table 3 show that any change in the demand scenario by varying in the value of β (cumulative probability of demand), leads to a different arrangement for facilities. This case reveals the sensitivity of the layout problems to variation in demand of products, and indicates that uncertainty in the product demand can lead to inefficiency in the layout of facilities.

Table 3 Computational results for the case

Test problem	β	Inter-cell layout		Intra-cell layout		Material handling cost
(n_p = 6)	0.3	Cells:	1-2-3	Cell 1:	1-2-3	891.8055
(n_m = 7)				Cell 2:	4-5	
(n_c = 3)				Cell 3:	7-6	
	0.5	Cells:	1-2-3	Cell 1:	1-2-3	974.7808
				Cell 2:	4-5	
				Cell 3:	6-7	
	0.7	Cells:	1-3-2	Cell 1:	1-2-3	1053.9890
				Cell 3:	7-6	
				Cell 2:	4-5	

5 Conclusion

In this paper, a triangular stochastic facility layout model was developed to determine the layout of facilities in a cellular manufacturing system. The model arranges the machine cells within the shop-floor, and the machines within each of the machine cells to minimize the material handling cost. To validate the model, a case was solved by two solution approaches, Lingo 12.0 optimization software, and an enumeration method. Solving the model for different demand scenarios, which lead to different layouts for facilities, shows that facility layout problems are sensitive to any change in the demand of products.

References

1. Padayachee, J., Bright, G., Masekamela, I.: Modular reconfigurable machine tools: design, control and evaluation. S. Afr. J. Ind. Eng. **20**, 127–143 (2009)
2. Ariafar, S., Ismail, N., Tang, S.H., Ariffin, M.K.A.M., Firoozi, Z.: Design of a facility layout model in hybrid cellular manufacturing systems under variable demand. Int. J. Ind. Syst. Eng. **9**, 373–387 (2011)
3. Rafiee, K., Rabbani, M., Rafiei, H., Rahimi-Vahed, A.: A new approach towards integrated cell formation and inventory lot sizing in an unreliable cellular manufacturing system. Appl. Math. Model. **35**, 1810–1819 (2011)
4. Ying, K.C., Lin, S.W., Lu, C.C.: Cell formation using a simulated annealing algorithm with variable neighbourhood. Eur. J. Ind. Eng. **5**, 22–42 (2011)
5. Kesen, S.E., Toksari, M.D., Güngör, Z., Güner, E.: Analyzing the behaviors of virtual cells (VCs) and traditional manufacturing systems: ant colony optimization (ACO)-based metamodels. Comput. Oper. Res. **36**, 2275–2285 (2009)
6. Ariafar, S., Ismail, N., Tang, S.H., Ariffin, M.K.A.M., Firoozi, Z.: Facility layout design for hybrid cellular manufacturing system. Int. J. Phys. Sci. **6**, 3551–3556 (2011)
7. Sangwan, K.S., Kodali, R.: FUGEN: a tool for the design of layouts for cellular manufacturing systems. Int. J. Serv. Oper. Manag. **5**, 595–616 (2009)
8. Singh S.P., Singh, V.K.: Three-level AHP-based heuristic approach for a multi-objective facility layout problem. Int. J. Prod. Res. **49**, 1105–1125 (2011)
9. Ahi, A., Aryanezhad, M.B., Ashtiani, B., Makui, A.: A novel approach to determine cell formation, intracellular machine layout and cell layout in the CMS problem based on TOPSIS method. Comput. Oper. Res. **36**, 1478–1496 (2009)

10. Ariafar, S., Ismail, N.: An improved algorithm for layout design in cellular manufacturing systems. J. Manuf. Sys. **28**, 132–139 (2009)
11. Rosenblatt, M.J.: Dynamics of plant layout. Manag. Sci. **32**, 76–86 (1986)
12. Balakrishnan, J., Cheng, C.H.: Dynamic layout algorithms: a state-of-the-art survey. Omega. **26**, 507–521 (1998)
13. Balakrishnan, J., Cheng, C.H.: Dynamic cellular manufacturing under multiperiod planning horizons. J. Manuf. Technol. Manag. **16**, 516–530 (2005)
14. Balakrishnan, J., Cheng, C.H.: Multi-period planning and uncertainty issues in cellular manufacturing: a review and future directions. Eur. J. Oper. Res. **177**, 281–309 (2007)
15. Aiello, G., Enea, M.: Fuzzy approach to the robust facility layout in uncertain production environments. Int. J. Prod. Res. **39**, 4089–4101 (2001)
16. Enea, M., Galante, G., Panascia, E.: The facility layout problem approached using a fuzzy model and a genetic search. J. Intell. Manuf. **16**, 303–316 (2005)
17. Palekar, U.S., Batta, R., Bosch, R.M., Elhence, S.: Modeling uncertainties in plant layout problems. Eur. J. Oper. Res. **63**, 347–359 (1992)
18. Wang, T.Y., Wu, K.B., Liu, Y.W.: A simulated annealing algorithm for facility layout problems under variable demand in cellular manufacturing systems. Comput. Ind. **46**, 181–188 (2001)
19. Tavakkoli-Moghaddam, R., Javadian, N., Javadi, B., Safaei, N.: Design of a facility layout problem in cellular manufacturing systems with stochastic demands. Appl. Math. Comput. **184**, 721–728 (2007)
20. Suer, G.A., Huang, J., Maddisetty, S.: Design of dedicated, shared and remainder cells in a probabilistic demand environment. Int. J. Prod. Res. **48**, 5613–5646 (2010)
21. Ariafar, S., Ismail, N., Tang, S.H., Ariffin, M.K.A.M., Firoozi, Z.: The reconfiguration issue of stochastic facility layout design in cellular manufacturing systems. Int. J. Serv. Oper. Manag. **11**, 255–266 (2012)
22. Ariafar, S., Ismail, N., Tang, S.H., Ariffin, M.K.A.M., Firoozi, Z.: A stochastic facility layout model in cellular manufacturing systems. Int. J. Phys. Sci. **6**, 3666–3670 (2011)

Constructing a Three-Stage Asymptotic Coverage Probability for the Mean Using Edgeworth Second-Order Approximation

Ali S. Yousef

Abstract In this paper we consider a three-stage procedure that was presented by Hall (Ann Stat 9(6):1229–1238, 1981) to yield a fixed-width confidence interval for the mean with a precise confidence level using Edgeworth second-order expansion assuming the underlying continuous distribution has finite but unknown six moments. The procedure is based on expanding an asymptotic second order approximation of a differentiable and bounded function of the final stage stopping rule found in Yousef et al. (J Stat Plan Inference 143(9):1606–1618, 2013) by Edgeworth expansion. The performance of the asymptotic coverage was shown to be controlled by the performance of the Edgeworth approximation for the standardized underlying density and thus sensitive to the skewness and kurtosis of the underlying standardized distribution. The impact of several parameters on the asymptotic coverage is explored under continuous classes of distributions; normal, student's t-distribution, uniform, beta and chi-squared. For brevity, simulation results are given for three types of underlying distributions: standard uniform, standard normal and standard exponential.

1 Introduction

Several sampling techniques were developed during and after the World War II to perform inference. These included Stein's [1] group sampling in two stages, Cox [2] double sampling and Anscombe [3] and Chow and Robbins's [4], one-by-one purely sequential sampling procedure. In Stein's [1] and Cox's [2] the sampling is made in two stages where observations are taken in batches over the sampling course to calculate the required estimates; hence an optimal stopping rule is checked to determine whether to stop or to continue sampling and thus update the estimates for the next stage. It is readily known that this sampling technique often leads to oversampling especially if the initial sample size chosen was much smaller than the optimal sample size required to perform inference had the population variance been known. To overcome this deficiency entailed in the two-stage sampling procedure, Anscombe [3] introduced the one-by-one purely sequential sampling procedure; Robbins [5] devised one-by-one purely sequential sampling procedure to construct a fixed-width

A. S. Yousef (✉)
College of Arts and Sciences, Middle East University, Amman, Jordan
e-mail: dr.ayousef@aol.com, Ayousef@meu.edu.jo

A. Kilicman et al. (eds.), *International Conference on Mathematical Sciences and Statistics 2013,* DOI 10.1007/978-981-4585-33-0_7,

confidence interval for the normal mean with a prescribed width and coverage. He also laid down the basic foundation of the purely sequential sampling. Chow and Robbins [4] gave a detailed account of the asymptotic characteristics of the one-by-one purely sequential sampling while constructing a fixed-width confidence interval for the normal mean with unknown variance. Although the procedure enjoys all the required asymptotic characteristics one desires, the nature of how the sampling is performed in one-by-one sequential sampling makes it impossible to implement in real-life applications, especially when the time and money are targeted design factors.

Hall [6] proposed a triple sampling sequential procedure to construct a fixed-width confidence interval for the normal mean with a prescribed width and coverage. The procedure is designed to satisfy both Anscombe [3] and Chow and Robbins [4] asymptotic efficiency and the operational savings made possible by sampling in bulks as in Stein's [1] and Cox [2] double sampling.

The idea of triple sampling fixed-width confidence interval for the mean was discussed by several authors; Vik and Mukhophadhyay [7] developed a general triple sampling procedure to construct a fixed-width confidence interval for the mean based on the characteristics of U-statistics. Mukhophadhyay and deSilva [8] discussed the fixed-width confidence interval for the difference of two independent normal means in two cases; when the variances are equal but unknown and under the Behrens–Fisher case. Hamdy [9] utilized the triple sampling procedure to treat the point estimation, confidence interval estimation and hypothesis testing in one framework for the normal mean.

To the best of our knowledge, none of the existing papers in the literature discussed the asymptotic coverage probability based on Edgeworth second-order approximation and examined its performance as the numerical values of both the skewness and kurtosis of the standardized distribution function vary. Our approach will be as follows: in Sect. 2 we state the problem and give a review about the three-stage sampling procedure as described by Hall [6], also we state Theorem 1 and Theorem 2 as mentioned in Yousef et al. [10] that facilate our main result in this paper. Finally we state Edgeworth second-order approximation formula and its validity region. In Sect. 3 we find Corollary 1, which finds the Edgeworth asymptotic coverage probability for the mean and give illustrative examples to explore the performance for both the Edgeworth approximation for the standardized density and the asymptotic coverage's as the width approaches zero. In Sect. 4, for brevity we give simulation results for only three types of underlying distributions—standard uniform, standard normal and standard exponential—and in Sect. 5 we state our conclusions.

2 Definitions and Statement of the Problem

Let $\{X_i, i \geq 1\}$ be a sequence of independent identically distributed random variables from an absolutely continuous distribution function F, with mean $\mu \in \mathbb{R}$, variance $\sigma^2 \in \mathbb{R}^+ = (0, \infty)$, skewness $\gamma \in \mathbb{R}$ and kurtosis $\beta \in (1, \infty)$, all are finite but unknown. Here, $\gamma = E((X-\mu)/\sigma)^3$ and $\beta = E((X-\mu)/\sigma)^4$. Moreover, we assume that $E|X_1|^6 < \infty$. Having recorded a sample of size $n, (X_1, \ldots, X_n)$, our objective is to construct a confidence interval for μ such that the width of the interval is $2d$ and $P(|\overline{X}_n - \mu| \leq d) \geq (1-\alpha)$, uniformly for all $(\mu, \sigma) \in \mathbb{R} \times \mathbb{R}^+, d > 0$ and $0 < \alpha < 1$, are known and fixed prior to sampling. If σ is known and n is sufficiently large, then this fixed-width confidence interval problem has a simple fixed sample size solution given by

$$n \geq n_d^* = \lambda\sigma^2 \tag{1}$$

where $\lambda = (a/d)^2$ and a is an upper $(\alpha/2)$ quantile point from $N(0,1)$. By taking $[n_d^*]+1$, (x is an integer part of x) observations, the desired confidence interval for μ is $\mu \in (\overline{X}_{n_d^*} - d, \overline{X}_{n_d^*} + d)$ has coverage probability of at least $(1-\alpha)$ and fixed-width $2d$. The quantity n_d^* is referred to as the optimal fixed sample size required constructing a fixed-width confidence interval for μ, given that σ^2 in known. Since σ in (1) is unknown, then n_d^* is unknown and it was shown by Dantzig [11] that there is exists no fixed sample size procedure which allows us to construct a size $(1-\alpha)$ confidence interval for μ with a fixed-width $2d$. Therefore, we resort to Hall's [6] three-stage procedure to estimate n_d^* via estimation of σ^2. In the following lines we give a rigorous account of the three-stage procedure as described by Hall [6].

Pilot Phase Take an initial sample of size $m \geq 2$, observations from F and calculate S_m^2, where $S_m^2 = (m-1)^{-1}\sum_{i=1}^{m}(X_i - \overline{X})^2$.

The Main Study Phase A fraction of n^* is estimated during that phase, say, δn^* where δ is a prescribed constant, $\delta \in (0,1)$. The sample size required to complete the main study phase is defined by the following stopping rule as

$$N_1 = \max\left\{\mathrm{m}, \left[\delta\lambda \mathrm{S}_{\mathrm{m}}^2\right] + 1\right\}. \tag{2}$$

If $m \geq N_1$, stop sampling at this stage; otherwise continue to observe an extra random sample of size $N_1 - m$ from F, say $X_{m+1}, X_{m+2}, \ldots, X_{N_1}$. Hence, we augment the $N_1 - m$ observations by the previous m observations and calculate $S_{N_1}^2$ as new estimates of σ^2.

The Fine Tuning Phase This is defined according to the following final stage stopping rule

$$N = \max\left\{\mathrm{N}_1, \left[\lambda \mathrm{S}_{\mathrm{N}_1}^2\right] + 1\right\}. \tag{3}$$

If $N_1 \geq N$, then stop at this stage; otherwise continue to sample $N - N_1$ more observations randomly from F, say $X_{N_1+1}, X_{N_1+2}, \ldots, X_N$. Whenever sampling is terminated and N is realized, $\overline{X}_N$ is a sequential point estimator of μ and $I_N(d) = \overline{X}_N \pm d$ is a desired confidence interval for μ. It was shown by Hall [6] that the asymptotic characteristics of the three-stage procedure was carried out under the condition

$$m = O\left(d^{-2/r}\right) \text{ as } d \to 0\,(r > 1). \tag{4}$$

This yields that:

(i) $P(\mu \in I_N(d)) = (1-\alpha) + o(d^2)$ and *(ii)* $E(N - n_d^*) = K < \infty$, K is a constant unrelated to n_d^*. Thus the three-stage procedure is asymptotically consistent and asymptotically second-order efficient. See Hall [6] and Hamdy [9]. Yousef et al. [10] generalized the results of Hall [6] to a wider class of underlying continuous distributions, where $E|X_1|^6 < \infty$. His results based on expanding a second-order differentiable and bounded function of the final stage sample size using second-order Taylor series approximation. The results are shown in Theorem 1.

Theorem 1 *Let $E|X_1|^6 < \infty$ and let N be defined as in* (3) *and assume condition* (4) *holds. Also, let $\rho_{N_1} = 1 - \{\lambda s_{N_1}^2 - [\lambda s_{N_1}^2]\}$ and $c = E(\rho_{N_1})$. Then as $d \to 0$, N is asymptotically normal with*

$$(i)\ E(N) = n^* - \delta^{-1}(\beta - 1) + c + o(1), \quad (ii)\ Var(N) = \delta^{-1}(\beta - 1)n^* + o(\lambda),$$

$$(iii)\ E(h(N)) = \left\{h(n^*) + h'(n^*)\left(c - (\beta - 1)\delta^{-1}\right) + (\beta - 1)(2\delta)^{-1} n^* h''(n^*) + o\left(\lambda^{-1}|h'''|\right)\right\},$$

where, $h(> 0)$ be a continuously differentiable real-valued function in a neighbourhood around n_d^* such that $\underset{n \geq m}{Sup} h(n) = O\left(h'''(n_d^*)\right)$. See, Yousef et al. [10], Theorems [5], [6] and [7] for the proof.

Hall [6] showed that under the normal distribution $\rho_{N_1} \to U(0,1)$ as $\lambda \to \infty$.

Theorem 2 *Let $E|X_1|^6 < \infty$ and let N be defined as in* (3) *and assume condition* (4) *holds. Then as $d \to 0$ we have*

$$(i)\ E\left(\overline{X}_N\right) = \mu - \gamma\sigma\left(n^*\right)^{-1} + o(d) \text{ and}$$

$$(ii)\ Var\left(\overline{X}_N\right) = \sigma^2/n^* - 2\left(\sigma^2/\delta n^{*2}\right)\{(\beta - 3)(\delta - 1) - 2\} + o\left(d^2\right)$$

(iii) $\overline{X}_N$ is asymptotically normal with mean and variance listed in *(i)* and *(ii)*.

Proof Parts *(i)* and *(ii)* follows from Yousef et al. [10], Theorem [3]. Part *(iii)* follows by taking the expectations over expanding a real-valued three times continuously differentiable function $h(\cdot)$ around μ using Taylor series. That is, $E\left\{h\left(\overline{X}_N\right)\right\} = h(\mu) + \sum_{k=1}^{2}\left(h^k(\mu)/k!\right)E\left(\overline{X}_N - \mu\right)^k + (1/6)E\left\{h'''(v)\left(\overline{X}_N - \mu\right)\right\}^3$, v lies between $\overline{X}_N$ and μ. Now, by taking $h(\overline{X}_N) = \exp(t(\overline{X}_N - \mu)/\sigma_{\overline{X}_N})$ and using *(i)* and *(ii)* we get $M_{\overline{X}_N}(t) \approx 1 + (t^2/2), d \to 0$. To show that $R = E\left\{h'''(v)\left(\overline{X}_N - \mu\right)\right\}^3$ vanishes as $d \to 0$, let $Z = X - \mu$ and consider the case $\overline{X}_N < v < \mu, t > 0 \Rightarrow |R| \leq t^3\sigma_{\overline{X}_N}^{-3} E\left|N^{-1}\sum_{i=1}^{N} Z_i\right|^3 \leq t^3\sigma_{\overline{X}_N}^{-3} m^{-3} E\left|\sum_{i=1}^{N} Z_i\right|^3$. But since, $N^{-1/2}\sum_{i=1}^{N} Z_i \to N(0, \sigma^2)$ as $d \to 0$ (Anscombe [12]), $\left\{\left|N^{-1/2}\sum_{i=1}^{N} Z_i\right|^3\right\}$ is uniformly integrable, Chow and

Yu [13], Lemmas [2], [4] and [5] then it follows $E\left|N^{-1/2}\sum_{i=1}^{N} Z_i\right|^3 = 2\sigma^3\sqrt{2/\pi}$. By using (4) and the fact that $N/n^* \to 1$ as $d \to 0$, we get the result. Similar justification can be performed for the other case. Proof is complete.

In the next lines we state Edgeworth second-order approximation formula and its validity region.

Let Φ and ϕ stand for the distribution and density function of $N(0, 1)$ then the Edgeworth second-order approximation for a standardized distribution F (DasGupta [14], Theorem [13].[1], p. 186) is

$$F(x) = \Phi(x) - \frac{1}{72}\phi x(x^4 - 10x^2 + 15)\gamma^2 - \frac{1}{6}\gamma\phi(x^2 - 1) \\ - \frac{1}{24}\phi x(x^2 - 3)(\beta - 3) + O(1). \tag{5}$$

If the distribution function F of an absolutely continuous random variable admits an Edgeworth expansion, then we can obtain an expansion of the density function heuristically by differentiating (5) with respect to x. Hence the probability density function of the Edgeworth expansion is

$$f(x) = \phi(x)\begin{Bmatrix} 1 + (\gamma/6)\, x\left(x^2 - 3\right) + ((\beta - 3)/24)\left(x^4 - 6x^2 + 3\right) \\ + \left(\gamma^2/72\right)\left(x^6 - 15x^4 + 45x^2 - 15\right) \end{Bmatrix}. \tag{6}$$

Draper and Tierney [15] found the numerical values for both γ and β (Note that γ and β are the skewness and kurtosis for the standardized distribution) where (6) behaves as a positive definite and unimodal probability density function, the numerical values are $\{(\gamma, \beta), |\gamma| \le 9/20, 3 \le \beta \le 107/20\}$. If the numerical values are outside the validity region then the results may be misleading (as we will see in the forthcoming examples). For more details about Edgeworth expansion sees, Bhattacharya and Ghosh [16], Barndorff-Nielsen and Cox [17] and Hall [18]. Now, we are ready to construct the Edgeworth three-stage asymptotic coverage probability for the mean under the asymptotic independence between $\overline{X}_N$ and N.

3 Three-Stage Asymptotic Coverage Probability for the Mean

The following Corollary is the main result in this paper.

Corollary 1 Let Theorem 1 hold and let F be an absolutely continuous distribution function that satisfies all regularity conditions to admit an Edgeworth expansion. Then if $\overline{X}_N$ and N are asymptotically independent as $d \to 0$, then the three-stage asymptotic coverage probability (ACP) for the mean under (1) as $d \to 0$, is

$$P(\mu \in I_N) = (1 - \alpha) - \tfrac{a\phi(a)}{4\delta n^*}\left\{v_1 + \tfrac{a^2 v_3}{72} + \tfrac{a^4 v_5}{36} - \tfrac{a^6 v_7}{72} + \tfrac{a^8}{72}\gamma^2(\beta - 1)\right\} + o(d^2),$$
$$v_1 = (5(\beta - 1) - 4\delta c) - (\beta - 3 - 5/3\gamma^2)((n^* + c/2)\delta - (5/8)(\beta - 1)).$$

$$v_3 = \left\{\begin{matrix} 72(\beta-1) - (10\gamma^2 - 3(\beta-3))(12\delta c + 8\delta n^* - 9(\beta-1)) \\ -6(\beta - 1 - 2\delta c)(3(\beta-3) - 5\gamma^2) \end{matrix}\right\}.$$
$$v_5 = \left\{\begin{matrix} (\beta - 1 + 2\delta c)(10\gamma^2 - 3(\beta-3)) - (3/2)(\beta-1)(3(\beta-3) - 5\gamma^2) \\ +4\gamma^2((n^* + 5c/2)\delta - 5/8(\beta-1)) \end{matrix}\right\}.$$
$$v_7 = (\beta-1)(10\gamma^2 - 3(\beta-3)) + 2\gamma^2(3(\beta-1) + 2\delta c).$$

Proof

$$\begin{aligned} P\left(\mu \in I_N\right) &= P\left(\left|\overline{X}_N - \mu\right| \le d\right) \\ &= \sum_{n=m}^{\infty} P\left(\left|\overline{X}_N - \mu\right| \le d,\ N = n\right) \\ &= \sum_{n=m}^{\infty} P\left(\left|\overline{X}_N - \mu\right| \le d \,|N = n\right) P\left(N = n\right) \end{aligned}$$

Since $\overline{X}_N$ and N are asymptotically independent. Then, it follows

$$\begin{aligned} P\left(\mu \in I_N\right) &= \sum_{n=m}^{\infty} P\left(\left|\overline{X}_n - \mu\right| \le d\right) P\left(N = n\right) \\ &= \sum_{n=m}^{\infty} P\left(-d \le \overline{X}_n - \mu \le d\right) P\left(N = n\right) \\ &= E_N\left(F_{Z_N}\left(d\sqrt{N}/\sigma\right)\right) - E_N\left(F_{Z_N}\left(-d\sqrt{N}/\sigma\right)\right) \end{aligned}$$

Let $u_N = u = \sqrt{N}/\sigma$. Then from (5) we have

$$\begin{aligned} F_{Z_N}(du_N) - F_{Z_N}(-du_N) &= \Phi(du_N) - \Phi(-du_N) - \frac{1}{72}\gamma^2(2\phi(du_N)d^5)u_N^5 \\ &\quad +(2\phi(du_N)d^3)u_N^3\left(-\frac{1}{24}(\beta-3) + \frac{5}{36}\gamma^2\right) \\ &\quad +(2\phi(du_N)d)u_N\left(\frac{1}{8}(\beta-3) - \frac{5}{24}\gamma^2\right) + O(1). \end{aligned}$$

Setting $\eta_k = (2\phi(du_N)d^k)u_N^k$ and taking the expectation we obtain

$$\begin{aligned} &E_N(F_{Z_N}(du_N) - F_{Z_N}(-du_N)) \\ &= E_N(\Phi(du_N) - \Phi(-du_N)) - \frac{1}{72}\gamma^2 E_N(\eta_5) + \left(-\frac{1}{24}(\beta-3)\right. \\ &\quad \left. + \frac{5}{36}\gamma^2\right) E_N(\eta_3) + \left(\frac{1}{8}(\beta-3) - \frac{5}{24}\gamma^2\right) E_N(\eta_1) + O(d^2). \end{aligned} \tag{7}$$

To find $E_N(\Phi(du_N) - \Phi(-du_N))$, take $h(N) = \Phi(du_N) - \Phi(-du_N)$ and apply Theorem 1 and then substituting for each $\sigma^2 = n^* d^2/a^2$ we got

$$E_N(\Phi(du_N) - \Phi(-du_N)) = (1-\alpha) - \frac{a\phi(a)}{4\delta n^*}\left\{(\beta-1)a^2 + (5(\beta-1) - 4\delta c)\right\}. \quad (8)$$

Similar procedure for the remaining terms yields the following results

$$E(\eta_1) = \frac{a\phi(a)}{4\delta n^*}\left\{a^4(\beta-1) + 2a^2(\beta-1-2\delta c) + 8\left(\left(n^*+c/2\right)\delta - 5/8(\beta-1)\right)\right\} \quad (9)$$

$$E(\eta_3) = \frac{a^3\phi(a)}{4\delta n^*}\left\{a^4(\beta-1) - 2a^2(\beta-1+2\delta c) + 8((n^*+3c/2)\delta - 9/8(\beta-1))\right\} \quad (10)$$

$$E(\eta_5) = \frac{a^5\phi(a)}{4\delta n^*}\left\{a^4(\beta-1) - 2a^2(3(\beta-1)+2\delta c) + 8((n^*+5c/2)\delta - 5/8(\beta-1))\right\}. \quad (11)$$

Substituting (8)–(11) in (7) we obtain our result. The proof is complete.

Clearly, the asymptotic coverage probability is sensitive to the values of γ, β, δ, c, and n_d^*, while it is robust as $d \to 0$. Note that (8) represents the asymptotic coverage probability for the normal mean.

To explore the performance of the asymptotic coverage probability found in Corollary 1 we consider several types of underlying distributions to cover symmetric and skewed distributions and light and heavy tailed distributions. The featured underlying distributions are: $N(0,1)$, $U(0,1)$, $t(r)$, with $r = 5, 10$ and 20, beta $(4,4)$ and Chi-squared distribution with $r = 2, 5, 10$ and 30.

3.1 *Illustrative Examples*

Let $Cg(\delta) \triangleq$ the asymptotic coverage probability calculated from Corollary 1 at $\delta = 0.5$. Let the optimal sample sizes $n^* = 24, 43, 61, 76, 96, 125, 171, 246$ and 500. These values of n^* are the same as those used by Hall [6] and represent small, medium and large sample sizes. Although our results are valid for all values of δ we shall concentrate only on the case $\delta = 0.5$ since this has been recommended previously for practical reasons (see Hall [6] and Mukhophadhyay and deSilva [19], p. 122).

3.1.1 Normal Distribution

The ACP for the mean of the normal distribution is

$$P(\mu \in I_N) \approx (1-\alpha) - \frac{a\phi(a)}{2\delta n^*}\left\{a^2 + (5-\delta)\right\} < (1-\alpha)$$

Table 1 The ACP under the normal distribution

Optimal sample size	*EN*	$1-\alpha=\mathbf{0.95}$	$1-\alpha=\mathbf{0.99}$
24	20.5	0.9102	0.9727
43	39.5	0.9278	0.9804
61	57.5	0.9343	0.9832
76	72.5	0.9374	0.9845
96	92.5	0.9401	0.9857
125	121.5	0.9424	0.9867
171	167.5	0.9444	0.9876
246	242.5	0.9461	0.9883
500	496.5	0.9481	0.9892

From Table 1 we see that $E(N) < n^*$ (early stopping on average) for all values of n^* and thus the quantity $|E(N) - n^*|$ decreases as n^* increases. The asymptotic coverage is less than $(1-\alpha)$ and attains it only asymptotically. To maintain a coverage probability of at least the nominal value, Hall [6] suggested taking extra sample of size $(2\delta)^{-1}(a^2 - \delta + 5)$ (the cost of ignorance σ^2, Simons [20]) after termination of the triple sampling procedure, but his procedure leads to be impractical in practice.

3.1.2 The Student t(r) Distribution

$$
\begin{aligned}
P(\mu \in I_N) &\approx (1-\alpha) - \frac{a\phi(a)}{8\delta n^*}\left\{k_0 + k_2 a^2 + k_4 a^4 + k_6 a^6\right\},\\
k_0 &= (-4\delta + 20)\, r^2 + [-85 + (29 - 12n)\,\delta]\, r + [65 + (48n - 52)\,\delta]\,.\\
k_2 &= 4r^2 + [(4n+6)\,\delta - 35]\, r + [31 + (-16n - 24)\,\delta]\,.\\
k_4 &= (-5 - \delta)\, r + (5 + 4\delta)\,.\\
k_6 &= (r-1)\,.
\end{aligned}
$$

From (6), $f(x) = \phi(x)(4(r-4))^{-1}(x^4 - 6x^2 + 4r - 13) + O(1), \forall r > 4$ and its standardized density is $g(x) = \Gamma((r+1)/2)(\Gamma(r/2)\sqrt{\pi})^{-1}(r-2)^{r/2}(r-2+x^2)^{-(r+1)/2}$, $x \in \Re, r > 2$.

Figure 1 below explore the performance of the Edgeworth series for the standardized $t(r)$at$r = 5, 10$ and 20, respectively. Clearly the performance at $r=5$ is bad since it has a sharp kurtosis $\beta = 9$ while the performance of the Edgeworth approximation for the standardized $t(20)$ is better than for $t(20)$.

Table 2 below shows the impact of increasing r on the performance of the asymptotic coverage at 90, 95 and 99 %. Here, we consider $r=5$, 10, 20, 50 and 100 and $n^* = 500$. Clearly as r increases the asymptotic results approach the nominal values from above at 90 % and from below at 95 and 99 %.

3.1.3 Uniform Distribution

$$
\begin{aligned}
P(\mu \in I_N) &\approx (1-\alpha)\\
&\quad - \frac{a\phi(a)}{288\delta n^*}\begin{Bmatrix} 72(3.4 - 1.7\delta + 1.2n^*) + a^2(100.8 - 43.2\delta - 28.8n^*\delta) \\ + a^4(14.4 + 7.2\delta) - a^6(2.88) \end{Bmatrix}.
\end{aligned}
$$

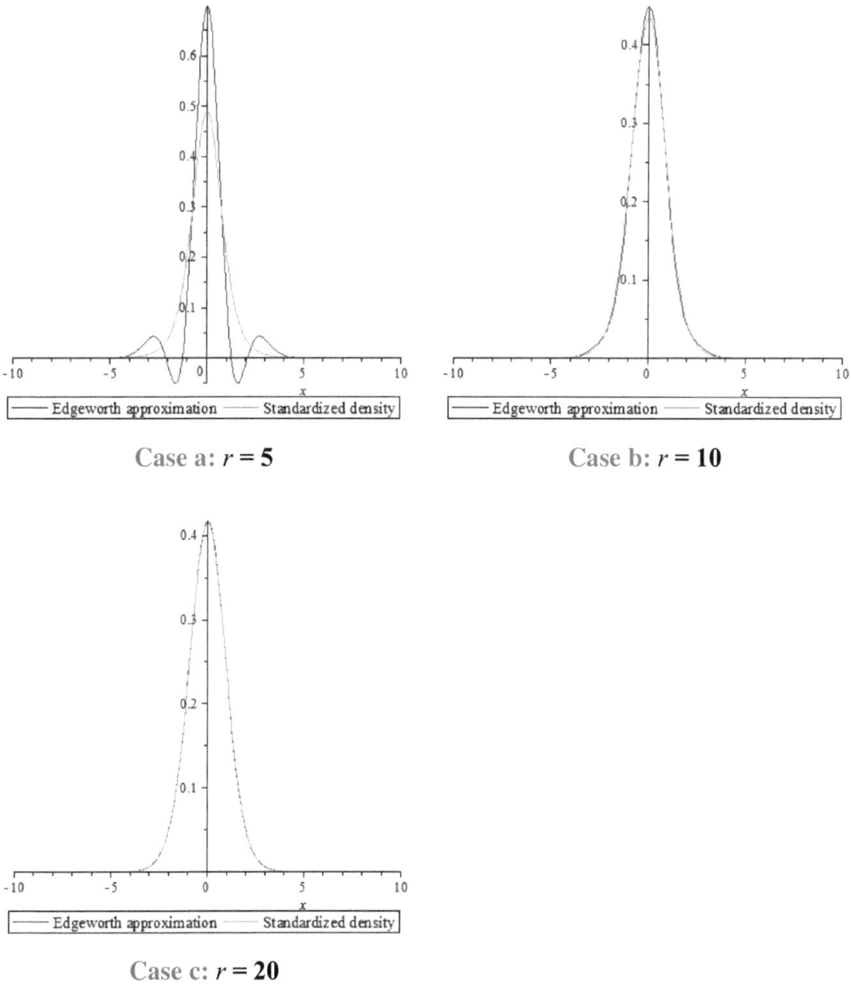

Fig. 1 The standardized $t(r)$ densities at $r = 5$, 10 and 20 and their Edgeworth approximations

From (6), $f(x) = \phi(x)(0.85 + 0.3x^2 - 0.05x^4) + O(1)$. Clearly, the performance is bad.

Table 3 below shows the asymptotic results under the uniform distribution. We noticed early stopping and the quantity $|E(N) - n^*|$ decreases as n^* increases. The asymptotic coverage's results should be treated with caution since the Edgeworth approximation is poor.

3.1.4 Beta (4,4) Distribution

From (6), $f(x) = \phi(x)\left\{1 - 0.02273(x^4 - 6x^2 + 3)\right\} + O(1)$ while its standardized density is $g(x) = 0.0005(9 - x^2)^3, |x| < 3$. Figure 2 below shows that the performance of the Edgeworth series for the standardized density is good (note the

Table 2 The ACP under t distribution as r increases

r	β	$1-\alpha=\mathbf{0.90}$	$1-\alpha=\mathbf{0.95}$	$1-\alpha=\mathbf{0.99}$
5	9	0.9249	0.9018	0.9223
10	4	0.9042	0.9420	0.9787
20	3.375	0.9016	0.9470	0.9858
50	3.130	0.9005	0.9490	0.9885
100	3.0625	0.9003	0.9495	0.9893

Table 3 The ACP under the uniform distribution

Optimal sample size	EN	$1-\alpha=\mathbf{0.95}$	$1-\alpha=\mathbf{0.99}$
24	22.9	0.94008	0.99642
43	41.9	0.94872	0.99957
61	59.9	0.95194	1.00074
76	74.9	0.95346	1.00129
96	94.9	0.95475	1.00176
125	123.9	0.95588	1.00217
171	169.9	0.95689	1.00254
246	244.9	0.95773	1.00284
500	498.9	0.95870	1.00320

Table 4 The ACP under beta (4,4)

Optimal sample size	EN	$1-\alpha=\mathbf{0.95}$	$1-\alpha=\mathbf{0.99}$
24	21.6	0.9217	0.9834
43	40.6	0.9361	0.9890
61	58.6	0.9415	0.9911
76	73.6	0.9441	0.9921
96	93.6	0.9462	0.9930
125	122.6	0.9481	0.9937
171	168.6	0.9498	0.9944
246	243.6	0.9512	0.9949
500	497.6	0.9528	0.9955

numerical value of $\beta = 27/11$ is close to its lower bound specified by its validity region). Table 4 below shows that as n^* increases the asymptotic coverage probability exceed the nominal values from below.

3.1.5 Chi-Squared Distribution

$$P(\mu \in I_N) \approx (1-\alpha) - \frac{a\phi(a)}{18r^2\delta n^*}\left\{k_0 + k_1 r + k_2 r^2\right\},$$

$$k_0 = 6a^8 - 69a^6 + 45a^4 + 315a^2 - 45.$$

$$k_1 = \left\{\begin{matrix} a^8 + a^6(-11.5-\delta) + a^4(7.5 + (4n^* + 10.5)\delta) \\ +a^2(106.5 + (-22n^* - 18)\delta) + (262.5 + (6n^* + 1.5)\delta) \end{matrix}\right\}.$$

$$k_2 = 9\left(a^2 - \delta + 5\right).$$

From (6), $f(x) = \phi(x)(18r)^{-1}(6\sqrt{2r}x(x^2-3) + 3(6r-1) + x^2(2x^4 - 21x^2 + 36)) + O(1)$ while $g(x) = \frac{\sqrt{r2^{1-r}}}{\Gamma(r/2)}\frac{(r+\sqrt{2r}x)^{r/2-1}}{e^{(r+\sqrt{2r}x)/2}}, x > -\sqrt{r/2}$.

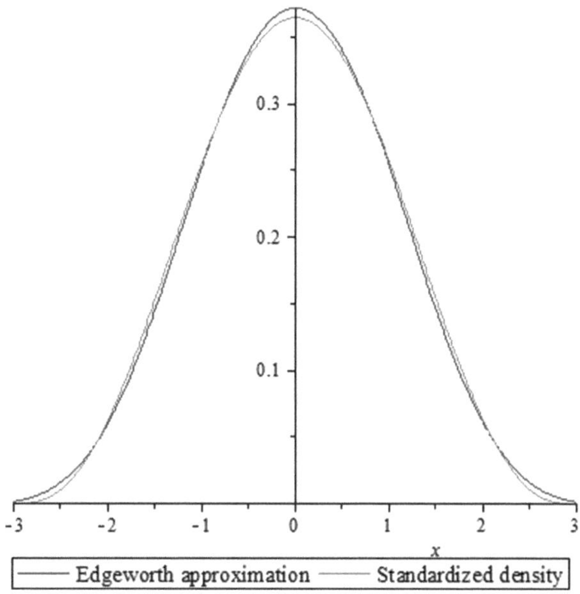

Fig. 2 The standardized beta (4,4) and its Edgeworth approximation

Table 5 The ACP for underlying chi-squared with $r = 5$, 10 and 30, $n^* = 500$ and $\delta = 0.5$

r	γ	β	$1 - \alpha = \mathbf{0.90}$	$1 - \alpha = \mathbf{0.95}$	$1 - \alpha = \mathbf{0.99}$
5	1.2650	5.4	0.9457	0.9748	0.9751
10	0.8944	4.2	0.9228	0.9624	0.9825
30	0.5164	3.4	0.9076	0.9541	0.9875

Table 6 The ACP under the standard exponential distribution

Optimal sample size	EN	$1 - \alpha = \mathbf{0.95}$
24	8.5	0.9787
43	27.5	0.9934
61	45.5	0.9989
76	60.5	1.0015
96	80.5	1.0037
125	109.5	1.0056
171	155.5	1.0073
246	230.5	1.0087
500	484.5	1.0104

From Fig. 3 above we noticed that as r increases, the performance of the Edgeworth approximation for the standardized chi-squared improved. Consequently, the performance of the asymptotic coverage improved (see, Table 5).

Table 6 below shows the asymptotic results under the standard exponential distribution. Clearly, the Edgeworth-based approach leads to poor results.

In the next section, we use simulation to study the performance of the three-stage asymptotic coverage probability for the mean under three types of underlying distributions; standard normal, standard uniform and standard exponential considering the same values of n^* and δ while we take $1 - \alpha = 0.95$ and $m = 10.0$

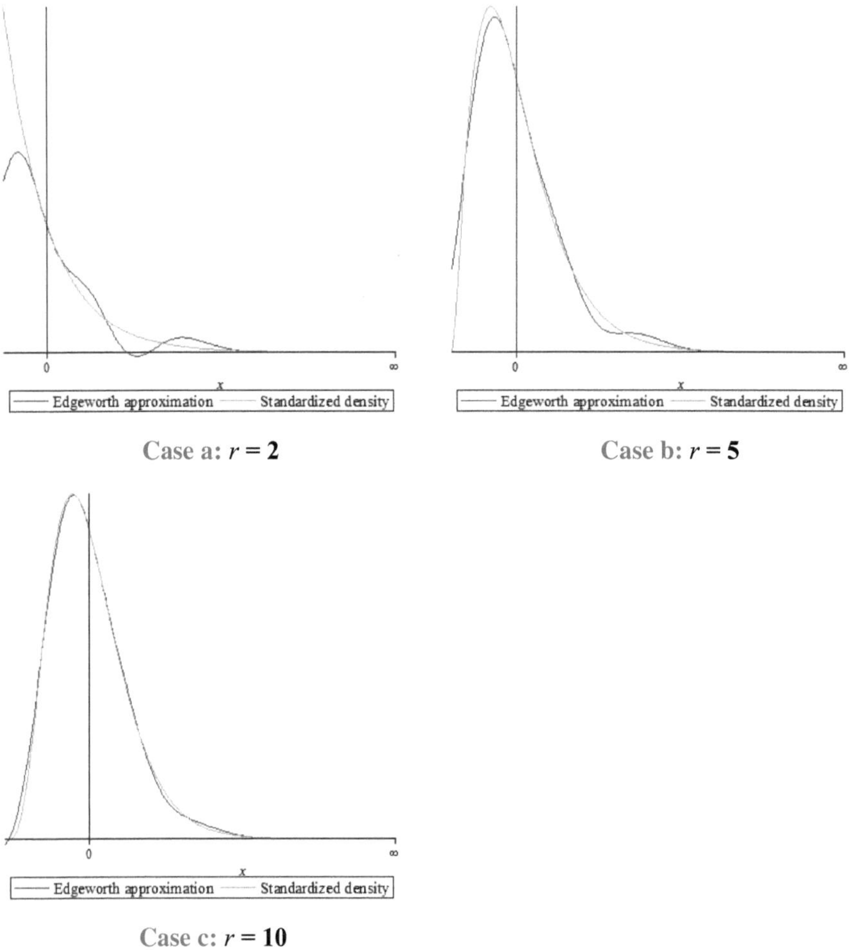

Fig. 3 The standardized chi-squared densities with $r = 2$, 5 and 10 and their Edgeworth approximations

4 Simulations and Results

A FORTRAN program is coded using Microsoft Developer Studio software to generate a series of simulations. For each experimental situation 50,000 replicate samples were used. Random samples from the above underlying distributions were generated and a three-stage sampling rule (2) and (3) was implemented to estimate the mean $\widehat{\mu}$ and its standard error S.E. $(\widehat{\mu})$, $\overline{N}$ the estimated values of n^* and its standard errors S.E. $(\overline{N})$ and finally, the estimated value of the coverage probability $(1-\widehat{\alpha})$. Now, for the i-th sample generated for a particular combination of$m$, δ and an underlying distribution we have

Table 7 Simulation results under the standard uniform

n^*	d	$\overline{N}$	$\widehat{\mu}$	$S.E\,(\mu)$	$CI = \widehat{\mu} \pm d$		$1-\widehat{\alpha}$
24	0.11549	21.29	0.50050	0.00034	0.38501	0.61599	0.8718
43	0.08628	40.70	0.50008	0.00024	0.41380	0.58636	0.9112
61	0.07244	58.70	0.50007	0.00019	0.42763	0.57251	0.9263
76	0.06490	73.70	0.49993	0.00017	0.43503	0.56483	0.9312
96	0.05775	93.95	0.49996	0.00014	0.44221	0.55771	0.9389
125	0.05061	123.13	0.49990	0.00012	0.44929	0.55051	0.9409
171	0.04327	169.09	0.49996	0.00010	0.45669	0.54323	0.9442
246	0.03607	244.27	0.49987	0.00008	0.46380	0.53594	0.9469
500	0.02530	498.43	0.50003	0.00006	0.47473	0.52533	0.9470

First Take an initial sample of size m, say $X_{1,i}, X_{2,i}, \ldots, X_{m,i}$.

Second Compute the sample mean and sample variance for the pilot sample.

Third Apply the triple sampling procedure as presented in (2) and (3) to determine the stopping sample size at this iteration whether in the first or second stage (say, N_i^*).

Fourth Record the resultant sample size and the sample mean $(N_i^*, \overline{X}_i^{\,*})$.

Hence, for each experimental combination we have two vectors of size 50,000 as follows:

Vector (1) contains $(N_1^*, N_2^*, \ldots, N_{50{,}000}^*)$, while Vector (2) contains $(\overline{X}_1^*, \overline{X}_2^*, \ldots, \overline{X}_{50{,}000}^*)$. Let $\overline{N} = \sum_{i=1}^{50{,}000} (N_i^*/50{,}000)$ and $\widehat{\mu} = \overline{X} = \sum_{i=1}^{50{,}000} (\overline{X}_i^*/50{,}000)$, where, $\overline{N}$ and $\overline{X}$ are respectively the estimated mean sample size and the estimated mean of the estimator of the population mean across replicates. Thus, $\widehat{\mu} = \overline{X}$ may be regarded as an estimate of the expected value of the estimator of the population mean μ. The standard errors are $S.E.(\widehat{\mu}) = s_1/\sqrt{50{,}000}$, where $s_1 = \sqrt{\sum_{i=1}^{50{,}000} (\overline{X}_i - \widehat{\mu})^2/49{,}999}$ and $S.E.(\overline{N}) = s_2/\sqrt{50{,}000}$, where $s_2 = \sqrt{\sum_{i=1}^{50{,}000} (\overline{N}_i - \overline{N})^2/49{,}999}$.

Table 7, Table 8 and Table 9 below reflect the simulation results for $U(0,1)$, $N(0,1)$ and $Exp(1)$, respectively. From Table 9 we noticed bad estimates for the simulated coverage's and the reason goes due to the sharp values of γ and β which delay the convergence of the coverage probability. Better performance expected as both n^* and m increase.

5 Conclusions

As noticed from the above tables regardless the underlying distribution, a three-stage procedure provides confidence intervals which are generally less than the nominal values. It is also evident that the confidence intervals attain the nominal values

Table 8 Simulation results under the standard normal

n^*	d	$\overline{N}$	$\widehat{\mu}$	$S.E\,(\mu)$	$CI = \widehat{\mu} \pm d$		$1-\widehat{\alpha}$
24	0.40008	20.45	0.00225	0.00113	−0.39783	0.40233	0.8878
43	0.29890	38.38	−0.00054	0.00082	−0.29944	0.29836	0.9033
61	0.25095	56.31	−0.00095	0.00066	−0.25190	0.25000	0.9167
76	0.22483	71.23	0.00108	0.00058	−0.22375	0.22591	0.9225
96	0.20004	91.35	−0.00033	0.00050	−0.20037	0.19971	0.9302
125	0.17531	120.67	0.00006	0.00043	−0.17525	0.17537	0.9369
171	0.14989	167.12	0.00005	0.00036	−0.14984	0.14994	0.9404
246	0.12497	242.77	0.00032	0.00029	−0.12465	0.12529	0.9458
500	0.08765	497.82	−0.00015	0.00020	−0.08780	0.08750	0.9475

Table 9 Simulation results under the standard exponential

n^*	d	$\overline{N}$	$\widehat{\mu}$	$S.E\,(\mu)$	$CI = \widehat{\mu} \pm d$		$1-\widehat{\alpha}$
24	0.40008	19.22	0.92597	0.00102	0.52589	1.3261	0.9004
43	0.29890	34.07	0.91981	0.00092	0.62091	1.2187	0.8254
61	0.25095	50.35	0.92849	0.00085	0.67754	1.1794	0.8111
76	0.22483	64.89	0.93804	0.00078	0.71321	1.1629	0.8255
96	0.20004	84.49	0.94790	0.00072	0.74786	1.1479	0.8391
125	0.17531	114.72	0.96120	0.00062	0.78589	1.1365	0.8643
171	0.14989	162.13	0.97330	0.00052	0.82342	1.1232	0.8871
246	0.12497	240.53	0.98371	0.00039	0.85875	1.1087	0.9084
500	0.08765	508.45	0.99405	0.00023	0.90640	1.0817	0.9318

asymptotically. We have seen that the results of the sequential triple sampling procedure depend mainly on the characteristics of the underlying distribution; in particular, the skewness and the kurtosis. Finally, the good or bad behaviour of the asymptotic coverage probability is controlled mainly by the behaviour of the Edgeworth approximation for the standardized underlying distribution. So in the cases where the underlying distribution is poorly approximated, the asymptotic coverage's values naturally differ considerably from the simulated coverage values.

References

1. Stein, C.: A two sample test for a linear hypothesis whose power is independent of the variance. Ann. Math. Stat. **16**, 243–258 (1945)
2. Cox, D.R.: Estimation by double sampling. Biometrika, **39**, 217–227 (1952)
3. Anscombe, F.: Sequential estimation. J. Roy. Stat. Soc. B. Met. **15**, 1–21 (1953)
4. Chow, Y.S., Robbins, H.: On the asymptotic theory of fixed width sequential confidence intervals for the mean. Ann. Math. Stat. **36**, 1203–1212 (1965)
5. Robbins, H.: Sequential estimation of the mean of a normal population. In: Probability and Statistics (Harold Cramer Volume), pp. 235–245. Almquist and Wiksell, Stockholm (1959)
6. Hall, P.: Asymptotic theory of triple sampling for sequential estimation of a mean. Ann. Stat. **9**(6), 1229–1238 (1981)
7. Vik, G., Mukhophadhyay, N.: Triple sampling to construct fixed-width confidence intervals for estimable parameters based on U-statistics. Metron **46**(1–4), 165–174 (1988)

8. Mukhophadhyay, N., deSilva, B.M.: Multistage fixed-width confidence intervals in the two-sample problem: the normal case. J. Stat. Res. **31**(1), 1–20 (1997)
9. Hamdy, H.I.: Remarks on the asymptotic theory of triple stage estimation of the normal mean. Scand. J. Stat. **15**, 303–310 (1988)
10. Yousef, A.S., Kimber, A.C., Hamdy, H.I.: Sensitivity of normal-based triple sampling sequential point estimation to the normality assumption. J. Stat. Plan. Inference **143**(9), 1606–1618 (2013)
11. Dantzig, G.B.: On the non-existence of tests of 'students' hypothesis having power functions independent of σ. Ann. Math. Stat. **11**, 186–192 (1940)
12. Anscombe, F.J.: Large-sample theory of sequential estimation. Math. Proc. Camb. Philos. Soc. **45**, 600–607 (1952)
13. Chow, Y.S., Yu, K.F.: The performance of a sequential procedure for the estimation of the mean. Ann. Stat. **9**, 184–188 (1981)
14. DasGupta, A.: Asymptotic Theory of Statistics and Probability. Springer, New York (2008)
15. Draper N.R., Tierney, D.E.: Regions of positive and unimodal series expansion of the Edgeworth and Gram-Charlier approximations. Biometrika **59**, 463–465 (1972)
16. Bhattacharya, R.N., Ghosh, J.K.: On the validity of the formal Edgeworth expansion. Ann. Stat. **6**, 434–451 (1978)
17. Barndorff-Nielsen, O.E., Cox, D.R.: Asymptotic Techniques for Use in Statistics. Chapman and Hall, London (1989)
18. Hall, P.: The Bootstrap and Edgeworth Expansion. Springer, New York (1992)
19. Mukhopadhyay, N., deSilva, B.M.: Sequential Methods and Their Application. CRC Press, Boca Raton (2009)
20. Simons, G.: On the cost of not knowing the variance when making a fixed-width interval estimation of the mean. Ann. Math. Stat. **39**, 1946–1952 (1968)

Mathematical and Numerical Modelling of the Thermoplastic Coupled Problem

Abduvali A. Khaldjigitov, Nik Mohd Asri Nik Long, Aziz Qalandarov and Zainidin K. Eshkuvatov

Abstract The coupled thermoplastic dynamic boundary problem is formulated using the deformation theory of plasticity for small deformations. The explicit and implicit schemes of finite difference equations in one-dimension case are constructed. The discreet equations are numerically solved using the explicit and implicit schemes. Comparison shows the coincidence of the numerical results received using two methods.

1 Introduction

Investigation of the plastic deformations of isotropic and anisotropic materials accounting the thermal properties and subjected to mechanical forces is important in analyzing the material structures. In recent years, in conjunction with the development of composite materials, researchers have proposed various approaches for constructing the constitutive relations for composite materials. Aboudi [1] proposed the thermomechanical continuum theory for the prediction of the average behavior of unidirectional fiber reinforced in studying the properties of the graphite–aluminum composite under various types of mechanical and thermal changes. Pobedria [2] and Khaldjigitov [3] considered the deformation theory of thermoplasticity for transversely isotropic and orthotropic materials, which is the modification of the Ilyushins deformation theory for isotropic materials [4]. Kawai et al. [5] considered a phenomenological thermoviscoplastic model for investigating the stress-strain field in

A. A. Khaldjigitov (✉) · A. Qalandarov
Faculty of Information technologies, Tashkent University of Information technology, Amir Temur Str.108, 100075 Tashkent, Uzbekistan
e-mail: khaldjigitov@gmail.com

A. Qalandarov
e-mail: abrorshox@mail.com

N. M. A. N. Long · Z. K. Eshkuvatov
Department of Mathematics, Faculty of Science, University Putra Malaysia, Institute for Mathematical Research, University Putra Malaysia, 43400 Serdang, Selangor, Malaysia
e-mail: nmasri@math.upm.edu.my

Z. K. Eshkuvatov
e-mail: ezaini@science.upm.edu.my

A. Kilicman et al. (eds.), *International Conference on Mathematical Sciences and Statistics 2013,* DOI 10.1007/978-981-4585-33-0_8,

carbon/epoxy composites at high temperatures. Green and Naghdi [6] proposed a rate-type thermoplasticity theory at finite deformations, whereas Miehe [7] examined the rate-independent case. Ulz [11] used a Green-Naghdi's approach for modeling anisotropic rate-independent and rate-dependent thermoplasticity in the logarithmic Lagrangean strain-entropy space at finite strains. Thermodynamical restrictions are discussed in [6], [9] and [7]. The thermomechanical coupling problems, in which the mechanical response of the structure depends upon its thermal behavior and vice-versa are considered by Sloderback and Pajak [10]. Recently, Vaz Jr. et al. [12] modeled the coupled effects between ductile damage and temperature evolution. In this paper, using the Ilyishins deformation theory, the coupled dynamic thermoplasticity boundary problem is formulated. The governing equations of the thermoplastic coupled boundary problems consist of the motion, modelling, and heat conduction equations [13]. Note that the motion equation and heat conduction equations are the corresponding hyperbolic and parabolic type of partial differential equations.

In Sect. 2, the coupled dynamic thermoplasticity boundary problem is formulated. For the 1D case, the equations of motion and heat conduction depend on the displacement and temperature, with the corresponding initial and boundary conditions given. In Sect. 3, using the finite difference method, discrete equations are constructed. In the discretization process, all the derivatives are approximated by their corresponding difference formulas and it turns out that two kinds of schemes appear, i.e., explicit and implicit schemes. The explicit scheme is solved with the help of recurrent formulas. For the solution of implicit scheme, the "consecutive" method is used [14], [3].

In Sect. 4, as an example the coupled dynamic thermoplastic problem has been solved. Using the recurrent formulas (explicit scheme) and "consecutive" method (implicit scheme), some coupled thermoplasticity numerical tests are solved. Comparison of the numerical results received aforementioned, two methods show a good coincidence.

2 Formulation of the Dynamic Thermoplastic Coupled Boundary Problem Using Deformation Theory of Plasticity

The coupled thermodynamic elastic plastic boundary value problem consists of the motion equations [13]

$$\sigma_{ij,j} + X_i = \rho \ddot{u}_i, \tag{1}$$

nonlinear constitutive relation between the strains and stresses tensors for isotropic or anisotropic materials [4]

$$\sigma_{ij} = K(\theta - 3\alpha\vartheta)\delta_{ij} + \frac{\sigma_u}{\varepsilon_u} e_{ij}, \tag{2}$$

the heat equation for isotropic materials [13]

$$\lambda_0 T_{,ii} - c_\varepsilon \dot{T} - T_0 \cdot \alpha(3\lambda + 2\mu) \cdot \dot{\varepsilon}_{ij} = 0, \tag{3}$$

Cauchy relations [13]

$$\varepsilon_{ij} = \frac{1}{2}\left(u_{i,j} + u_{j,i}\right) \tag{4}$$

with the corresponding initial

$$u_i|_{t=t_0} = \phi_i,\ \dot{u}_i|_{t=t_0} = \psi_i,\ T|_{t=t_0} = T_0, \tag{5}$$

and boundary conditions

$$u_i|_{\sum_1} = u_i^0,\ T|_{\sum_1} = \bar{T}_0,\ \sigma_{ij}n_j\big|_{\sum_2} = S_i^0 \tag{6}$$

where, c_ε is heat at a constant temperature, α is thermal expansion coefficient, and λ_0 is the heat flow coefficient.

Equations (1)–(6) in one-dimension case take the form

$$\frac{\partial \sigma_{11}}{\partial x} + X_1 = \rho\frac{\partial^2 u}{\partial t^2} \tag{7}$$

$$\sigma_{11} = (\lambda + 2\mu - \frac{4}{3}(\mu - \mu')(1 - \frac{\varepsilon_u^*}{\varepsilon_u}))\varepsilon_{11} - \alpha\left(3\lambda + 2\mu\right)\left(T - T_0\right) \tag{8}$$

$$\varepsilon_{11} = \frac{\partial u}{\partial x} \tag{9}$$

Substituting the Eq. (9) into Eq. (8),

$$\sigma_{11} = (\lambda + 2\mu - \frac{4}{3}(\mu - \mu')(1 - \frac{\varepsilon_u^*}{\varepsilon_u}))\frac{\partial u}{\partial x} - \alpha\left(3\lambda + 2\mu\right)\left(T - T_0\right) \tag{10}$$

and expression obtained therein into Eq. (7) allows us to write the motion equation in terms of displacements:

$$(\lambda + 2\mu - \frac{4}{3}(\mu - \mu'))\frac{\partial^2 u}{\partial x^2} - \alpha\gamma(T - T_0) = \rho\frac{\partial^2 u}{\partial t^2}, \gamma = 3\lambda + 2\mu. \tag{11}$$

The heat equation in one-dimension form takes the form

$$\lambda_0\frac{\partial^2 T}{\partial x^2} - c_\varepsilon\frac{\partial T}{\partial t} - T_0\alpha\gamma\frac{\partial^2 u}{\partial x \partial t} = 0, \tag{12}$$

with appropriate initial and boundary conditions

$$u\left(x,t\right)|_{t=0} = \phi(x_i),\ \left.\frac{\partial u}{\partial t}\right|_{t=0} = \psi(x_i),\ T\left(x,t\right)|_{t=0} = T_0 \tag{13}$$

$$u\left(x,\,t\right)|_{x=0} = u_0;\ u\left(x,\,t\right)|_{x=\ell} = \bar{u}_0;\ T\left(x,t\right)|_{x=0} = T_1\left(t\right);\ T\left(x,t\right)|_{x=\ell} = T_2\left(t\right) \tag{14}$$

where, $\lambda, \mu, \mu', \alpha, C_\varepsilon, \lambda_0$ are the known values, l is the length of the rod, and φ, ψ, T_0, T_1, and T_2 are the specified amounts.

3 The Construction of the Explicit and Implicit Finite Difference Equations

Considering in the area $t \geq 0,\ 0 \leq x \leq l$ two sets of parallel lines $x = ih_1 (i = \overline{0, n})$, $t = k\tau\ (k = 0, 1, 2, \ldots)$ replace the derivatives in Eqs. (11) and (12) by difference quotients, we obtain

$$(\lambda + 2\mu - \frac{4}{3}(\mu - \mu'))\frac{u_{i+1}^{j} - 2u_{i}^{j} + u_{i-1}^{j}}{h^2} - \alpha\gamma\frac{T_{i+1}^{j} - T_{i-1}^{j}}{2h} = \rho\frac{u_{i}^{j+1} - 2u_{i}^{j} + u_{i}^{j-1}}{\tau^2} \tag{15}$$

$$\lambda_0\frac{T_{i+1}^{j} - 2T_{i}^{j} + T_{i-1}^{j}}{h^2} - C_{\varepsilon}\frac{T_{i}^{j+1} - T_{i}^{j-1}}{2\tau} - \alpha\gamma T_0\frac{u_{i+1}^{j+1} - u_{i-1}^{j+1} - u_{i+1}^{j-1} + u_{i-1}^{j-1}}{4h\tau} = 0 \tag{16}$$

Solving the differential Eqs. (15) and (16) about u_i^{j+1} and T_i^{j+1} respectively, we get

$$u_i^{j+1} = \frac{\tau^2}{\rho}\left((\lambda + 2\mu - \frac{4}{3}(\mu - \mu'))\frac{u_{i+1}^{j} - 2u_{i}^{j} + u_{i-1}^{j}}{h^2} - \alpha(3\lambda + 2\mu)\frac{T_{i+1}^{j} - T_{i-1}^{j}}{2h}\right) + 2u_i^j - u_i^{j-1} \tag{17}$$

$$T_i^{j+1} = -\frac{2\tau}{C_{\varepsilon}}\left(\alpha\gamma T_i^j\frac{u_{i+1}^{j+1} - u_{i-1}^{j+1} - u_{i+1}^{j-1} + u_{i-1}^{j-1}}{4h\tau} - \lambda'\frac{T_{i+1}^{j} - 2T_{i}^{j} + T_{i-1}^{j}}{h^2}\right) - T_i^{j-1} \tag{18}$$

As can be seen, Eqs. (17) and (18) allow us to find the function values $u(x,t)$ and $T(x,t)$ at the layer t^{j+1} using the given values of these functions at the two previous layers. We can find the values of $u(x,t)$ on two primary layers from the initial conditions and for the values of $T(x,t)$ and also find the following difference relations

$$u_i^1 = \frac{1}{2}\left(\frac{\tau^2}{\rho}\left((\lambda + 2\mu - \frac{4}{3}(\mu - \mu'))\frac{u_{i+1}^{0} - 2u_{i}^{0} + u_{i-1}^{0}}{h^2} - \alpha\gamma\frac{T_{i+1}^{0} - T_{i-1}^{0}}{2h}\right) + 2u_i^0 + 2\psi\tau\right) \tag{19}$$

$$T_i^1 = -\frac{2\tau}{C_{\varepsilon}}\left(\alpha\gamma T_0\frac{u_{i+1}^{1} - u_{i-1}^{1} - u_{i+1}^{0} + u_{i-1}^{0}}{2h\tau} - \lambda'\frac{T_{i+1}^{0} - 2T_{i}^{0} + T_{i-1}^{0}}{h^2}\right) - T_i^0 \tag{20}$$

The difference equation, Eq. (11) can be reduced to

$$a_i u_{i+1}^{j+1} + b_i u_i^{j+1} + c_i u_{i-1}^{j+1} = f_{ij}, \tag{21}$$

where

$$a_i = \frac{\lambda + 2\mu - \frac{4}{3}(\mu - \mu')}{h^2}, b_i = -\frac{2(\lambda + 2\mu - \frac{4}{3}(\mu - \mu'))}{h^2} - \frac{\rho}{\tau^2}$$

$$c_i = \frac{\lambda + 2\mu - \frac{4}{3}(\mu - \mu')}{h^2}, f_{ij} = \alpha(3\lambda + 2\mu)\frac{T_{i+1}^j - T_{i-1}^j}{2h} + \rho\frac{u_i^{j-1} - 2u_i^j}{\tau^2} \quad (22)$$

Similarly, we can bring the difference equation Eq. (12) into the form

$$a_i T_{i+1}^{j+1} + b_i T_i^{j+1} + c_i T_{i-1}^{j+1} = f_{ij}, \quad (23)$$

where

$$a_i = \frac{\lambda'}{h^2}, b_i = -\frac{2\lambda'}{h^2} - \frac{\rho}{\tau^2}, c_i = \frac{\lambda'}{h^2}$$

$$f_{ij} = \beta T_i^j \frac{u_{i+1}^{j+1} - u_{i-1}^{j+1} - u_{i+1}^{j-1} + u_{i-1}^{j-1}}{4h\tau} - c_\varepsilon \frac{T_i^{j-1}}{2\tau}.$$

Calculate the values of the functions $u(x,t)$ and $T(x,t)$ on two primary layers at $j = 0$ from the initial conditions and at $j = 0$ from Eqs. (15) and (16) respectively. The values of these functions on the other layers can be calculated from Eqs. (21) and (23) using the boundary conditions of the "consecutive" method [14].

The joint solution of the thermoplasticity equations with the heat equation may describe more adequately the process of thermoplastic deformations under mechanical and thermal effects.

4 Numerical Test

As an example, the coupled dynamic thermoplastic problem has been solved (Eq. 7)–Eq. (10) by an explicit grid method and "consecutive" method, with the following initial and boundary conditions and constants:

$$u\,(x,t)|_{t=0} = \sin\left(\frac{\pi x}{\ell}\right), \left.\frac{\partial u(x,t)}{\partial t}\right|_{t=0} = 0, T\,(x,t)|_{t=0} = T_0$$

$$u\,(x,t)|_{x=0} = 0, u\,(x,t)|_{x=1} = 0, T\,(x,t)|_{x=0} = T_0, T\,(x,t)|_{x=1} = T_0$$

$\lambda = 1.2$, $\lambda_0 = 0.8$, $\alpha = 0.05$, $\mu = 0.5$, $\rho = 0.9$, $C_\varepsilon = 3.5$, $T_0 = 90$, $h = 0.1$, $\tau = 0.01$, $l = 1$

In the Tables 1–4, the values of $u(x,t)$ and $T(x,t)$ which depend on the position - x and time - t are shown. It can be seen from the tables that the numerical results obtained using the explicit grid and the "sweep" methods are close enough.

Table 1 The value of the displacement $u(x,t)$ calculated by the "sweep" method

x t	0	0,1	0,2	0,3	0,4	0,5	0,6	0,7	0,8	0,9	1
0	0	0,3090	0,5877	0,8090	0,9510	1	0,9510	0,8090	0,5877	0,3090	0
0,01	0	0,3086	0,5870	0,8080	0,9499	0,9988	0,9499	0,8080	0,5870	0,3086	0
0,02	0	0,3075	0,5849	0,8051	0,9465	0,9952	0,9465	0,8051	0,5849	0,3075	0
0,03	0	0,3084	0,5750	0,8003	0,9408	0,9892	0,9408	0,8003	0,5703	0,3020	0
0,04	0	0,3043	0,5674	0,7936	0,9329	0,9809	0,9329	0,7936	0,5628	0,2979	0
0,05	0	0,2970	0,5647	0,7850	0,9229	0,9704	0,9229	0,7850	0,5647	0,2969	0
0,06	0	0,2929	0,5571	0,7746	0,9106	0,9575	0,9106	0,7746	0,5571	0,2928	0
0,07	0	0,2904	0,5424	0,7624	0,8963	0,9424	0,8963	0,7624	0,5379	0,2842	0
0,08	0	0,2834	0,5297	0,7485	0,8799	0,9252	0,8799	0,7484	0,5252	0,2772	0
0,09	0	0,2736	0,5214	0,7328	0,8616	0,9059	0,8615	0,7327	0,5213	0,2734	0
0,1	0	0,2666	0,5087	0,7156	0,8413	0,8845	0,8412	0,7154	0,5086	0,2663	0

Table 2 The value of the displacement $u(x,t)$ calculated by the grid method

x t	0	0,1	0,2	0,3	0,4	0,5	0,6	0,7	0,8	0,9	1
0	0	0,309	0,5877	0,809	0,951	1	0,951	0,809	0,5877	0,309	0
0,01	0	0,3086	0,5871	0,808	0,9499	0,9988	0,9499	0,808	0,5871	0,3086	0
0,02	0	0,3076	0,5852	0,8051	0,9465	0,9952	0,9465	0,8051	0,5852	0,3076	0
0,03	0	0,3059	0,582	0,8003	0,9408	0,9892	0,9408	0,8003	0,582	0,3059	0
0,04	0	0,3036	0,5775	0,7936	0,9329	0,9809	0,9329	0,7936	0,5775	0,3036	0
0,05	0	0,3005	0,5718	0,785	0,9228	0,9703	0,9228	0,785	0,5718	0,3005	0
0,06	0	0,2967	0,5647	0,7745	0,9105	0,9574	0,9105	0,7745	0,5647	0,2967	0
0,07	0	0,2922	0,5564	0,7622	0,8961	0,9422	0,8961	0,7622	0,5564	0,2922	0
0,08	0	0,287	0,5468	0,7482	0,8796	0,9249	0,8796	0,7482	0,5468	0,287	0
0,09	0	0,2811	0,5359	0,7325	0,8611	0,9055	0,8611	0,7325	0,5359	0,2811	0
0,1	0	0,2745	0,5238	0,715	0,8407	0,884	0,8407	0,715	0,5238	0,2745	0

Table 3 The temperature (x, t) values found using the "sweep" method

x t	0	0,1	0,2	0,3	0,4	0,5	0,6	0,7	0,8	0,9	1
0	90	90	90	90	90	90	90	90	90	90	90
0,01	90	90,01	90,01	90	90	89,99	89,99	89,98	89,98	89,98	90
0,02	90	90,06	90,05	90,03	90,01	89,98	89,96	89,93	89,92	89,93	90
0,03	90	90,12	90,12	90,08	90,02	89,97	89,91	89,86	89,83	89,85	90
0,04	90	90,21	90,21	90,14	90,05	89,94	89,84	89,76	89,71	89,75	90
0,05	90	90,31	90,33	90,22	90,07	89,91	89,76	89,64	89,57	89,62	90
0,06	90	90,43	90,45	90,31	90,11	89,88	89,66	89,49	89,4	89,48	90
0,07	90	90,56	90,6	90,42	90,14	89,84	89,55	89,32	89,21	89,32	90
0,08	90	90,69	90,76	90,53	90,18	89,79	89,42	89,13	89	89,15	90
0,09	90	90,83	90,92	90,66	90,23	89,75	89,29	88,93	88,77	88,97	90
0,1	90	90,97	91,1	90,79	90,28	89,69	89,14	88,71	88,53	88,78	90

Table 4 The temperature Ò(x,t) values found using the grid method

x t	0	0,1	0,2	0,3	0,4	0,5	0,6	0,7	0,8	0,9	1
0	90	90	90	90	90	90	90	90	90	90	90
0,01	90	90,02	90,017	90,012	90,006	90	89,993	89,987	89,982	89,979	90
0,02	90	90,077	90,07	90,051	90,026	90	89,973	89,948	89,929	89,922	90
0,03	90	90,161	90,155	90,113	90,059	90	89,94	89,886	89,844	89,838	90
0,04	90	90,267	90,269	90,2	90,105	90	89,894	89,799	89,73	89,732	90
0,05	90	90,392	90,407	90,308	90,163	90	89,836	89,691	89,592	89,607	90
0,06	90	90,531	90,568	90,435	90,231	90	89,768	89,564	89,431	89,468	90
0,07	90	90,683	90,747	90,58	90,31	90	89,689	89,419	89,252	89,316	90
0,08	90	90,845	90,941	90,74	90,398	90	89,601	89,259	89,058	89,154	90
0,09	90	91,016	91,149	90,912	90,494	90	89,505	89,087	88,85	88,983	90
0,1	90	91,194	91,367	91,095	90,596	90	89,403	88,904	88,632	88,805	90

Acknowledgements The first author was the visiting scientist at the "Institute for Mathematical Research", University Putra Malaysia in 2011. This article appears to thank the scientific collaboration between scientists from UPM and Tashkent University of Information technologies.

References

1. Aboudi, J.: The effective thermomechanical behavior of inelastic fiber-reinforced materials. Int. J. Engng. sci. **23**(7), 773–787 (1985)
2. B.E. Pobedria, Mechanics of composite materials, MGU, 336 p., 1984
3. A.A. Khaldjigitovs: On thermoplasticity relations, Problems of mechanics, 1, 1998, (in Russsian)
4. Ilyushin, A.A.: Plasticity, Part 1: The elastic-plastic deformation. GITTL, Moscow (1948) (in Russian)
5. Kawai, M., Kazama, T., Masuko, Y.: Stress relaxation behavior of unidirectional carbon/epoxy composites at elevated temperature and analysis using viscoplasticity model. JSME Int. J., Series A, **47**(1), 8–16 (2004)
6. Green, A.E., Naghdi, P.M.: A general theory of an elastic-plastic continuum. Arch. Rat. Mech. Analysis. **18**(4), 251–281 (1965)
7. Miehe, C.: A theory of large-strain isotropic thermoplasticity based on metric transformation tensors. Arch. Appl. Mech. **66**(4), 45–64 (1995)
8. Naghdi C.J.: Stress-strain relations in plasticity and thermoplasticity, Plasticity, Proc. 2nd Symp. Naval Struct. Mechanics, Lee, E.H., Symonds, P.S. (eds.), pp. 121–167. Pergamon Prss, New York (1960)
9. Casey, J.: On elastic-thermo-plastic materials at finite deformations. Int. J. Plasticity. **14**, 85–95 (1998)
10. Sloderbach, Z., Pajak, J.: Generalized coupled thermoplasticity taking into account large strains: Part I. Conditions of uniqueness of the solution of boundary-value problem and bifurcation criteria. Math. Mech. Solids. **15**, 308–327 (2010)
11. Ulz, M.H.: A Green-Naghdi approarch to finite anisotropic rate-independent and rate dependent thermo-plasticity in logarithmic Lagrangean strain-entropy space. Comput. Methods Appl. Mech. Engrg. **198**, 3262–3277 (2009)
12. Vaz Jr., M., Munoz-Rojas, B.A., Lange, M.N.: Damage evolution and thermal coupled effects in inelstic solids. Int. J. Mech. Sciences. **53**, 387–398 (2011)
13. V. Novatskiy: Theory of elasticity. M.Mir, 872 p 1975
14. Samarskii, A., Nikolaev, E.: Numerical methods of grid equation, p. 657. Birkhauser Verlag, Berlin (1989)

New Modification of Laplace–Adomian Decomposition Method for the Fifth-Order KdV Equation

H. O. Bakodah and B. S. Kashkari

Abstract In this paper, for obtaining accurate solutions of the fifth-order KdV equation with initial condition, a new modification of Laplace–Adomian decomposition method is proposed. To see the accuracy of the proposed method, L_2 and L_∞ error norms are calculated in test problems. The numerical results are found to be in a good agreement with exact solutions and with the literature.

1 Introduction

The fifth-order KdV (FKdV) equation which can be shown in the form

$$u_t + au^2 u_x + bu_x u_{xx} + cuu_{xxx} + du_{xxxxx} = 0, \tag{1}$$

where a, b, c and d are constants, plays an important role in describing motions of long waves in shallow water under gravity, one-dimensional nonlinear lattice fluid mechanics, quantum mechanics, plasma physics, nonlinear optics and other areas. This equation has been known as the general form of the fifth-order KdV equation [1]. The Adomian decomposition method (ADM) was applied to an FKdV equation [2]–[5]. In this work, the new modification of Laplace ADM [6]–[8] is implemented in the FKdV equation. The proposed method is seen to be an efficient technique in solving initial value problems accurately.

2 Analysis of the Method

In this section, we outline the new modification of Laplace ADM to obtain explicit solution of Eq. (1) with the initial condition $u(x,0) = f(x)$.

H. O. Bakodah (✉) · B. S. Kashkari
Department of Mathematics, Faculty of Science for Girls,
King Abdulaziz University, Jeddah, Saudi Arabia
e-mail: hbakodah@kau.edu.sa

B. S. Kashkari
e-mail: b.habiballah@gmail.com

A. Kilicman et al. (eds.), *International Conference on Mathematical Sciences and Statistics 2013,* DOI 10.1007/978-981-4585-33-0_9,

Let us consider the standard form of a FKDV Eq. (1) in an operator form

$$L_t(u) + a(Ku) + b(Mu) + c(Nu) + dL_x(u) = 0 \tag{2}$$

where the notation $Ku = u^2 u_x$, $Mu = u_x u_{xx}$ and $Nu = uu_{xxx}$ symbolize the nonlinear term, respectively. The notation $L_t = \frac{\delta}{\delta t}$ and $L_x = \frac{\delta^5}{\delta x^5}$ symbolize the linear differential operators. We represent the solution as an infinite series given below,

$$u(x,t) = \sum_{n=0}^{\infty} u_n(x,t). \tag{3}$$

The nonlinear terms Ku, Mu and Nu can be decomposed into infinite series of polynomial given by:

$$Ku = u^2 u_x = \sum_{n=0}^{\infty} A_n;$$

$$Mu = u_x u_{xx} = \sum_{n=0}^{\infty} B_n;$$

$$Nu = uu_{xxx} = \sum_{n=0}^{\infty} C_n, \tag{4}$$

where A_n, B_n and C_n are Adomian polynomials [9] of $u_0, u_1, \cdots\cdots, u_n$ and can be calculated by formula given below

$$A_n = \frac{1}{n!}\frac{d^n}{d\lambda^n}\left[K\left(\sum_{i=0}^{\infty}\lambda^i u_i\right)\right]_{\lambda=0}, n = 0, 1, 2, \cdots\cdots.;$$

$$B_n = \frac{1}{n!}\frac{d^n}{d\lambda^n}\left[M\left(\sum_{i=0}^{\infty}\lambda^i u_i\right)\right]_{\lambda=0}, n = 0, 1, 2, \cdots\cdots.;$$

$$C_n = \frac{1}{n!}\frac{d^n}{d\lambda^n}\left[N\left(\sum_{i=0}^{\infty}\lambda^i u_i\right)\right]_{\lambda=0}, n = 0, 1, 2, \cdots\cdots. \tag{5}$$

By applying the Laplace transform with respect to t to both sides of Eq. (2), we obtain

$$\pounds\{L_t(u)\} = -a\pounds\{Ku\} - b\pounds\{Mu\} - c\pounds\{Nu\} - d\pounds\{L_x(u)\}. \tag{6}$$

Thus,

$$\pounds\{u(x,t)\} = \frac{1}{s}u(x,0) - \frac{1}{s}(a\pounds\{Ku\} + b\pounds\{Mu\} + c\pounds\{Nu\} + d\pounds\{L_x(u)\}). \tag{7}$$

A sum of components defined by the decomposition series is given by

$$\pounds\{u(x,t)\} = U(x,s) = \sum_{n=0}^{\infty} U_n(x,s). \tag{8}$$

In the new modification [10], the initial condition $u(x,0)$ can be expressed as a series of infinite components i.e.

$$u(x,0) = \sum_{n=0}^{\infty} u_0^n(x). \tag{9}$$

and the new recursive relationship can be expressed in the form

$$\begin{aligned} U_0(x,s) &= \frac{1}{s}u_0^0(x), \\ U_{n+1}(x,s) &= \frac{1}{s}u_0^{n+1}(x) - \frac{1}{s}(a\pounds\{A_n\} + b\pounds\{B_n\} + c\pounds\{C_n\} \\ &\quad + d\pounds\{L_x(u_n(x,t))\}), n \geq 0. \end{aligned} \tag{10}$$

Finally, by applying inverse Laplace transformation we get

$$\begin{aligned} u_0(x,t) &= \pounds^{-1}\{U_0(x,s)\}, \\ u_{n+1}(x,t) &= \pounds^{-1}\{U_{n+1}(x,s)\}, n \geq 0. \end{aligned} \tag{11}$$

Using (9), the series solution follows immediately. In some cases, the exact solution in the closed form may be obtained. In practice, all terms of the series u_i cannot be determined, so we have an approximation of the solution by the following series

$$\varphi_k(x,t) = \sum_{m=0}^{k-1} u_m(x,t). \tag{12}$$

$$\lim_{k\to\infty} \varphi_k(x,t) = u(x,t). \tag{13}$$

3 Numerical Results

3.1 Lax's FKdV Equation

When we put $a = 30, b = 20, c = 10$ and $d = 1$ in (1) we get Lax's equation [11]:

$$u_t + 30u^2u_x + 20u_xu_{xx} + 10uu_{xxx} + u_{xxxxx} = 0 \tag{14}$$

Consider the initial condition

$$u(x,0) = -\frac{1}{6} \times 10^{-2}\left(1 - 3sech^2\left(\frac{x}{20}\right)\right). \tag{15}$$

By applying Laplace transform and using given initial condition we get

$$£\{u(x,t)\} = \frac{1}{s}u(x,0) - \frac{1}{s}(30£\{Ku\} + 20£\{Mu\} + 10£\{Nu\} + £\{L_x(u)\}). \quad (16)$$

The initial condition $u(x,0)$ can be expressed as a series of infinite components i.e.

$$\begin{aligned} u(x,0) = &0.003333 - 0.0000125x^2 + 2.083333 \times 10^{-8}x^4 - 2.951388 \times 10^{-11}x^6 \\ &+ 3.844246 \times 10^{-14}x^8 - 4.760527 \times 10^{-17}x^{10} + \cdots. \end{aligned} \quad (17)$$

Using recursive relation (10) with Adomian polynomials yields

$$u_0(x,t) = £^{-1}\{U_0(x,s)\} = £^{-1}\left\{\frac{1}{3s} \times 10^{-2}\right\} = \frac{1}{3} \times 10^{-2} \quad (18)$$

$$\begin{aligned} u_1(x,t) &= £^{-1}\{U_1(x,s)\} \\ &= £^{-1}\left\{\frac{1}{s}u_0^1(x) - \frac{1}{s}(30£\{A_0\} + 20£\{B_0\} + 10£\{C_0\} + £\{L_x(u_0(x,t))\})\right\} \\ &= £^{-1}\left\{-\frac{x^2}{8s} \times 10^{-4}\right\} = -\frac{x^2}{8} \times 10^{-4} \end{aligned} \quad (19)$$

$$\begin{aligned} u_2(x,t) &= £^{-1}\{U_2(x,s)\} \\ &= £^{-1}\left\{\frac{1}{s}u_0^2(x) - \frac{1}{s}(30£\{A_1\} + 20£\{B_1\} + 10£\{C_1\} + £\{L_x(u_1(x,t))\}\right\} \\ &= £^{-1}\left\{\frac{x}{12s^2} \times 10^{-7} + \frac{2.083333x^4}{s} \times 10^{-8}\right\} \\ &= 8.333333 \times 10^{-9}tx + 2.083333 \times 10^{-8}x^4 \end{aligned} \quad (20)$$

and so on. In this manner, the rest of the components of the decomposition series were obtained. Substituting (18)–(20) into (3) gives the solution $u(x,t)$ in a series form and in a close form as

$$u(x,t) = -\frac{1}{6} \times 10^{-2}\left(1 - 3sech^2\left(\frac{1}{20}\left(-\frac{t}{6} + x\right)\right)\right). \quad (21)$$

This result can be verified through substitution. From Table 1, we find that Laplace ADM has a small error. The L_2 and L_∞ errors are contained in Table 2. In Fig. 1 the surface shows the exact solution (a) and the approximation solution (b) of Lax's equation for $0 \leq t \leq 5$ and $-10 \leq x \leq 10$.

Table 1 Absolute error between the exact solution and approximate solution

$t\backslash x$	0.2	0.4	0.6	0.8	1
1	3.38898×10^{-9}	6.72855×10^{-9}	1.00777×10^{-8}	1.34383×10^{-8}	1.6809×10^{-8}
2	6.88738×10^{-9}	1.35664×10^{-8}	2.02644×10^{-8}	2.69853×10^{-8}	3.37262×10^{-8}
3	1.04952×10^{-8}	2.05134×10^{-8}	3.0560×10^{-8}	4.00409×10^{-8}	5.07515×10^{-8}
4	1.42124×10^{-8}	2.75698×10^{-8}	4.09648×10^{-8}	5.44052×10^{-8}	6.78851×10^{-8}
5	1.80391×10^{-8}	3.47354×10^{-8}	5.14785×10^{-8}	6.82781×10^{-8}	8.51268×10^{-8}

Table 2 L_2 and L_∞ errors

t	L_2	L_∞
1	7.38478×10^{-8}	1.6809×10^{-8}
2	1.47885×10^{-7}	3.37262×10^{-8}
3	2.22115×10^{-7}	5.07515×10^{-8}
4	2.96536×10^{-7}	6.78851×10^{-8}
5	3.71149×10^{-7}	8.51268×10^{-8}

3.2 Sawada–Kotera Equation (SK Equation)

When we put $a = 5, b = 5, c = 5$ and $d = 1$ in (1), we get the Sawada–Kotera equation [11]

$$u_t + 5u^2u_x + 5u_xu_{xx} + 5uu_{xxx} + u_{xxxxx} = 0. \tag{22}$$

Consider the initial condition

$$u(x,0) = -\frac{1}{200}\left(1 - 3sech^2\left(\frac{x}{20}\right)\right) \tag{23}$$

$$£\{u(x,t)\} = \frac{1}{s}u(x,0) - \frac{1}{s}(5£\{Ku\} + 5£\{Mu\} + 5£\{Nu\} + £\{L_x(u)\}) \tag{24}$$

The initial condition $u(x,0)$ can be expressed as a series of infinite components i.e.

$$\begin{aligned} u(x,0) &= 0.01 - 0.0000375x^2 + 6.25\times10^{-8}x^4 - 8.854167\times10^{-11}x^6 \\ &= 1.153274\times10^{-13}x^8 - 1.4281581\times10^{-16}x^{10} + \cdots. \end{aligned} \tag{25}$$

Using recursive relation (10) with Adomian polynomials yields

$$u_0(x,t) = £^{-1}\{U_0(x,s)\} = £^{-1}\left\{\frac{0.01}{s}\right\} = 0.01 \tag{26}$$

$$u_1(x,t) = £^{-1}\{U_1(x,s)\} = £^{-1}\left\{-\frac{3x^2}{8s}\times10^{-4}\right\} = -\frac{3x^2}{8}\times10^{-4} \tag{27}$$

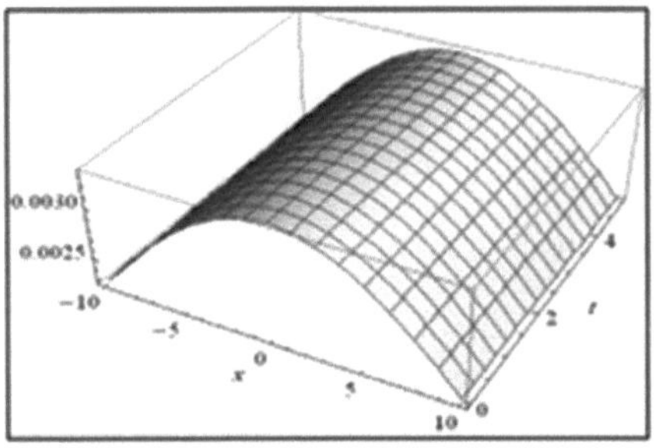

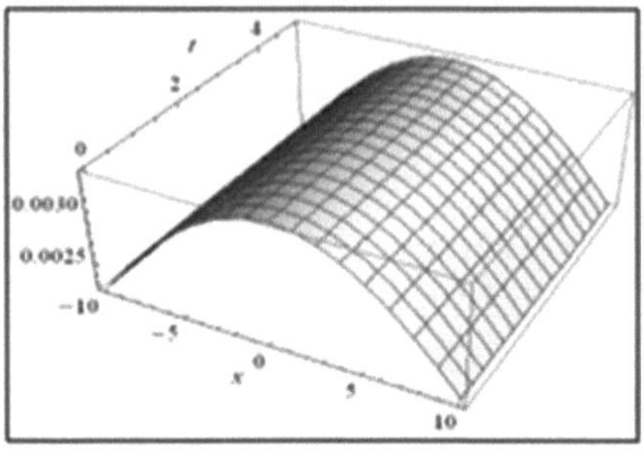

Fig. 1 The surface shows the exact and the approximate solution of Lax's equation for $0 \leq t \leq 5$ and $-10 \leq x \leq 10$

Table 3 Absolute error between the exact solution and approximate solution

$t\backslash x$	0.2	0.4	0.6	0.8	1
1	1.51855×10^{-8}	3.0215×10^{-8}	4.5293×10^{-8}	6.04399×10^{-8}	7.567088×10^{-8}
2	3.07337×10^{-8}	6.07927×10^{-8}	9.09491×10^{-8}	1.21243×10^{-7}	1.51706×10^{-7}
3	4.66444×10^{-8}	9.17333×10^{-8}	1.36968×10^{-7}	1.8241×10^{-7}	2.28105×10^{-7}
4	6.29177×10^{-8}	1.23037×10^{-7}	1.83351×10^{-7}	2.43941×10^{-7}	3.04868×10^{-7}
5	7.95537×10^{-8}	1.54703×10^{-7}	2.30096×10^{-7}	3.05835×10^{-7}	3.81996×10^{-7}

Table 4 L_2 and L_∞ errors

t	L_2	L_∞
1	5.11664×10^{-7}	7.56708×10^{-8}
2	1.02411×10^{-6}	1.51706×10^{-7}
3	1.53735×10^{-6}	2.28105×10^{-7}
4	2.05141×10^{-6}	3.04868×10^{-7}
5	2.56627×10^{-6}	3.81996×10^{-7}

$$
\begin{aligned}
u_2(x,t) &= £^{-1}\{U_2(x,s)\} \\
&= £^{-1}\left\{\frac{1}{s}u_0^2(x) - \frac{1}{s}(5£\{A_1\} + 5£\{B_1\} + 5£\{C_1\} + £\{L_x(u_1(x,t))\}\right\} \\
&= £^{-1}\left\{\frac{3.75\times10^{-8}x}{s^2} + \frac{6.25\times10^{-8}x^4}{s}\right\} \\
&= 3.75\times10^{-8}tx + 6.25\times10^{-8}x^4
\end{aligned} \tag{28}
$$

and so on. In this manner, the rest of the components of the decomposition series were obtained. Substituting (26)–(28) into (3) gives the solution $u(x,t)$ in a series form and in a close form as

$$
u(x,t) = -\frac{1}{200}\left(1 - 3sech^2\left(\frac{1}{20}\left(\frac{t}{4}\times10^{-4} + x\right)\right)\right). \tag{29}
$$

This result can be verified through substitution. From Table 3, we find that Laplace ADM has a small error. The L_2 and L_∞ errors are contained in Table 4. In Fig. 2

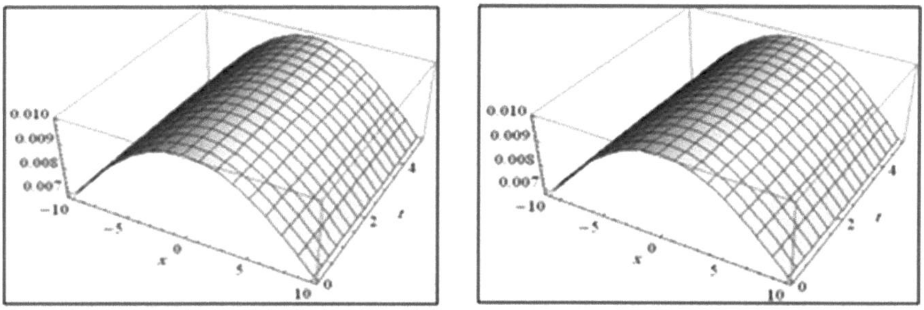

(a) The surface of the exact solution **(b) The surface of the approximation solution**

Fig. 2 The surface shows the exact and the approximate solution of SK equation for $0 \le t \le 5$ and $-10 \le x \le 10$

the surface shows the exact solution (a) and the approximation solution (b) of SK equation for $0 \le t \le 5$ and $-10 \le x \le 10$.

3.3 *Kaup–Kupershmidt Equation (KK Equation)*

When we put $a = 20, b = 25, c = 10$ and $d = 1$ in (1) we get Kaup–Kupershmidt equation [11]

$$u_t + 20u^2 u_x + 25u_x u_{xx} + 10uu_{xxx} + u_{xxxxx} = 0 \tag{30}$$

Consider the initial condition

$$u(x,0) = -\frac{1}{800}\left(1 - 3sech^2\left(\frac{x}{20}\right)\right) \tag{31}$$

$$£\{u(x,t)\} = \frac{1}{s}u(x,0) - \frac{1}{s}(20£\{Ku\} + 25£\{Mu\} + 10£\{Nu\} + £\{L_x(u)\}). \tag{32}$$

The initial condition $u(x,0)$ can be expressed as a series of infinite components i.e.

$$u(x,0) = 0.0025 - 0.000009375x^2 + 1.5625 \times 10^{-8}x^4 - 2.213542 \times 10^{-11}x^6 + 2.883185 \times 10^{-14}x^8 - 3.5703952 \times 10^{-17}x^{10} + \cdots. \tag{33}$$

Using recursive relation (10) with Adomian polynomials yields

$$u_0(x,t) = £^{-1}\{U_0(x,s)\} = £^{-1}\left\{\frac{0.0025}{s}\right\} = 0.0025 \tag{34}$$

Table 5 Absolute error between the exact solution and approximate solution

$t\backslash x$	0.2	0.4	0.6	0.8	1
1	1.89833×10^{-9}	3.77694×10^{-9}	5.66132×10^{-9}	7.55332×10^{-9}	9.45357×10^{-9}
2	3.84226×10^{-9}	7.59949×10^{-9}	1.13682×10^{-8}	1.51522×10^{-8}	1.89527×10^{-8}
3	5.8318×10^{-9}	1.14676×10^{-9}	1.71208×10^{-8}	2.27968×10^{-8}	2.84975×10^{-8}
4	7.86694×10^{-9}	1.53814×10^{-8}	2.29189×10^{-8}	3.04869×10^{-8}	3.80879×10^{-8}
5	9.94768×10^{-9}	1.93407×10^{-8}	2.87626×10^{-8}	3.8226×10^{-8}	4.77239×10^{-8}

Table 6 L_2 and L_∞ errors

t	L_2	L_∞
1	5.13321×10^{-8}	9.45357×10^{-9}
2	1.02756×10^{-7}	1.89527×10^{-8}
3	1.54277×10^{-7}	2.84975×10^{-8}
4	2.05894×10^{-7}	3.80879×10^{-8}
5	2.57609×10^{-7}	4.77239×10^{-8}

$$\begin{aligned} u_1(x,t) &= £^{-1}\{U_1(x,s)\} \\ &= £^{-1}\left\{\frac{1}{s}u_0^1(x) - \frac{1}{s}(20£\{A_0\} + 25£\{B_0\} + 10£\{C_0\} + £\{L_x(u_0(x,t))\})\right\} \\ &= £^{-1}\left\{-\frac{0.9375\times10^{-7}x^2}{s}\right\} = -0.9375\times10^{-7}x^2 \end{aligned} \tag{35}$$

$$\begin{aligned} u_2(x,t) &= £^{-1}\{U_2(x,s)\} \\ &= £^{-1}\left\{\frac{1}{s}u_0^2(x) - \frac{1}{s}(20£\{A_1\} + 25£\{B_1\} + 10£\{C_1\} + £\{L_x(u_1(x,t))\}\right\} \\ &= £^{-1}\left\{\frac{2.34375\times10^{-9}x}{s^2} + \frac{1.5625\times10^{-8}x^4}{s}\right\} \\ &= 2.34375\times10^{-9}tx + 1.5625\times10^{-8}x^4 \end{aligned} \tag{36}$$

and so on. In this manner, the rest of components of the decomposition series were obtained. Substituting (34)–(36) into (3) gives the solution $u(x,t)$ in a series form and in a close form as

$$u(x,t) = -\frac{1}{800}\left(1 - 3sech^2\left(\frac{1}{20}\left(-\frac{t}{16}\times10^{-4} + x\right)\right)\right). \tag{37}$$

This result can be verified through substitution. From Table 5, we find that Laplace ADM has a small error. The L_2 and L_∞ errors are contained in Table 6. In Fig. 3 the surface shows the exact solution (a) and the approximation solution (b) of KK equation for $0 \le t \le 5$ and $-10 \le x \le 10$.

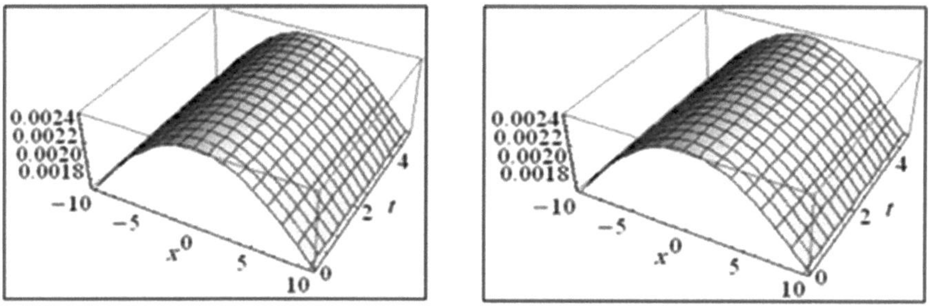

(a) The surface of the exact solution **(b)** The surface of the approximation solution

Fig. 3 The surface shows the exact and the approximate solution of KK equation for $0 \leq t \leq 5$ and $-10 \leq x \leq 10$

3.4 Caudrey–Dodd–Gibbon Equation (CDG Equation)

When we put $a = 180, b = 30, c = 30$ and $d = 1$ in (1) we get the Caudrey–Dodd–Gibbon equation [11]

$$u_t + 180u^2u_x + 30u_xu_{xx} + 30uu_{xxx} + u_{xxxxx} = 0 \tag{38}$$

Consider the initial condition

$$u(x,0) = -\frac{1}{1200}\left(1 - 3sech^2\left(\frac{x}{20}\right)\right) \tag{39}$$

$$£\{u(x,t)\} = \frac{1}{s}u(x,0) - \frac{1}{s}(180£\{Ku\} + 30£\{Mu\} + 30£\{Nu\} + £\{L_x(u)\}). \tag{40}$$

The initial condition $u(x,0)$ can be expressed as a series of infinite components i.e.

$$\begin{aligned} u(x,0) =& 0.001667 - 0.00000625x^2 + 1.041667 \times 10^{-8}x^4 - 1.475694 \times 10^{-11}x^6 \\ &+ 1.922123 \times 10^{-14}x^8 - 2.380263 \times 10^{-17}x^{10} + \cdots. \end{aligned} \tag{41}$$

Using recursive relation (10) with Adomian polynomials yields

$$u_0(x,t) = £^{-1}\{U_0(x,s)\} = £^{-1}\left\{\frac{1}{6} \times 10^{-2}\right\} = \frac{1}{6s} \times 10^{-2} \tag{42}$$

$$\begin{aligned} u_1(x,t) &= £^{-1}\{U_1(x,s)\} \\ &= £^{-1}\left\{\frac{1}{s}u_0^1(x) - \frac{1}{s}(180£\{A_0\} + 30£\{B_0\} + 30£\{C_0\} + £\{L_x(u_0(x,t))\}\right\} \\ &= £^{-1}\left\{-\frac{6.25 \times 10^{-6}x^2}{s}\right\} = -6.25 \times 10^{-6}x^2 \end{aligned} \tag{43}$$

Table 7 Absolute error between the exact solution and approximate solution

$t\backslash x$	0.2	0.4	0.6	0.8	1
1	2.53793×10^{-9}	5.04201×10^{-9}	7.55181×10^{-9}	1.00657×10^{-8}	1.25769×10^{-8}
2	5.14993×10^{-9}	1.01576×10^{-8}	1.51762×10^{-8}	2.02025×10^{-8}	2.5223×10^{-8}
3	7.83629×10^{-9}	1.53469×10^{-8}	2.28731×10^{-8}	3.04103×10^{-8}	3.79382×10^{-8}
4	1.05969×10^{-8}	2.06098×10^{-8}	3.06426×10^{-8}	4.06892×10^{-8}	5.07225×10^{-8}
5	1.34318×10^{-8}	2.59462×10^{-8}	3.84846×10^{-8}	5.10397×10^{-8}	6.35759×10^{-8}

Table 8 L_2 and L_∞ errors

t	L_2	L_∞
1	8.13987×10^{-8}	1.25769×10^{-8}
2	1.62888×10^{-7}	2.5223×10^{-8}
3	2.44464×10^{-7}	3.79382×10^{-8}
4	3.26124×10^{-7}	5.07225×10^{-8}
5	4.07870×10^{-7}	6.35759×10^{-8}

$$\begin{aligned} u_2(x,t) &= \pounds^{-1}\{U_2(x,s)\} \\ &= \pounds^{-1}\left\{\frac{1}{s}u_0^2(x) - \frac{1}{s}(180\pounds\{A_1\} + 30\pounds\{B_1\} + 30\pounds\{C_1\} + \pounds\{L_x(u_1(x,t))\}\right\} \\ &= \pounds^{-1}\left\{\frac{6.25\times10^{-9}x}{s^2} + \frac{1.041667\times10^{-8}x^4}{s}\right\} \\ &= 6.25\times10^{-9}tx + 1.041667\times10^{-8}x^4 \end{aligned} \tag{44}$$

and so on. In this manner, the rest of components of the decomposition series were obtained. Substituting (42)–(44) into (3) gives the solution $u(x,t)$ in a series form and in a close form as

$$u(x,t) = -\frac{1}{1200}\left(1 - 3sech^2\left(\frac{1}{20}\left(\frac{t}{4}\times10^{-4} + x\right)\right)\right). \tag{45}$$

This result can be verified through substitution. From Table 7, we find that Laplace ADM has a small error. The L_2 and L_∞ errors are contained in Table 8. In Fig. 4 the surface shows the exact solution (a) and the approximation solution (b) of CDG equation for $0 \le t \le 5$ and $-10 \le x \le 10$.

3.5 *Ito Equation*

When we put $a = 2, b = 6, c = 3$ and $d = 1$ in (1) we get the Ito equation [11]

$$u_t + 2u^2u_x + 6u_xu_{xx} + 3uu_{xxx} + u_{xxxxx} = 0 \tag{46}$$

Consider the initial condition

$$u(x,0) = -\frac{1}{40}\left(1 - 3sech^2\left(\frac{x}{20}\right)\right) \tag{47}$$

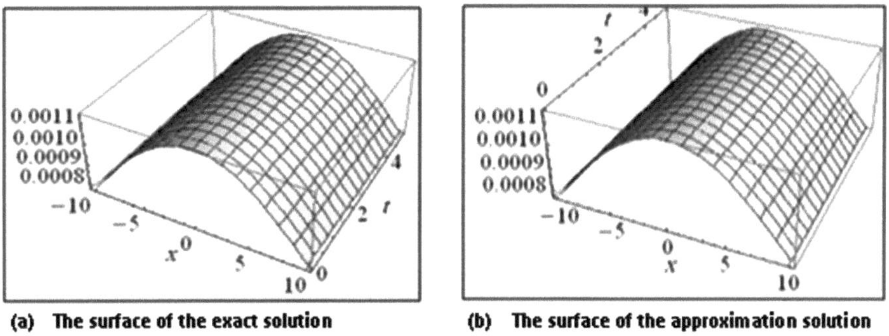

Fig. 4 The surface shows the exact and the approximate solution of CDG equation for $0 \le t \le 5$ and $-10 \le x \le 10$

$$£\{u(x,t)\} = \frac{1}{s}u(x,0) - \frac{1}{s}(2£\{Ku\} + 6£\{Mu\} + 3£\{Nu\} + £\{L_x(u)\}). \quad (48)$$

The initial condition $u(x,0)$ can expressed as a series of infinite components i.e.

$$\begin{aligned} u(x,0) = &0.05 - 0.0001875x^2 + 3.125 \times 10^{-4}x^4 - 4.427083 \times 10^{-10}x^6. \\ &+ 5.766369 \times 10^{-13}x^8 - 7.140790 \times 10^{-16}x^{10} + \cdots. \end{aligned} \quad (49)$$

Using recursive relation (10) with Adomian polynomials yields

$$u_0(x,t) = £^{-1}\{U_0(x,s)\} = £^{-1}\left\{\frac{1}{20s}\right\} = \frac{1}{20} \quad (50)$$

$$\begin{aligned} u_1(x,t) &= £^{-1}\{U_1(x,s)\} \\ &= £^{-1}\left\{\frac{1}{s}u_0^1(x) - \frac{1}{s}(2£\{A_0\} + 6£\{B_0\} + 3£\{C_0\} + £\{L_x(u_0(x,t))\})\right\} \\ &= £^{-1}\left\{-\frac{0.0001875x^2}{s}\right\} = -0.0001875x^2 \end{aligned} \quad (51)$$

$$\begin{aligned} u_2(x,t) &= £^{-1}\{U_2(x,s)\} \\ &= £^{-1}\left\{\frac{1}{s}u_0^2(x) - \frac{1}{s}(2£\{A_1\} + 6£\{B_1\} + 3£\{C_1\} + £\{L_x(u_1(x,t))\}\right\} \\ &= £^{-1}\left\{\frac{1.875 \times 10^{-6}x}{s^2} + \frac{3.125 \times 10^{-7}x^4}{s}\right\} \\ &= 1.875 \times 10^{-6}tx + 3.125 \times 10^{-7}x^4 \end{aligned} \quad (52)$$

Table 9 Absolute error between the exact solution and approximate solution

$t\backslash x$	0.2	0.4	0.6	0.8	1
1	2.36426×10^{-7}	4.61916×10^{-7}	6.88176×10^{-7}	9.15535×10^{-7}	1.14427×10^{-6}
2	4.95579×10^{-7}	9.46654×10^{-7}	1.39934×10^{-6}	1.85428×10^{-6}	2.31204×10^{-6}
3	7.77459×10^{-7}	1.45422×10^{-6}	2.13348×10^{-6}	2.81624×10^{-6}	3.50331×10^{-6}
4	1.08207×10^{-6}	1.98461×10^{-6}	2.89062×10^{-6}	3.80141×10^{-6}	4.71809×10^{-6}
5	1.40941×10^{-6}	2.53783×10^{-6}	3.67075×10^{-6}	4.80981×10^{-6}	5.95640×10^{-6}

Table 10 L_2 and L_∞ errors

t	L_2	L_∞
1	8.22362×10^{-6}	1.14427×10^{-6}
2	1.65218×10^{-5}	2.31204×10^{-6}
3	2.48947×10^{-5}	3.50331×10^{-6}
4	3.33425×10^{-5}	4.71809×10^{-6}
5	4.18653×10^{-5}	5.95640×10^{-6}

and so on. In this manner, the rest of the components of the decomposition series were obtained. Substituting (50)–(52) into (3) gives the solution $u(x,t)$ in a series form and in a close form as

$$u(x,t) = -\frac{1}{40}\left(1 - 3sech^2\left(\frac{1}{20}\left(-6\times10^{-4}t + x\right)\right)\right). \tag{53}$$

This result can be verified through substitution. From Table 9, we find that Laplace ADM has a small error. The L_2 and L_∞ errors are contained in Table 10. In Fig. 5 the surface shows the exact solution (a) and the approximation solution (b) of Ito equation for $0 \le t \le 5$ and $-10 \le x \le 10$.

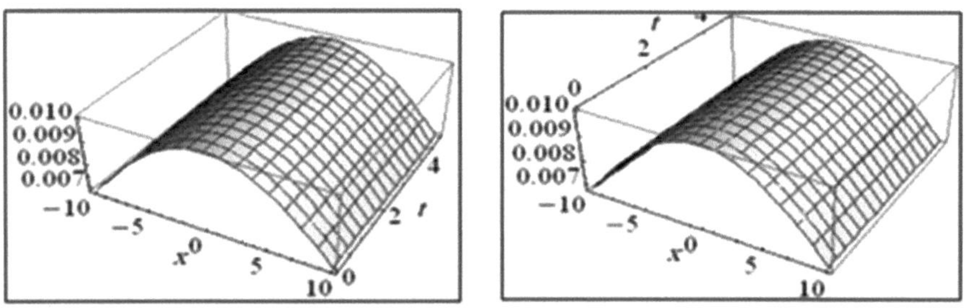

(a) The surface of the exact solution (b) The surface of the approximation solution

Fig. 5 The surface shows the exact and the approximate solution of Ito equation for $0 \le t \le 5$ and $-10 \le x \le 10$

4 Conclusions

In this paper, the new modification of Laplace ADM method has been proposed to obtain numerical solution of the FKdV equation. Numerical results showed that much smaller error with high spatial steps has been achieved by the current method. It is concluded that the new modification of Laplace ADM method has the capability of accurately solving the physical problem represented by the FKdV equation.

References

1. Wazwaz, A.M.: Partial Differential Equation and Solitary Waves Theory. Springer, Berlin (2009)
2. Adomian, G.: The fifth-order Korteweg–De Vries equation. Int. J. Math. Math. Sci. **19**(2), 415 (1996)
3. Kaya, D.: An explicit and numerical solutions of some fifth-order KdV equation by decomposition method. Appl. Math. Comput. **144**(2–3), 353–363 (2003)
4. Kaya, D.: An application for the higher order modified KdV equation by decomposition method. Commun. Nonlinear Sci. Numer. Simul. **10**(6), 693–702 (2005)
5. Kaya, D., El-Sayed, S.M.: On a generalized fifth order KdV equations. Phys. Lett. A. **310**(1), 44–51 (2003)
6. Yusfoglu, E.: Numerical solution of Duffing equation by the Laplace decomposition algorithm. Appl. Math. Comput. **177**, 572–580 (2006)
7. Kiymaz, O.: An algorithm for solving initial value problems using Laplace Adomian decomposition method. Appl. Math. Sci. **3**(30), 1453–1459 (2009)
8. Jafari, H., Khalique, C.M., Khan, M., Ghasemi, M.: A two-step Laplace decomposition method for solving nonlinear partial differential equations. Int. J. Phys. Sci. **6**(16), 4102–4109 (2011)
9. Adomian, G.: Solving Frontier problems of Physics: The Decomposition Method. Kluwer Academic, Boston (1994)
10. Wazwaz, A.M., El-Sayed, S.M.: A new modification of the Adomian decomposition method for linear and nonlinear operators. Appl. Math. Comput. **122**, 393–405 (2001)
11. Alvaro, H.S., Cesar, A.S., Jairo, E., Castillo, H.: Symbolic Computation of Solutions for the General Fifth-order KdV Equation. International Journal of Nonlinear Science. **9**(4), 394-401 (2010)

Extended Simpson Rule for Solving First-Order Fuzzy Differential Equations Using Generalized Differentiability

Reza Afsharinafar, Fudziah Ismail, Mohamed Suleiman and Ali Ahmadian Hosseini

Abstract In this present paper, extended Milne–Simpson rule is applied to find the fuzzy solutions of first order fuzzy differential equations (FDEs) under generalized Hukuhara differentiability concept. Based on the work done by Chalco-Cano we show how FDEs can be transformed to a system of ordinary differential equations. Then to find the solutions of fuzzy initial value problems (FIVPs), the Milne–Simpson formula is generalized. The errors, which guarantee pointwise convergence, are compared with other established methods which clearly show the advantage of our method for solving FIVPs.

1 Introduction

Fuzzy differential equations (FDEs) are studied as suitable setting for the modeling problems in science and engineering in which uncertainties or vagueness are permeated. Several approaches have been introduced for the solution of these problems [7], [8], [14], [18], [20] in which the Hukuhara differentiability, introduced by Puri and Ralescu [19], is used for fuzzy-number-valued mappings. However, their approaches have a disadvantage that the solutions have an increasing length of their support, which is figured out through interpreting FDEs as a family of differential inclusions by Hüllermeier [13]. Using differential inclusion has its turn shortcoming that the derivative of fuzzy-number-valued functions doesn't exist.

R. Afsharinafar (✉)
Institute for Mathematical Research, Universiti Putra Malaysia,
43400 Serdang, Selangor, Malaysia
e-mail: afshari@inspem.upm.edu.my

F. Ismail
Mathematics Department, Science Faculty, Universiti Putra Malaysia,
43400 Serdang, Selangor, Malaysia
e-mail: fudziah_i@yahoo.com

M. Suleiman · A. A. Hosseini
Universiti Putra Malaysia, 43400 Serdang, Selangor, Malaysia
e-mail: suleiman@fsas.upm.edu.my

A. A. Hosseini
e-mail: ahmadian.hosseini@gmail.com

A. Kilicman et al. (eds.), *International Conference on Mathematical Sciences and Statistics 2013*, DOI 10.1007/978-981-4585-33-0_10,

To solve this shortcoming the authors in [4] defined a new concept of derivative (*Generalized Hukuhara differentiability*) as a generalization of Hukuhara difference and studied in [2], [5], [6]. Using strongly generalized differentiability, the concept of derivative exists for more number of fuzzy-number-valued functions. Indeed, while the uniqueness is lost, the fuzzy initial value problem has two solutions with decreasing length. Therefore, this kind of derivative is used in this paper.

Different approaches such as Euler method, four-stage four-order Runge Kutta method, Predictor–Corrector method, Taylor method and Nyström method [16] presented for the solution of fuzzy differential equations. All the approaches use the Hukuhara derivative in which the fuzzy differential equations replaced by its parametric form. Another equivalent system introduced by authors in [3], [8]. Bede proved a characterization theorem to obtain the equivalent ODE systems under Hukuhara differentiability. In the current paper, we present the generalization of characterization theorem and, using strongly generalized differentiability, a numerical method is extended to solve numerically the ODE systems.

After preliminary system, we discuss the proved characterization theorems with its generalization to generalized Hukuhara differentiability. Then we extend Milne–Simpson rule to find the solutions of first-order fuzzy initial value problem under strongly generalized differentiability. At the end, some numerical examples are presented to illustrate our method with some conclusions for further research.

2 Preliminaries

First, we give some notations and present the necessary definitions, which will be used throughout the paper.

Consider, the initial value problem (1) for a single first-order differential equation

$$x' = f(t, x), \qquad x(a) = x_0 \tag{1}$$

The general linear multistep method for solving the initial value problem (1) can be written as

$$\sum_{j=0}^{k} \alpha_j x_{n+j} = h \sum_{j=0}^{k} \beta_j f_{n+j} \tag{2}$$

where, α_j and β_j are constant, $\alpha_k \neq 0$ and not both α_0 and β_0 are zero (it can be simplified by taking $\alpha_k = 1$). Moreover, x_n be an approximation to the theoretical solution at the mesh point t_n and $f_n = f(t_n, x_n)$.

We say that the method (2) is explicit if $\beta_k = 0$ and implicit otherwise. Several multistep methods can be derived from (2) such as the generalization of Milne–Simpson method called *Simpson's Rule*

$$x_{n+2} - x_n = \frac{h}{3}(f_{n+2} + 4f_{n+1} + f_n) \tag{3}$$

Definition 1 Let X be a nonempty set. A fuzzy set u in X is characterized by its membership function u: $X \to [0,1]$. Then $u(x)$ is interpreted as the degree of membership function of an element x in the fuzzy set u for each $x \in X$ [9].

Denote the class of fuzzy subsets of the real axis (i.e. u: $\mathbb{R} \to [0,1]$) by $\mathbb{R}_F$ such that:

i. u is normal, i.e., there exists $s_0 \in \mathbb{R}$ such that $u(s_0) = 1$,
ii. u is a convex fuzzy set (i.e. $u(ts)+(1-t)r) \geq \min\{u(s), u(r)\}, \forall t \in [0,1], s, r \in \mathbb{R}$,
iii. u is upper semicontinuous on $\mathbb{R}$,
iv. $cl\{s \in \mathbb{R}|u(s) > 0\}$ is compact, where cl denotes the closure of a subset.

So $\mathbb{R}_F$ is called the space of fuzzy numbers. Then for $0 < \alpha \leq 1$ if u belongs to $\mathbb{R}_F$, the α-level set $[u]^\alpha = \{s \in \mathbb{R}|u(s) \geq \alpha\}$ is a non-empty closed bounded interval. Also, the α-level set of u denoted by $[u^\alpha] = [\underline{u}^\alpha, \overline{u}^\alpha]$.

We refer to $\underline{u}$ and $\overline{u}$ as the lower and upper branches on u, respectively.

For $u, v \in \mathbb{R}_F$ and $\lambda \in \mathbb{R}$ the operations of sum and the scalar multiplication are defined as

$$[u+v]^\alpha = [u]^\alpha + [v]^\alpha$$

$$[\lambda \odot u]^\alpha = \lambda[u]^\alpha, \qquad \forall\alpha \in [0,1]$$

Also, we can extend the Hausdorff metric h to $\mathbb{R}_F$ by means

$$D(u, v) = \sup_{\alpha \in [0,1]} h([u]^\alpha, [v]^\alpha), \forall u, v \in \mathbb{R}_F.$$

$(\mathbb{R}_F, D)$ is a complete metric space and the following properties are well known:

$$D(u+w, v+w) = D(u, v), \forall u, v, w \in \mathbb{R}_F,$$

$$D(k \odot u, k \odot v) = |k|\, D(u, v), \ \forall k \in \mathbb{R}, u, v \in \mathbb{R}_F$$

$$D(u+v, w+e) \leq D(u, w) + D(v, e), \forall u, v, w, e \in \mathbb{R}_F.$$

Definition 2 Let $x, y \in \mathbb{R}_F$. If there exists $z \in \mathbb{R}_F$ such that $x = y + z$, then z is called the H-difference of x which is denoted by $x \ominus y$.

Definition 3 Let $F: I \to \mathbb{R}_F$ be a fuzzy function and $t_0 \in I$. We say that F is differentiable at t_0 if:

- An element $F'(t_0) \in \mathbb{R}_F$ exist such that, for all h > 0 sufficiently near 0, there are $F(t_0+h) \ominus F(t_0)$, $F(t_0) \ominus F(t_0-h)$ and the limits (in D-metric)

$$lim_{h \to 0^+} \frac{F(t_0+h) \ominus F(t_0)}{h} = lim_{h \to 0^+} \frac{F(t_0) \ominus F(t_0-h)}{h} = F'(t_0)$$

or

- There is an element $F^{'}(t_0) \in \mathbb{R}_F$ such that, for all $h < 0$ sufficiently near to 0, there are $F(t_0 + h) \ominus F(t_0)$, $F(t_0) \ominus F(t_0 - h)$ and the limits

$$lim_{h \to 0^-} \frac{F\,(t_0 + h) \ominus F\,(t_0)}{h} = lim_{h \to 0^-} \frac{F\,(t_0) \ominus F\,(t_0 - h)}{h} = F^{'}\,(t_0)$$

Note that if F is differentiable in the first form of Definition 3, it is not differentiable in the second form and vice versa.

Remark 4 In the previous definition, case (1) corresponds to the H-derivative introduced in [8], so this differentiability concept is a generalized of the H-derivative.

Remark 5 At the end of the paper [4], the authors' defined 4 cases to the generalized concepts of differentiability that in case 3 and 4 the derivative is reduced to a crisp element (Theorem 7 in [5]) so it is trivial. Therefore, we consider only the first two conditions of the definition.

Definition 6 Let F: $I \to \mathbb{R}_F$ be differentiable on I. We say that f is (1)-differentiable (D_1F) on I if F is differentiable according to case (1) and (2)-differentiable (D_2F) in the case (2) of Definition 3.

We can see more details of defined derivatives in [4], [8], [15]. Here, we make use of the following Theorem [2].

Theorem 7 If $F : I \to \mathbb{R}_F$ be a differentiable function and $[F(t)]^\alpha = [f^{'}_\alpha(t), g^{'}_\alpha(t)]$ denotes the α-level of F for each $\alpha \in [0,1]$ [8]. Then

1. If F is (1)-differentiable then f_α and g_α are differentiable functions and

$$[D_1F(t)]^\alpha = \left[f^{'}_\alpha(t), g^{'}_\alpha(t)\right].$$

2. If F is (2)-differentiable then f_α and g_α are differentiable functions and

$$[D_2F(t)]^\alpha = \left[g^{'}_\alpha(t), f^{'}_\alpha(t)\right].$$

3 Characterization Theorem for FDE Under Generalized Differentiability

Following Kaleva [15], Theorem 7 shows us a useful procedure to solve the fuzzy differential equation. Consider the fuzzy initial value problem

$$x^{'}\,(t) = f\,(t, x\,(t)), \qquad x\,(0) = x_0 \tag{4}$$

where f: $I \times \mathbb{R}_F \to \mathbb{R}_F$ is a continuous fuzzy mapping and x_0 is a fuzzy number. The interval I can be $[0, A]$ for $A > 0$ or can be $[0, \infty)$.

In the following we discuss the existence and uniqueness of the solutions.

Theorem 9 Let $f\colon I \times \mathbb{R}_F \to \mathbb{R}_F$ be a continuous fuzzy mapping and there exist a $k > 0$, such that $D(f(t,x), f(t,z)) \leq kD(x,z)$ for all $t \in I, x, z \in \mathbb{R}_F$. The problem (3) has two unique solutions on I (one solution for each derivative).

Then for solving FIVP (4), using theorem 7 we can translate it into a system of ODEs.

Let $[x(t)]^\alpha = [\underline{x}_\alpha(t), \overline{x}_\alpha(t)]$ and $[f(t,x(t))]^\alpha = [\underline{f}^\alpha(t,\underline{x}_\alpha,\overline{x}_\alpha), \overline{f}^\alpha(t,\underline{x}_\alpha,\overline{x}_\alpha)]$. Then $[x'(t)]^\alpha = [\underline{x}'_\alpha(t), \overline{x}'_\alpha(t)]$ for $x(t)$ be (1)-differentiable and FIVP (3) translates into the following system of ODEs:

$$\begin{cases} \underline{x}'(t) = \underline{f_\alpha}\left(t, \underline{x}_\alpha, \overline{x}_\alpha\right), \underline{x}\,(0) = \underline{x_0}, \\ \overline{x}'(t) = \overline{f}_\alpha\left(t, \underline{x}_\alpha, \overline{x}_\alpha\right), \overline{x}\,(0) = \overline{x_0}. \end{cases} \tag{5}$$

Also, if we consider $x'(t)$ by using the derivative in the second form ((2)-differentiability), then $[x'(t)]^\alpha = [\underline{x}'_\alpha(t), \overline{x}'_\alpha(t)]$ and, consequently, (3) substitute by following ODEs system:

$$\begin{cases} \underline{x}'(t) = \underline{f_\alpha}\left(t, \underline{x}_\alpha, \overline{x}_\alpha\right), \underline{x}\,(0) = \underline{x_0}, \\ \overline{x}'(t) = \overline{f}_\alpha\left(t, \underline{x}_\alpha, \overline{x}_\alpha\right), \overline{x}\,(0) = \overline{x_0}. \end{cases} \tag{6}$$

For solving FIVP (3), as we can see in [2], [15], after ensuring that the solution $\left[\underline{x}_\alpha(t), \overline{x}_\alpha(t)\right]$ and $\left[\underline{x}'_\alpha(t), \overline{x}'_\alpha(t)\right]$ are valid level sets of a fuzzy number valued function, by using stacking theorem [20], the fuzzy solution $x(t)$ can be constructed in both (1)-differentiability and (2)-differentiability forms.

The characterization theorem shows us how to deal with numerical solutions of FDEs (see [8], [15], [18]). The FDE can be translated equivalently to a system of ODEs under certain conditions. So any numerical method can be considered for solving the system of ODEs.

4 Extended Simpson's Rule of FDE Under Generalized Differentiability

In this section we replace the fuzzy differential equation equivalently with ODEs system under characterization theorems and then to approximate the two fuzzy solutions and solve numerically two ODE systems consist of four classic ordinary differential equations we extend the Generalized Milne–Simpson method (3) and give its error analysis.

Lemma 10 For $x_0 \in I$, the fuzzy differential equation (4) where $f\colon I \times \mathbb{R}_F \to \mathbb{R}_F$ is continuous, is equivalent to one of the integral equations [12]:

$$x(t) = x_0 + \int_0^t f(s, x(s))\, ds, \qquad \forall t \in I$$

or

$$x_0 = x(t) + (-1) \odot \int_0^t f(s, x(s))\, ds, \qquad \forall t \in I,$$

depending on the strongly differentiability considered, (1)-differentiability or (2)-differentiability, respectively.

Now, to extend the implicit Simpson's Rule (3) under generalized differentiability for solving fuzzy initial value problem (4), using Eqs. (5) and (6), let a set of discrete equally spaced grid points $0 = t_0 < t_1 < t_2 < \cdots < t_N = A$ in interval $I = [0, A]$ such that at which two exact solutions $[Y_1(t)]^\alpha = [\underline{Y_1}(t,\alpha), \overline{Y_1}(t,\alpha)]$ and $[Y_2(t)]^\alpha = [\underline{Y_2}(t,\alpha), \overline{Y_2}(t,\alpha)]$ are approximated by $[y_1(t)]^\alpha = [\underline{y_1}(t,\alpha), \overline{y_1}(t,\alpha)]$ and $[y_2(t)]^\alpha = [\underline{y_2}(t,\alpha), \overline{y_2}(t,\alpha)]$ respectively.

Here, also the grid points at which the solutions are calculated are $t_n = t_0 + nh$, $h = \frac{A}{N}$; and also, the exact and approximate solutions are denoted at t_n, $0 \le n \le N$ by $Y_{1n}(\alpha), Y_{2n}(\alpha)$, $y_{1n}(\alpha)$ and $y_{2n}(\alpha)$, respectively. The extended Adams-Moulton three-step method on the first-order approximation of $\underline{Y_1}'(t,\alpha), \overline{Y_1}'(t,\alpha), \underline{Y_2}'(t,\alpha)$, $\overline{Y_2}'(t,\alpha)$ is attached as follows:

$$\begin{cases} \underline{y_1}_{n+1}(\alpha) = \underline{y_1}_n(\alpha) + h\mathrm{F}(t_{n+1}, \underline{y_1}_{n+1\alpha}, \overline{y_1}_{n+1\alpha}), \\ \overline{y_1}_{n+1}(\alpha) = \overline{y_1}_n(\alpha) + h\mathrm{G}(\mathrm{t_{n+1}}, \underline{\mathrm{y_1}}_{\mathrm{n+1}\alpha}, \overline{\mathrm{y_1}}_{\mathrm{n+1}\alpha}), \\ \underline{y_1}_0(\alpha) = \underline{y_0}(\alpha), \\ \overline{y_1}_0(\alpha) = \overline{y_0}(\alpha), \end{cases} \tag{7}$$

$$\begin{cases} \underline{y_2}_{n+1}(\alpha) = \underline{y_2}_n(\alpha) + h\mathrm{G}(t_{n+1}, \underline{y_1}_{n+1\alpha}, \overline{y_1}_{n+1\alpha}), \\ \overline{y_2}_{n+1}(\alpha) = \overline{y_2}_n(\alpha) + h\mathrm{F}(t_{n+1}, \underline{y_1}_{n+1\alpha}, \overline{y_1}_{n+1\alpha}), \\ \underline{y_2}_0(\alpha) = \underline{y_0}(\alpha), \\ \overline{y_2}_0(\alpha) = \overline{y_0}(\alpha). \end{cases} \tag{8}$$

where

$$\begin{cases} F\left(t_n, \underline{y_1}_{n\alpha}, \overline{y_1}_{n\alpha}\right) = \frac{1}{3}[\underline{f_\alpha}(t_{n+1}, y(t_{n+1})) \\ +4\underline{f_\alpha}(t_n, y(t_n)) - \underline{f_\alpha}(t_{n-1}, y(t_{n-1}))] \\ G(t_n, \underline{y_1}_{n\alpha}, \overline{y_1}_{n\alpha}) = \frac{1}{3}[\overline{f}_\alpha(t_{n+1}, y(t_{n+1})) \\ +4\overline{f}_\alpha(t_n, y(t_n)) - \overline{f}_\alpha(t_{n-1}, y(t_{n-1}))] \end{cases} \tag{9}$$

5 Convergence

The exact and approximated solutions at t_n, $0 \le n \le N$, are denoted by $[Y_i(t)]^\alpha = [\underline{Y_i}(t,\alpha), \overline{Y_i}(t,\alpha)]$ and $[y_i(t)]^\alpha = [\underline{y_i}(t,\alpha), \overline{y_i}(t,\alpha)]$ for $i = 1{,}2$, respectively. The solutions are calculated at grid points in previous section. From Eqs. (7)–(9), we have

$$\underline{y}\,(t_{n+1},\alpha) = \underline{y}\,(t_n,\alpha) + hF(t_{n+1},\underline{y}_{n+1}\,(\alpha),\overline{y}_{n+1}\,(\alpha)\,)$$

$$\overline{y}(t_{n+1},\alpha) = \overline{y}(t_n,\alpha) + hG(t_{n+1},\underline{y}_{n+1}(\alpha),\overline{y}_{n+1}(\alpha)),$$

also,

$$\underline{Y}\,(t_{n+1},\alpha) \approx \underline{Y}\,(t_n,\alpha) + h\mathrm{F}(t_{n+1},\underline{Y}_{n+1}\,(\alpha),\overline{Y}_{n+1}\,(\alpha)\,)$$

$$\overline{Y}\,(t_{n+1},\alpha) \approx \overline{Y}\,(t_n,\alpha) + h\mathrm{G}(t_{n+1},\underline{Y}_{n+1}(\alpha),\overline{Y}_{n+1}(\alpha)).$$

Theorem 13 The implicit Simpson's approximate of case (1)-differentiability and (2)-differentiability (Eq. (7) and (8)) converge to the exact solutions $\underline{\mathrm{Y}}(\mathrm{t},\alpha),\overline{\mathrm{Y}}(\mathrm{t},\alpha)$ for arbitrary fixed α, where $\underline{Y},\overline{Y} \in C^1(K)$.

Proof Simpson's rule is constructed from two simpler approximations: the midpoint method and the trapezoidal method. The errors and the convergence of these methods considered in [1] in case (1)-differentiability and also, for (2)-differentiability.

6 Examples

Example 1 Let us consider the initial value problem

$$\mathrm{y}'\,(\mathrm{t}) = -\mathrm{y}\,(\mathrm{t})\,, \qquad \mathrm{y}\,(0) = (\alpha - 1{,}1 - \alpha). \tag{10}$$

For solving the problem by presented method, divide I into $N = 10$ equally spaced subintervals.

Using systems (5) and (6), the exact solutions for case (1)-differentiability and case (2)-differentiability are as follows, respectively:

$$\begin{cases} x\,(t,\alpha) = \left[(\alpha - 1)\,e^{\mathrm{t}},\ (1 - \alpha)\,e^{t}\right] \\ x\,(t,\alpha) = \left[(\alpha - 1)\,e^{-t},\ (1 - \alpha)\,e^{-t}\right] \end{cases}$$

The results of the exact and approximate solutions and a comparison between the errors of presented method with the numerical methods applied by Allahviranloo [2] and Nieto [17] in case (1)-differentiability and (2)-differentiability are shown in Tables 1 and 2, and Figs. 1 and 2.

Table 1 The comparison between Simpson and existing methods errors in case (1)-differentiability

α	Simpson errors		Predictor–Corrector errors		Euler errors	
0	0.6101e − 11	0.6101e − 11	0.6522e − 2	0.4298e − 2	0.4553e − 3	0.4553e − 3
0.1	0.5491e − 11	0.5491e − 11	0.5836e − 2	0.3868e − 2	0.4098e − 3	0.4098e − 3
0.2	0.4881e − 11	0.4881e − 11	0.5150e − 2	0.3438e − 2	0.3643e − 3	0.3643e − 3
0.3	0.4271e − 11	0.4271e − 11	0.4464e − 2	0.3008e − 2	0.3187e − 3	0.3187e − 3
0.4	0.3661e − 11	0.3661e − 11	0.3778e − 2	0.2578e − 2	0.2732e − 3	0.2732e − 3
0.5	0.3050e − 11	0.3050e − 11	0.3092e − 2	0.2149e − 2	0.2277e − 3	0.2277e − 3
0.6	0.2440e − 11	0.2440e − 11	0.2406e − 2	0.1719e − 2	0.1821e − 3	0.1821e − 3
0.7	0.1830e − 11	0.1830e − 11	0.1720e − 2	0.1289e − 2	0.1366e − 3	0.1366e − 3
0.8	0.1220e − 11	0.1220e − 11	0.1034e − 2	0.8595e − 3	0.0911e − 3	0.0911e − 3
0.9	0.0610e − 11	0.0610e − 11	0.3470e − 3	0.4297e − 3	0.0455e − 3	0.0455e − 3
1	0	0	0	0	0	0

Table 2 The comparison between Simpson and existing methods errors in case (2)-differentiability

α	Simpson errors		Predictor–Corrector errors		Euler errors	
0	0.5062e − 11	0.5062e − 11	0.6522e − 2	0.4298e − 2	0.4553e − 3	0.4553e − 3
0.1	0.4556e − 11	0.4556e − 11	0.5836e − 2	0.3868e − 2	0.4098e − 3	0.4098e − 3
0.2	0.4050e − 11	0.4050e − 11	0.5150e − 2	0.3438e − 2	0.3643e − 3	0.3643e − 3
0.3	0.3544e − 11	0.3544e − 11	0.4464e − 2	0.3008e − 2	0.3187e − 3	0.3187e − 3
0.4	0.3037e − 11	0.3037e − 11	0.3778e − 2	0.2578e − 2	0.2732e − 3	0.2732e − 3
0.5	0.2531e − 11	0.2531e − 11	0.3092e − 2	0.2149e − 2	0.2277e − 3	0.2277e − 3
0.6	0.2025e − 11	0.2025e − 11	0.2406e − 2	0.1719e − 2	0.1821e − 3	0.1821e − 3
0.7	0.1519e − 11	0.1519e − 11	0.1720e − 2	0.1289e − 2	0.1366e − 3	0.1366e − 3
0.8	0.1012e − 11	0.1012e − 11	0.1034e − 2	0.8595e − 3	0.0911e − 3	0.0911e − 3
0.9	0.0506e − 11	0.0506e − 11	0.3470e − 3	0.4297e − 3	0.0455e − 3	0.0455e − 3
1	0	0	0	0	0	0

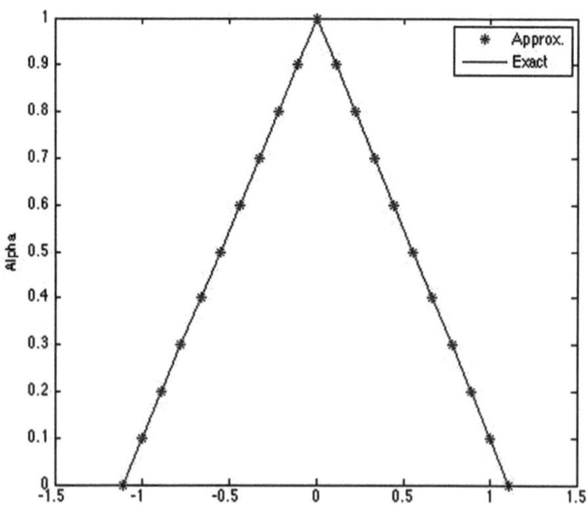

Fig. 1 The comparison between the exact and approximate solutions in case (1)-differentiability

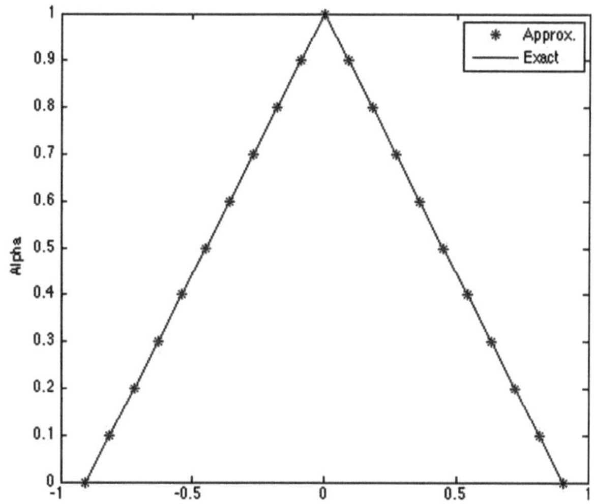

Fig. 2 The comparison between the exact and approximate solutions in case (2)-differentiability

Example 2 Consider the problem

$$y'(t) = -y(t) + t + 1, \qquad y(0) = y_0 \tag{11}$$

where y_0 is a fuzzy number. Let I $= [0,0.1]$ and $y_0 = (0.96 + 0.04\alpha, 1.01 - 0.01\alpha)$.

To solve problem (11) we divide I into $N = 10$ equally spaced subintervals. We can get the exact and approximate solutions related to case (1)-differentiability and case (2)-differentiability solutions of the problem. But to compare with other existing method, only the results of case (1)-differentiability are presented and compared with Neural Network method [10]:

The exact solution at $t = 0.1$ is given by

$$\begin{cases} \underline{Y}(t,\alpha) = (0.985 + 0.015\alpha)\, e^{-t} - 0.025\,(1-\alpha)\, e^{t} + t \\ \overline{Y}(t,\alpha) = (0.985 + 0.015\alpha)\, e^{-t} + 0.025\,(1-\alpha)\, e^{t} + t \end{cases}$$

and the Table 3 and Fig. 3 will show the comparison between the errors of presented method for case (1)-differentiability in compare with Neural Network method.

7 Conclusion

A generalized representation theorem has been presented for the approximate solution of FDEs under generalized differentiability. As a matter of fact by stacking theorem a fuzzy differential equation can be transferred into two systems of ODEs, which are solved by suitable numerical methods and then by stacking theorem we can bunch the fuzzy solutions of FDE. Plus, we discussed Simpson's rule as an implicit method under generalized differentiability that shows the more accuracy than Hukuhara derivative. As a future work, we will extend numerical methods for second order fuzzy differential equations.

Table 3 The comparison between the exact and approximate solutions (case (1)-differentiability)

α	Simpson errors		Neural network errors	
0	0.5139111e − 11	0.4833689e − 11	5.067853e − 5	3.753089e − 5
0.1	0.5131117e − 11	0.4857003e − 11	3.873561e − 5	4.153197e − 5
0.2	0.5123679e − 11	0.4879430e − 11	2.067133e − 5	7.614799e − 6
0.3	0.5115796e − 11	0.4902300e − 11	6.021995e − 5	3.922612e − 5
0.4	0.5108358e − 11	0.4925615e − 11	7.410150e − 6	4.092031e − 5
0.5	0.5100475e − 11	0.4948264e − 11	6.225859e − 6	3.571873e − 5
0.6	0.5092926e − 11	0.4970912e − 11	3.472099e − 5	2.494235e − 5
0.7	0.5085376e − 11	0.4994227e − 11	4.320305e − 5	4.355865e − 5
0.8	0.5077605e − 11	0.5016653e − 11	1.614909e − 5	2.703282e − 5
0.9	0.5070166e − 11	0.5039524e − 11	1.253534e − 5	3.883441e − 5
1	0	0	0	0

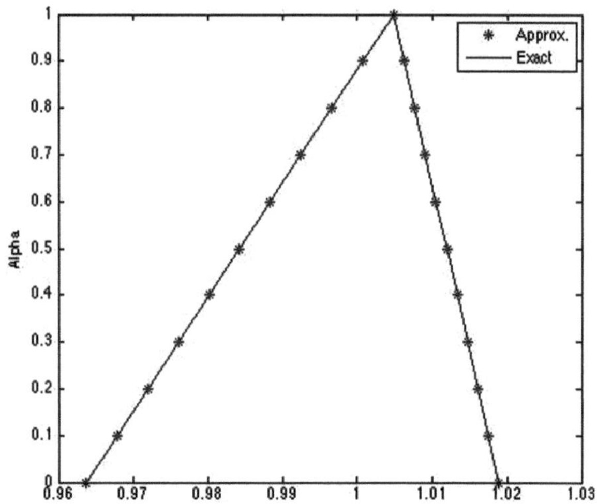

Fig. 3 The comparison between the exact and approximate solutions case (1)-differentiability

References

1. Afsharinafar, R., Ismail, F., Ahmadian, A.: Extended midpoint method for solving fuzzy differential equations. IEEE Symposium on humanities, Science and Engineering Research (2012)
2. Allahviranloo, T., Abbasbandy, S., Ahmady, N., Ahmady, E.: Improved Predictor–Corrector method for solving fuzzy initial value problems. Info. Sci. **179**, 945–955 (2009)
3. Bede, B.: Note on "Numerical solutions of fuzzy differential equations by predictor corrector method". Info. Sci. **178**, 1917–1922 (2008)
4. Bede, B., Gal, S.G.: Almost periodic fuzzy-number-valued functions. Fuzzy Set. Syst. **147**, 385–403 (2004)
5. Bede, B., Gal, S.G.: Generalizations of the differentiability of fuzzy-number-valued functions with applications to fuzzy differential equations. Fuzzy Set. Syst. **151**, 581–599 (2005)
6. Bede, B., Rudas, I.J., Bencsik, A.L.: First order linear fuzzy differential equations under generalized differentiability. Info. Sci. **177**, 1648–1662 (2007)
7. Buckley, J.J., Feuring, T.: Fuzzy differential equations. Fuzzy Set. Syst. **110**, 43–54 (2000)

8. Chalco-Cano, Y., Román-Flores, H.: On new solutions of fuzzy differential equations, Chaos, Solitons & Fractals **38**, 112–119 (2008)
9. Diamond, P., Kloden, P.: Metric Spaces of Fuzzy Sets. World Scientific, Singapore (1994)
10. Effati, S., Pakdaman, M.: Artificial neural network approach for solving fuzzy differential equations. Info. Sci. **180**, 1434–1457 (2010)
11. Friedman, M., Ma, M., Kandel, A.: Numerical solution of fuzzy differential and integral equations. Fuzzy Set. Syst. **106**, 35–48 (1999)
12. Goeken, D., Johnson, O.: Runge–Kutta with higher derivative approximations, Appl. Numer. Math. **39**, 249–257 (2000)
13. Hullermeier, E.: An approach to modeling and simulation of uncertain dynamical systems, Int. J. Uncertain. Fuzz. **5**,117–137 (1997)
14. Kaleva, O.: Fuzzy differential equations. Fuzzy Set. Syst. **24**, 301–317 (1987)
15. Kaleva, O.: A note on fuzzy differential equations. Nonlinear Anal. **64**, 895–900 (2006)
16. Khastan, A., Ivaz, K.: Numerical solution of fuzzy differential equations by Nyström method. Chaos, Solitons & Fractals **41**, 859–868 (2009)
17. Nieto, J.J., Rodriguez-Lopez, R.: Hybrid metric dynamical systems with impulses. Nonlinear Anal. **64**, 368–380 (2006)
18. Nieto, J.J., Khastan, A., Ivaz, K.: Numerical solution of fuzzy differential equations under generalized differentiability, Nonlinear Anal. Hybrid Syst. **3**, 700–707 (2009)
19. Puri, M., Ralescu, D.: Differential and fuzzy functions. J. Math. Anal. Appl. **91**, 552–558 (1983)
20. Seikkala, S.: On the fuzzy initial value problem. Fuzzy Set. Syst. **24**, 319–330 (1987)

Modified Decomposition Method for Solving Nonlinear Volterra–Fredholm Integral Equations

F. S. Zulkarnain, Z. K. Eshkuvatov, Z. Muminov and N. M. A. Nik Long

Abstract In this note, an approximate solution of nonlinear Volterra–Fredholm integral equation is obtained by using modified decomposition method. General cases of nonlinear terms in the equations are considered. Finally, some numerical examples are presented to validate the accuracy and efficiency of the method.

1 Introduction

Adomian decomposition method [1], [2] has been focus on many stochastic and deterministic problems in physics, biology, and chemical reactions. In mathematical problems, this method has been applied for solving a large class of linear and nonlinear equations including differential and integral equations, integro-differential equations, and so on.

The Volterra–Fredholm integral equation arises from parabolic boundary value problems, the mathematical modeling of the spatio-temporal development of an epidemic and biological problems. The essential features of these models are of wide applicability. Detailed descriptions and formulation of these models can be found in [3], [4] and references therein.

Particularly, Bildik and Inc [5] have solved the following nonlinear Volterra–Fredholm integral equation by using modified decomposition method,

$$u(x) = f(x) + \lambda_1 \int_a^x K_1(x,t)[u(t)]^p dt + \lambda_2 \int_a^b K_2(x,t)[u(t)]^q dt \tag{1}$$

where p is a positive, $q = 1, f(x)$ is the given function, $K_1(x,t)$ and $K_2(x,t)$ are square integrable kernels, λ_1, λ_2, a and b are constants. Yalçinbaş [6] have obtained an approximation solution of Eq. (1) by using Kanwal and Liu method.

F. S. Zulkarnain (✉) · Z. Muminov
Department of Mathematics, Faculty of Science, Universiti Putra Malaysia (UPM), Serdang, Selangor, Malaysia
e-mail: fsamihah88@gmail.com

Z. K. Eshkuvatov · N. M. A. N. Long
Institute for Mathematical Research (INSPEM), Universiti Putra Malaysia (UPM), Serdang, Selangor, Malaysia

A. Kilicman et al. (eds.), *International Conference on Mathematical Sciences and Statistics 2013*, DOI 10.1007/978-981-4585-33-0_11,

Maleknejad and Hadizadeh [7] have presented the solution for following nonlinear mixed Volterra–Fredholm integral equation by using the standard decomposition method

$$u(x,t) = f(x,t) + \int_0^t \int_\Omega F(x,t,\xi,\tau,u(\xi,\tau))\,d\xi\,d\tau, (x,t) \in \Omega \times [0,T], \quad (2)$$

where $(x,t) \in \Omega \times [0,T]$ and $u(x,t)$ is an unknown function, $f(x,t)$ and $F(x,t,\xi,\tau,u(\xi,\tau))$ are analytic functions on $D = \Omega \times [0,T]$, where Ω is a closed subset of R^n, $n = 1,2,3$. Wazwaz [9] has treated Eq. (2) using modified decomposition method.

In this paper, we applied the modified decomposition method (MDM) to obtain approximate solution of nonlinear Volterra–Fredholm integral equation for any $p, q \geq 1$. Moreover, we have shown four examples to validate the method proposed.

2 Modified Decomposition Method (MDM)

2.1 *General Scheme for Operator Equation*

Let us consider the operator equation of the form

$$Lu + Ru + Nu = g\,(x), \quad (3)$$

where L is the operator of highest order derivative which is assumed to be invertible, R is the remaining linear operator collecting the lower order derivatives, Nu represents the nonlinear terms and g is a source term. Multiplying inverse function L^{-1} to both sides of Eq. (3) we get

$$u = f\,(x) - L^{-1}\,(Ru) - L^{-1}\,(Nu), \quad (4)$$

where $f(x) = L^{-1}(g(x))$. In standard decomposition method, the unknown function $u(x)$ is searched as a series of the form

$$u(x) = \sum_{n=0}^{\infty} u_n(x), \quad (5)$$

where the components u_0, u_1, $u_2, \ldots$ are usually determined recursively by

$$u_0(x) = f(x), \quad u_{n+1}(x) = -L^{-1}(Ru_n(x)) - L^{-1}(Nu_n(x)), \quad n \geq 0. \quad (6)$$

It is noted that the decomposition method suggests the initial component $u_0(x)$ as the function $f(x)$. To apply modified decomposition method, the function $f(x)$ on the right side of the Eq. (2) is divided into two parts

$$f(x) = f_1(x) + f_2(x). \quad (7)$$

Based on the expression above, Wazwaz [9] proposed a slight variation only on the components $u_0(x)$ and $u_1(x)$. He proposed that $f_1(x)$ contain the zeroth component u_0 while $f_2(x)$ is the remainder terms. Thus, the following recursive relations for the modified decomposition method are formulated as

$$\begin{aligned} u_0(x) &= f_1(x), \\ u_1(x) &= f_2(x) - L^{-1}(Ru_0(x)) - L^{-1}(Nu_0(x)), \\ u_{n+1}(x) &= -L^{-1}(Ru_n(x)) - L^{-1}(Nu_n(x)), \quad n \geq 1. \end{aligned} \tag{8}$$

We can conclude that the zeroth component of standard decomposition method in Eq. (4) is defined by the function f while in modified decomposition method, the zeroth component is defined by selecting some terms $f_1(x)$ of function $f(x)$ The remaining part will be added to the next component u_1. Easy calculations and fast solutions in many cases depend on our proper choice of the parts f_1 and f_2 of f.

2.2 Application of MDM for Nonlinear Volterra–Fredholm Integral Equations

We consider nonlinear Volterra–Fredholm equation

$$u(x) = f(x) + \lambda_1 \int_a^b K_1(x,t)\, F_1(u(t))\, dt + \lambda_2 \int_a^x K_2(x,t)\, F_2(u(t))\, dt, \tag{9}$$

where $f(x)$ is the given function, $K_1(x,t)$ and $K_2(x,t)$ are square integrable kernels of the equation, λ_1 and λ_2 are parameters, while $F_1(u(t))$ and $F_2(u(t))$ are nonlinear terms, and $u(x)$ is unknown function to be determined.

The nonlinear terms $F_1(u(t))$ and $F_2(u(t))$ are represented by the Adomian polynomial series. Wazwaz [9] introduced the procedure to calculate the Adomian polynomials

$$F_1(u(t)) = \sum_{n=0}^{\infty} A_n(t),\ F_2(u(t)) = \sum_{n=0}^{\infty} B_n(t), \tag{10}$$

where $A_n(t)$ and $B_n(t)$ are Adomian polynomials defined by

$$\begin{aligned} C_0 &= F(u_0), \\ C_1 &= u_1 F'(u_0), \\ C_2 &= u_2 F'(u_0) + \frac{1}{2!} u_1^2 F''(u_0), \\ C_3 &= u_3 F'(u_0) + u_1 u_2 F''(u_0) + \frac{1}{3!} u_1^3 F'''(u_0), \\ &\vdots \end{aligned} \tag{11}$$

Here, $A_i = C_i$ if $F = F_1$ and $B_i = C_i$ if $F = F_2$. The nonlinear terms can be considered in different cases. For example, for the case of $F(u) = (u(x))^p$, the first few Adomian polynomials C_n are given by

$$
\begin{aligned}
C_0 &= u_0^p, \\
C_1 &= p u_0^{p-1} u_1, \\
C_2 &= p u_0^{p-1} u_2 + \frac{p(p-1)}{2!} u_0^{p-2} {u_1}^2, \\
C_3 &= p u_0^{p-1} u_3 + p(p-1) u_0^{p-2} u_1 u_2 + \frac{p(p-1)(p-2)}{3!} u_0^{p-3} u_1^3, \\
&\vdots
\end{aligned}
\tag{12}
$$

For the case of $F(u) = \sin u(x)$, few terms of Adomian polynomials A_n are defined by

$$
\begin{aligned}
C_0 &= \sin u_0, \\
C_1 &= u_1 \cos u_0, \\
C_2 &= u_2 \cos u_0 - \frac{1}{2!} u_1^2 \sin u_0, \\
C_3 &= u_3 \cos u_0 - u_1 u_2 \sin u_0 - \frac{1}{3!} u_1^3 \cos u_0, \\
&\vdots
\end{aligned}
\tag{13}
$$

For the case of $F(u) = e^{u(t)}$, the first few Adomian polynomials C_n are given in the form

$$
\begin{aligned}
C_0 &= e^{u_0}, \\
C_1 &= u_1 e^{u_0}, \\
C_2 &= \left(u_2 + \frac{1}{2!} u_1^2 \right) e^{u_0}, \\
C_3 &= \left(u_3 + u_1 u_2 + \frac{1}{3!} u_1^3 \right) e^{u_0}, \\
&\vdots
\end{aligned}
\tag{14}
$$

Thus, the components $u_0, u_1, u_2, \ldots$ of $u(x)$ substitute into series in Eq. (5) and each component is defined by

$$
\begin{aligned}
u_0(x) &= f_1(x), \\
u_1(x) &= f_2(x) + \lambda_1 \int_a^x K_1(x,t)\, A_0(t)\, dt + \lambda_2 \int_a^b K_2(x,t)\, B_0(t) dt, \\
u_{n+1}(x) &= \lambda_1 \int_a^x K_1(x,t)\, A_n(t)\, dt + \lambda_2 \int_a^b K_2(x,t)\, B_n(t)\, dt, \quad n \geq 1
\end{aligned}
\tag{15}
$$

3 Examples

Example 1 Consider the nonlinear Fredholm integral equation of the second kind

$$u(x) = 1 - \frac{5}{4}x + \int_0^1 3xt\ u^2(t)\,dt, \tag{16}$$

Solution It is clear that $f(x) = 1 - \frac{5}{4}x$, $\lambda_1 = 3$, $\lambda_2 = 0$ and $K_1(x,t) = xt$. Since $\lambda_2 = 0$ the Adomian polynomials $B_n = 0$, $n = 0,1,\ldots$ and A_n for nonlinear term $F_1(u(t)) = u^2(t)$ are given in (12) with $p = 2$.

$$\begin{aligned} A_0 &= u_0^2, \\ A_1 &= 2u_0u_1, \\ A_2 &= 2u_0u_2 + u_1^2, \\ A_3 &= 2u_0u_3 + 2u_1u_2, \\ &\vdots \end{aligned} \tag{17}$$

Now, we divided *f(x)* into two parts such as $f_1(x) = 1 - x$ and $f_2(x) = -\frac{1}{4}x$. Using the modified recursive relation (15) we obtain

$$u_1(x) = -\frac{1}{4}x + \int_0^1 3xt\ A_0(t)\,dt = 0,$$

consequently

$$u_k(x) = 0, \quad k \geq 1.$$

Then, the solution is

$$u(x) = 1 - x.$$

Example 2 Consider the nonlinear Volterra integral equation of the second kind

$$u(x) = 4x^2 + \frac{x}{2}(\cos(4x^2) - 1) + 4\int_0^x xt\ \sin u(t)\,dt, \tag{18}$$

Solution From Eq. (9) we found that

$$\begin{aligned} &f(x) = 4x^2 + \frac{x}{2}(\cos(4x^2) - 1), \quad \lambda_1 = 0, \quad \lambda_2 = 4, \\ &K_1(x,t) = 0, K_2(x,t) = xt. \end{aligned} \tag{19}$$

The nonlinear term $F_2(u) = \sin u(x)$ are given in Eq. (13). Again, we divide *f(x)* into two parts by selecting $f_1(x) = 4x^2$ and $f_2(x) = \frac{x}{2}(\cos(4x^2) - 1)$. Then, it follows that

$$u_0(x) = 4x^2.$$

Consequently, $u_1(x)$ is defined by

$$u_1(x) = \frac{x}{2}(\cos(4x^2) - 1) + 4\int_0^x xt\ A_0\, dt, = 0.$$

so that

$$u_k(x) = 0, \quad k \geq 1.$$

Thus, the solution is

$$u(x) = 4x^2.$$

Example 3 Consider the nonlinear Volterra–Fredholm integral equation,

$$u(x) = \frac{5}{4} + \frac{2}{3}x - e^x + \int_0^1 (x-t)u^2(t)dt + \int_0^x e^{u(t)}dt \tag{20}$$

Solution Note that

$$f(x) = \frac{5}{4} + \frac{2}{3}x - e^x, \quad \lambda_1 = 1, \quad \lambda_2 = 1,$$
$$K_1 = x - t, \quad K_2 = 1.$$

The Adomian polynomials for $F_1(t) = u^2(t)$ are given in (17) while for $F_2(t) = e^{u(t)}$ are given in (14). We choose $f_1(x) = x$ as the initial value. Then it follows from Eq. (15) that

$$u_0(x) = x,$$
$$u_1(x) = \frac{5}{4} + \frac{1}{3}x - e^x + \int_0^1 (x-t)t^2 dt + \int_0^x e^t dt = 0.$$

Consequently,

$$u_k(x) = 0, \quad k \geq 1.$$

Then, the solution is

$$u(x) = x.$$

Example 4 Consider the nonlinear Volterra–Fredholm integral equation,

$$u(x) = \frac{1}{15}(20x - 12x^6) + \frac{1}{2}\int_0^1 xu^2(t)dt + \frac{1}{4}\int_0^x xu^4(t)dt. \tag{21}$$

Its exact solution is $u(x) = 2x$.

Table 1 Numerical results for Example 4

x	Exact solution $u(x)$	MDM (five iterations) with initial assumption $u_0(x) = \frac{4}{3}x$	MDM (five iterations) with initial assumption $u_0(x) = \frac{999}{500}x$
0	0	0	0
0.1	0.2	0.17106581	0.19970755
0.2	0.4	0.34210702	0.39941485
0.3	0.6	0.51290891	0.59911980
0.4	0.8	0.68263034	0.79881408
0.5	1.0	0.84902789	0.99847536
0.6	1.2	1.00712624	1.19805306
0.7	1.4	1.14667287	1.39743720
0.8	1.6	1.24685921	1.59637529
0.9	1.8	1.26622664	1.79420977
1.0	2.0	1.12955769	1.98889004

Solution Note that

$$f(x) = \frac{1}{15}\left(20x - 12x^6\right), \quad \lambda_1 = \frac{1}{2}, \quad \lambda_1 = \frac{1}{4},$$
$$K_1(x,t) = x, \quad K_2(x,t) = x.$$

The Adomian polynomials for $F_1(t) = u^2(t)$ are given in (17) and $F_2(t) = u^4(t)$ are given in (12) with $p = 4$. To use the MDM we compare two initial assumptions: $u_0(x) = \frac{4}{3}x$ and $u_0(x) = \frac{999}{500}x$. By using the recursive relation in Eq. (15) we present the numerical results in Table 1. The results in Table 1 are obtained by substituting the five terms into the series Eq. (5).

Based on Table 1, initial value show $u_0(x) = \frac{999}{500}x$ that MDM gives more approximate to the exact solution than initial value $u_0(x) = \frac{4}{3}x$.

4 Conclusion

In this paper, we have solved general nonlinear Volterra–Fredholm integral equations by using modified decomposition method. From the numerical Examples 1–3 we can see that MDM is identical with the exact solution for the given problems considered. In Example 4 (Table 1) developed MDM is not exact but rate of convergence of MDM can be accelerated by the choice of initial guess.

Acknowledgements This work was supported by University Putra Malaysia under Fundamental Research GrandScheme (FRGS). Project code is 01–12-10-989FR.

References

1. Adomian, G.: Solving Frontier Problems of Physics: The Decomposition Method. Kluwer, Boston (2004)
2. Adomian, G.: Comp. Math. Appl. **21**, 101–127 (1991)
3. Diekmann, O.: Thresholds and travelling waves for the geographical spread infection. J. Math. Biol. **6**, 109–130 (1978)
4. Thieme, H.R.: A model for the spatial spread of an epidemic. J. Math. Biol. **4**, 337–351 (1977)
5. Bildik, N., Inc, M.: Modified decomposition method for nonlinear method Volterra–Fredholm integral equations. Chao. Solit. Fract. **33**, 308–313 (2007)
6. Yalçinbaş, S.: Taylor polynomial solutions of nonlinear Volterra–Fredholm integral equations. Appl. Math. Comp. **127**, 195–206 (2002)
7. Maleknejad, K., Hadizadeh, M.: A new computational method for Volterra–Fredholm integral equations. Comp. Math. Appl. **37**, 1–8 (1999)
8. Wazwaz, A. M.: A new algorithm for calculating Adomian polynomials for nonlinear operators. Appl. Math. Comp. **111**, 53–69 (2000)
9. Wazwaz, A. M.: A reliable treatment for mixed Volterra–Fredholm integral equations. Appl. Math. Comp. **127**, 405–414 (2002)
10. Wazwaz, A. M.: A reliable modification of Adomian decomposition method. Appl. Math. Comp. **102**, 77–86 (1999)
11. Shah, A. W., Shaikh, S., Sandilo, S., H.: Modified decomposition method for nonlinear Volterra–Fredholm integro-differential equation. J. Basic Appl. Sci. **6**, 13–16 (2010)

Solution of Second Order Ordinary Differential Equations by Direct Diagonally Implicit Block Methods

Noorini Zainuddin, Zarina Bibi Ibrahim, Mohamed Suleiman, Khairil Iskandar Othman and Yong Faezah Rahim

Abstract This paper presents a diagonally implicit method for directly solving stiff second-order ordinary differential equations (ODEs) which produce a block of two new values at each step of application. The formulation of the method is described. Numerical results are given to demonstrate the efficiency of the proposed method.

1 Introduction

The problems of Ordinary Differential Equations (ODEs) arise in many fields in applied sciences. These include particle tracking problems, structural mechanic problems and network analysis problems [1]. These ODE problems can be categorized into stiff and non-stiff.

Stiff ODEs are defined as the presence of one or more fast-decay processes in time, with time constants that are short compared to the time span of interest [1]. Many papers have been published on the studies of stiff ODEs [1]–[6]. In Hindmarsh's paper, several solvers of large ODEs problems based on implicit Adams for the case of non-stiff problems and Backward Differentiation Formulas (BDFs) method for the case of stiff problems are discussed. These solvers include the Livermore Solver for ODEs (LSODE) and the variant of it Livermore Solver for ODEs (implicit form) (LSODI).

Block method for solving ODEs has been discussed by many researchers such as Shampine and Watts [7], Ibrahim et al. [5] and Olabode [8]. Olabode derived a five-step block method which solves third-order ODEs directly. Ibrahim discussed the two- and three-point block methods for solving first-order stiff ODEs.

Z. B. Ibrahim (✉) · N. Zainuddin · M. Suleiman · Y. F. Rahim
Institute for Mathematical Research, Department of Mathematics, Faculty of Science, Universiti Putra Malaysia, 43400 Serdang, Selangor, Malaysia
e-mail: zarinabb@science.upm.edu.my

K. I. Othman
Department of Mathematics, Faculty of Computer and Mathematical Sciences, UiTM Malaysia, 40450 Shah Alam, Selangor, Malaysia

N. Zainuddin
Fundamental and Applied Sciences Department, Universiti Teknologi Petronas, 31750 Tronoh, Perak Darul Ridzuan, Malaysia

A. Kilicman et al. (eds.), *International Conference on Mathematical Sciences and Statistics 2013,* DOI 10.1007/978-981-4585-33-0_12,

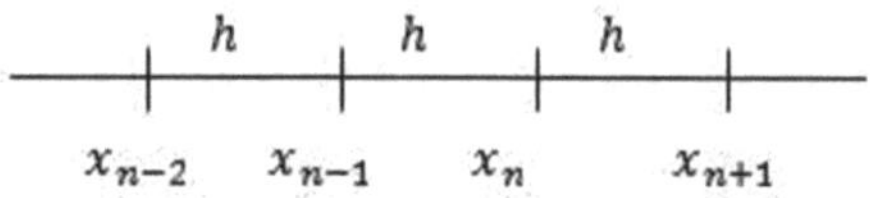

Fig. 1 2DBBDF for y_{n+1}

Direct solving of higher order ODEs has been suggested by several researchers. Among them are Lambert [9], Suleiman [10], Fatunla [11] and Merkanoon [12]. Fatunla [11] solved special-second order ODEs directly by using block approach. Merkanoon proposed a method which solves third-order ODEs directly. The method is based on Adams' method which produces multiple solutions in a block. In Suleiman's paper [10], the convergence of direct ODEs solvers has been discussed.

This paper proposed a direct solver for stiff ODEs based on the BDFs method. The method derived is in lower triangular matrix form which is called diagonally implicit [13] and approximates the numerical solutions in block term. The method enhances the works done by Ibrahim [5] to become direct solver of second-order ODEs.

2 Formulations of 2DBBDF method

This section discusses the derivation of the two-point diagonally implicit block method (2DBBDF). The method produces approximations at y_{n+1} and y_{n+2} simultaneously for solving second-order ODEs in the form,

$$y'' = f(x, y, y'), \quad y(a) = y_0, \quad y'(a) = y'_0, \quad x \in [a, b]. \tag{1}$$

The general definition of block method for second order ODEs can be found in Fatunla [11]. The solution of y_{n+1} at the point x_{n+1} is derived by using three back values which are x_n, x_{n-1}, and x_{n-2} with step size h (Fig. 1).

The backward difference interpolating polynomial P_{n+1} which interpolates the set of points $\{x_{n-2}, \cdots, x_{n+1}\}$ is as follows:

$$P_{n+1}(x) = \sum_{m=0}^{3} (-1)^m \binom{-s}{m} \nabla^m y_{n+1} \tag{2}$$

and $s = \frac{x - x_{n+1}}{h}$.

Equation (2) is differentiated $j = 1, 2$ times and therefore,

$$h^j P_{n+1}^{(j)}(x) = \sum_{m=0}^{3} (-1)^m \frac{d^j}{ds^j} \binom{-s}{m} \bigg|_{s=0} \nabla^m y_{n+1} = \sum_{m=0}^{3} \delta_{j,m} \nabla^m y_{n+1} \tag{3}$$

where $j = 1,2$.

The technique of generating function is used in order to obtain the general form of $\delta_{j,m}$. Given below are the equations of $\delta_{j,m}$ after differentiating once and twice.

Fig. 2 2DBBDF for y_{n+2}

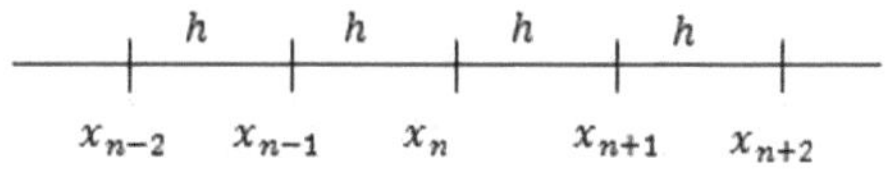

- $j = 1$:

$$\sum_{m=0}^{\infty} \delta_{1,m} t^m = -\log(1-t) = t + \frac{t^2}{2} + \frac{t^3}{3} + \frac{t^4}{4} + \cdots \tag{4}$$

- $j = 2$:

$$\sum_{m=0}^{\infty} \delta_{2,m} t^m = [-\log(1-t)][-\log(1-t)]$$

$$= \left(t + \frac{t^2}{2} + \frac{t^3}{3} + \cdots\right)\left(t + \frac{t^2}{2} + \frac{t^3}{3} + \cdots\right) \tag{5}$$

Substituting the respective values of $\delta_{1,m}$ and $\delta_{2,m}$ into (3) produce the following formulas of 2DBBDF at the point x_{n+1}.

$$y'_{n+1} = \frac{1}{h}\left(\frac{11}{6}y_{n+1} - 3y_n + \frac{3}{2}y_{n-1} - \frac{1}{3}y_{n-2}\right),$$

$$y_{n+1} = \frac{5}{2}y_n - 2y_{n-1} + \frac{1}{2}y_{n-2} + \frac{1}{2}h^2 f_{n+1}.$$

For the second point x_{n+2}, the backward difference interpolating polynomial which interpolates the points $\{x_{n-2}, \cdots, x_{n+2}\}$ is (Fig. 2).

$$P_{n+2}(x) = \sum_{m=0}^{4} (-1)^m \binom{-s}{m} \nabla^m y_{n+2}. \tag{6}$$

Following the step from obtaining (3) and finally producing the corrector formulas of y_{n+1} and y'_{n+1}, the formulas of y_{n+2} and y'_{n+2} are as follows:

$$y'_{n+2} = \frac{1}{h}\left(\frac{25}{12}y_{n+2} - 4y_{n+1} + 3y_n - \frac{4}{3}y_{n-1} + \frac{1}{4}y_{n-2}\right),$$

$$y_{n+2} = \frac{104}{35}y_{n+1} - \frac{114}{35}y_n + \frac{8}{5}y_{n-1} - \frac{11}{35}y_{n-2} + \frac{12}{35}h^2 f_{n+2}.$$

The next section gives the details for the implementation of this new method.

3 Implementation of 2DBBDF

To approximate the solutions at y_{n+1} and y_{n+2} simultaneously, the application of Newton-type scheme is considered. The general form for 2DBBDF method is as follows,

$$hy'_{n+1} = \beta_1 y_{n+1} + G_{n+1},$$

$$\begin{aligned} y_{n+1} &= \alpha_1 h^2 f_{n+1} + D_{n+1}, \\ hy'_{n+2} &= \hat{\beta}_1 y_{n+1} + \hat{\beta}_2 y_{n+2} + G_{n+2}, \\ y_{n+2} &= \theta_2 y_{n+1} + \alpha_2 h^2 f_{n+2} + D_{n+2}. \end{aligned} \tag{7}$$

The difference of y'_{n+1} and y'_{n+2} from (i)-th to $(i+1)$-th iterations, denoted as $e_{n+1}^{'(i+1)}$ and $e_{n+2}^{'(i+1)}$, respectively are equivalent to

$$\begin{pmatrix} e'_{n+1} \\ e'_{n+2} \end{pmatrix}^{(i+1)} = h \begin{pmatrix} \beta_1 & 0 \\ \hat{\beta}_1 & \hat{\beta}_2 \end{pmatrix} \begin{pmatrix} e_{n+1} \\ e_{n+2} \end{pmatrix}^{(i+1)}. \tag{8}$$

The values of $e_{n+1}^{'(i+1)}$ and $e_{n+2}^{'(i+1)}$ of y_{n+1} and y_{n+2}, are approximated by solving the Newton-type scheme.

The $(i+1)$-th iterations of y_{n+1} is

$$y_{n+1}^{(i+1)} = \alpha_1 h^2 f_{n+1}^{(i+1)} + D_{n+1}. \tag{9}$$

By equating $y_{n+1}^{(i+1)} = y_{n+1}^{(i)} + y_{n+1}^{(i+1)}$ and $y_{n+1}^{'(i+1)} = y_{n+1}^{'(i)} + e_{n+1}^{'(i+1)}$, and subsequently expanding the function $f_{n+1}^{(i+1)}$ by Taylor expansion, generates the following:

$$e_{n+1}^{(i+1)} + y_{n+1}^{(i)} = \alpha_1 h^2 \left(f_{n+1}^{(i)} + \frac{\partial f_{n+1}}{\partial y_{n+1}} e_{n+1}^{(i+1)} + \frac{\partial f_{n+1}}{\partial y'_{n+1}} \left(\frac{1}{h} \beta_1 e_{n+1}^{(i+1)} \right) \right) + D_{n+1}. \tag{10}$$

Simplify for $e_{n+1}^{(i+1)}$ gives,

$$\left(1 - \alpha_1 h^2 \frac{\partial f_{n+1}}{\partial y_{n+1}} + \alpha_1 \beta_1 h \frac{\partial f_{n+1}}{\partial y'_{n+1}} \right) e_{n+1}^{(i+1)} = -y_{n+1}^{(i)} + \alpha_1 h^2 f_{n+1}^{(i)} + D_{n+1} \tag{11}$$

The $(i+1)$-th iterations of y_{n+2} is as follows,

$$y_{n+2}^{(i+1)} = \alpha_2 h^2 f_{n+2}^{(i+1)} + \theta_2 y_{n+1}^{(i+1)} + D_2, \tag{12}$$

and by following the same strategy from (10) to (11), the equation for $e_{n+1}^{(i+1)}$ is obtained. $e_{n+1}^{(i+1)}$ and $e_{n+2}^{(i+1)}$ in matrix form is equivalent to

$$\begin{pmatrix} 1 - \alpha_1 h^2 J_{n+1} - \alpha_1 \beta_1 h J'_{n+1} & 0 \\ -\theta_2 - \alpha_2 \hat{\beta}_1 h J'_{n+2} & 1 - \alpha_2 h^2 J_{n+2} - \alpha_2 \hat{\beta}_2 h J'_{n+2} \end{pmatrix}$$

$$\begin{pmatrix} e_{n+1} \\ e_{n+2} \end{pmatrix}^{(i+1)} = \begin{pmatrix} -y_{n+1}^{(i)} + \alpha_1 h^2 f_{n+1}^{(i)} + D_{n+1} \\ -y_{n+2}^{(i)} + \theta_2 y_{n+1}^{(i)} + \alpha_2 h^2 f_{n+2}^{(i)} + D_{n+2} \end{pmatrix}. \tag{13}$$

By taking $i = 0, 1$ and based on evaluating $J_{n+1,n+2} = \left(\frac{\partial F}{\partial Y}\right)\left(Y_{n+1,n+2}\right)$ and $J'_{n+1,n+2} = \left(\frac{\partial F}{\partial Y\prime}\right)(Y_{n+1,n+2})$, the Jacobian matrix of F with respect to Y and Y' respectively, (13) is solved through LU decomposition for $e_{n+1}^{(i+1)}$ and $e_{n+2}^{(i+1)}$. Subsequently, the values $e_{n+1}^{(i+1)}$ and $e_{n+2}^{(i+1)}$ are used to approximate the $(i+1)$-th iterations of 2DBBDF formulas.

4 Numerical Results

In this section, we will present two numerical examples taken from Ibrahim et al. [6] to illustrate the accuracy of the new method. The performance of the proposed 2DBBDF method is compared with the Variable Step 2-Point Block Backward Differentiation Formulas (BBDF2) developed by Ibrahim et al. [6], which solves second-order ODEs directly. See Ibrahim et al. [6] for the details. We consider the comparison of these two methods in terms of maximum error. The problems are listed below.

Problem 1

$$\begin{aligned} &y'' = -1000y - 70y', \\ &y(0) = 2, \quad y'(0) = -70, \quad x \in [0,10], \\ &y(x) = e^{-20x} + e^{-50x}. \end{aligned}$$

Problem 2

$$\begin{aligned} &y'' = -16y - 8y', \\ &y(0) = 1 \quad y'(0) = -12, \quad x \in [0,10], \\ &y(x) = e^{-4x}(1 - 8x). \end{aligned}$$

The notations used in the following tables are as follows:

H step size
MAXE maximum error
TIME total execution time (seconds)
MTD method used

The maximum errors are defined as

$$\text{MAXE} = \max_{i \le i \le STEP}(\max_{i \le i \le N}(e_{it})), \tag{14}$$

and

$$(e_i)_t = \left| \frac{(y_i)_t - (y(x_i))_t}{A + B(y(x_i))_t} \right|. \tag{15}$$

$(y(x_1))_t$ is the t-th component of the exact solution and $(y_i)_t$ is the t-th component of the computed solution at x_i. For $(A = 1, B = 1), (A = 1, B = 0)$ and $(A = 0, B = 1)$, (15) is associated with mixed error test, absolute error test and relative error test respectively.

From Table 1, the method of 2DBBDF outperformed the method of BBDF2 at step size 10^{-4} and from Table 2, the 2DBBDF method gives the least maximum error at the step sizes 10^{-4} and 10^{-6}.

Table 1 Numerical Result for Problem 1

H	MTD	MAXE
10^{-2}	2DBBDF	1.3623e−001
	BBDF2	1.8840e−003
10^{-4}	2DBBDF	8.4620e−008
	BBDF2	1.1381e−004
10^{-6}	2DBBDF	1.4534e−005
	BBDF2	4.5819e−006

Table 2 Numerical Result for Problem 2

H	MTD	MAXE
10^{-2}	2DBBDF	4.2786e−003
	BBDF2	1.9115e−003
10^{-4}	2DBBDF	2.7781e−006
	BBDF2	1.1411e−004
10^{-6}	2DBBDF	4.3702e−007
	BBDF2	4.6212e−006

5 Conclusion

From Tables 1 and 2, the method of 2DBBDF has managed to give competitive result when compared with the BBDF2 method. It shows that the 2DBBDF method can be an alternative solver in solving the problems of stiff second-order ODEs.

Acknowledgement The authors would like to thank the editor and referee for their suggestion for improving this manuscript.

References

1. Hindmarsh, A.C.: Large ordinary differential equation systems and software. Control Syst. Mag. **2**(4), 24–30 (1982)
2. Gear, C.W.: The automatic integration of ordinary differential equations. Commun. ACM **14**(3), 176–179 (1971)
3. Visnak, K., Kubicek, M.: A class of numerical methods of higher order for stiff problems in ordinary differential equations. IMA J. Appl. Math. **21**(3), 251–264 (1978)
4. Ababneh, O.Y., Ahmad, R., Ismail, E.S.: Design of new diagonally implicit Runge–Kutta methods for stiff problems. Appl. Math. Sci. **3**(45), 2241–2252 (2009)
5. Ibrahim, Z.B., Othman, K.I., Suleiman, M.B.: Implicit *r*-point block backward differentiation formula for solving first-order stiff ODEs. Appl. Math. Comput. **186**(1), 558–565 (2007)
6. Ibrahim, Z.B., Suleiman, M.B., Othman, K.I.: Direct block backward differentiation formulas for solving second order ordinary differential equations. Int. J. Comp. Math. Sci. **3**(3), 120–122 (2009)
7. Shampine, L.F., Watts, H.A.: Block implicit one-step methods. Math. Comp. **23**, 731–740 (1969)
8. Olabode, B.T.: An accurate scheme by block method for third order ordinary differential equations. Pac. J. Sci. Technol. **10**(1), 136–142 (2009)
9. Lambert, J.D.: Numerical Methods for Ordinary Differential Systems: The Initial Value Problem. Wiley, New York (1993)

10. Suleiman, M.B.: Convergence of the variable order and variable stepsize direct integration methods for the solution of the higher order ordinary differential equations. Pertanika **8**(1), 59–66 (1985)
11. Fatunla, S.O.: Block methods for second order ODEs. Int. J. Comput. Math. **41**(1–2), 55–63 (1991)
12. Mehrkanoon, S.: A direct variable step block multistep method for solving general third-order ODEs. Numer. Algorithm **57**, 53–66 (2011)
13. Majid, Z.A., Suleiman, M.B.: Performance of 4-point diagonally implicit block method for solving ordinary differential equations. Matematika, **22**(2), 137–146 (2006)

Group Algebra Codes Define Over Extra-Special p-Group

Denis C. K. Wong

Abstract In this paper, group algebra code defined over any extra-special p-group G is constructed. If $char(F) \nmid |G|$, then FG is semisimple and hence $FG = \bigoplus_{e_i \in M} FGe_j$, where e_j is an idempotent of FG and M is the set consisting of all idempotents of FG. Any idea I of FG is a direct sum of some FGe_J, say $I = \bigoplus_{k=1}^{t} FGe_{j_k}$, for some t such that $1 \leq t \leq |G|$. Let $\beta = \{e_{j_k}\}_{k=1}^{t}$ and $\mu = M \backslash \beta$, then I is generated by β and for technical reason, I denotes $I_\mu = \{u \in FG \mid ue_{j_r} = 0, \forall e_{j_r} \in \mu\}$. The idempotent e_j provides useful information to determine the minimum distance for this family of group algebra code. Our primary task is to identify all such idempotents and thus construct a family of MDS group algebra code by choosing a suitable subset of μ in order to maximize the minimum distance.

1 Introduction

Coding theory is important in modern digital communication; however noises might occur during the transmission of digital data across a communication channel. This may cause the received data to be different from the transmitted data. Therefore, error correcting and detecting codes are used in modern digital communication system. The study of group codes as an ideal in a group algebra FG has been developed long time ago, refer [5] and [6].

Let $\mathcal{F}_q$ denote a finite field with q elements such that q is a prime. Given a finite group G of order n, the group algebra $\mathcal{F}_q G$ is a vector space over $\mathcal{F}_q$, with basis G and so, is isomorphic to $\mathcal{F}_q^n$ as a vector space. A group algebra code is defined as an ideal of the group algebra $\mathcal{F}_q G$. In particular, if G is a cyclic group then the ideal in $\mathcal{F}_q G$ is a cyclic code and if G is an abelian group then the ideal in $\mathcal{F}_q G$ is an abelian code. It is well known that if $q \nmid n$, it follows from Maschke's Theorem (Theorem 1.9 in [2]) that the group algebra $\mathcal{F}_q G$ is semisimple and hence $\mathcal{F}_q G$ is a direct sum of minimal ideals, $\mathcal{F}_q G = I_1 \bigoplus I_2 \bigoplus \cdots \bigoplus I_s$, where $I_j = \mathcal{F}_q Ge_j$ is the principal ideal of $\mathcal{F}_q G$ generated by e_j where e_j is an idempotent in $\mathcal{F}_q G$ for

D. C. K. Wong (✉)
Department of Applied Mathematics and Actuarial Science, Faculty of Engineering and Science, Universiti Tunku Abdul Rahman, Off Jalan Genting Kelang, 53300 Setapak, Kuala Lumpur, Malaysia
e-mail: deniswong@utar.edu.my

A. Kilicman et al. (eds.), *International Conference on Mathematical Sciences and Statistics 2013*, DOI 10.1007/978-981-4585-33-0_13,

$j = 1, 2, \ldots, s$. Let $M = \{e_j\}_{j=1}^{s}$ be the set of all pairwise orthogonal idempotents. Every ideal I of $\mathcal{F}_q G$ is a direct sum $I = I_{i_1} \bigoplus I_{i_2} \bigoplus \cdots \bigoplus I_{i_t}$, where $t \leq s$. Now, write $\mathcal{F}_q G = I \bigoplus J$, where $J = I_{j_1} \bigoplus I_{j_2} \bigoplus \cdots \bigoplus I_{j_{s-t}}$ is the direct sum of minimal ideals such that $I_{i_l} \neq I_{j_m}$ for all $1 \leq l \leq t$ and $1 \leq m \leq s - t$.

Lemma 1.1 *Let $\mu = \{e_{j_1}, e_{j_2}, \ldots, e_{j_{s-t}}\}$. Then $I = I_\mu$, where $I_\mu = \{u \in \mathcal{F}_q G \mid ue_{j_m} = 0, \forall e_{j_m} \in \mu\}$.*

Proof If $u \in I$, then $u = \sum_{e_i \in M \setminus \mu} a_i e_i, a_i \in \mathcal{F}_q G$. For all $e_j \in \mu$, $ue_j = \sum_{e_i \in M \setminus \mu} a_i (e_i e_j) = 0$. Thus, $u \in I_\mu$, and so $I \subseteq I_\mu$.

Conversely, for any $u \in \mathcal{F}_q G, u = \sum_{e_i \in M} a_i e_i$. If $u \in I_\mu$, then $ue_j = 0$ for all $e_j \in \mu$. Thus, $\left(\sum_{e_i \in M} a_i e_i\right) e_j = 0$ for all $e_j \in \mu$ if and only if $a_j e_j = 0$ for all $e_j \in \mu$ if and only if $a_j = 0$ for all j. Therefore, $u = \sum_{e_i \in M \setminus \mu} a_i e_i \in I$, and so $I_\mu \subseteq I$. Q.E.D.

Since I_μ is an ideal of $\mathcal{F}_q G$ then I_μ is a subspace of $\mathcal{F}_q G$ which implies that I_μ is a linear code over $\mathcal{F}_q$. It is well known that a linear code can be defined by using a parity check matrix. In this paper, the usage of idempotents in $\mathcal{F}_q G$ as parity checks is equivalent to the definition of a linear code by its parity check matrix. As a parity check matrix of a linear code can tell us the minimum distance of the code, the set of idempotents μ should also provide some information to determine the minimum distance of I_μ.

In [1], we study the group algebra codes defined over extra special p-group of order p^3 by constructing two families of group algebra codes generated by linear idempotents and nonlinear idempotents separately. In this paper, we will generalized our study to any extra special p-group of order p^{2r+1} and thus construct a family of group algebra codes which is generated by arbitrary idempotents.

We now give some basic definitions. Let us consider I_μ as defined in Lemma 1.1. The element in $\mathcal{F}_q G$ is called a word, the element in I_μ is called a codeword, and $N = dim(\mathcal{F}_q G) =$ length of codeword in I_μ. Furthermore, define the weight of u as $wt(u) = |\{a_g | a_g \neq 0\}|$ and the minimum distance of I_μ, $d(I_\mu) = min\{wt(u) | 0 \neq u \in I_\mu\}$. Therefore, I_μ with parameters $N, K = dim(I_\mu)$ and $d = d(I_\mu)$ is called an $[N, K, d -]$ group algebra code. In this paper, our main task is to determine the parameters N, K, and d for I_μ. The length and dimension of a code can be derived directly from the construction. However, to determine the minimum distance of a code, it may be a difficult task. The following lemma is needed later.

Lemma 1.2 *Let μ_1, μ_2 be the sets consisting of idempotents in $\mathcal{F}_q G$ such that $\mu_1 \subseteq \mu_2$, then $d(I_{\mu_2}) \geq d(I_{\mu_1})$.*

Proof For all $u \in I_{\mu_2}, ue = 0$ for all $e \in \mu_2$ and so $ue = 0$ for all $e \in \mu_1$. Thus, $u \in I_{\mu_1}$. Therefore, $I_{\mu_2} \subseteq I_{\mu_1}$. Next, let $u \in I_{\mu_1}$ with minimum weight, that is, $ue = 0$ for all $e \in \mu_1$. If $ue \neq 0$ for all $e \in \mu_2 \setminus \mu_1$, then $u \notin I_{\mu_2}$, so $d(I_{\mu_2}) > d(I_{\mu_1})$. On the other hand, if $ue = 0$ for all $e \in \mu_2 \setminus \mu_1$, then $u \in I_{\mu_2}$ and so the equility holds. Q.E.D.

2 Extra Special p-Group and Characters

Let p be a prime (distinct from q). A p-group G is called extra-special if $G' = Z(G)$, $|G'| = p$ and G/G' is elementary abelian. From Theorem 2.17 in [4], there is an integer $r \geq 1$ such that $|G| = p^{2r+1}$ and G has normal subgroups $N_1, N_2, \ldots, N_r$ such that

(a) N_i is a nonabelian group of order p^3 for all i, $1 \leq i \leq r$.
(b) $G = N_1 N_2 \ldots N_r$.
(c) $[N_i, N_j] = 1$ for all $i \neq j$.
(d) $N_i \cap N_1 \ldots N_{i-1} N_{i+1} \ldots N_r = Z(G)$ for all i.

Lemma 2.1 *Let G be the extra special p-group of order p^{2r+1}. Then each N_i is also an extra special p-group of order p^3 for $i = 1, 2, \ldots, r$.*

Proof Since each N_i is a p-group, $Z(N_i) > 1$. Thus, $|N_i/Z(N_i)| \leq p^2$. As N_i is nonabelian, $N_i/Z(N_i)$ is not cyclic and so $|N_i/Z(N_i)| = p^2$ which then implies $|Z(N_i)| = p$ and $N_i/Z(N_i) \cong Z_p \times Z_p$. Finally, as $N_i' \subseteq Z(N_i)$, $Z(N_i) = N_i'$.
Q.E.D.

Since $|G'| = |Z(G)| = p$, then we may write $G' = \langle g \mid g^p = 1 \rangle$. As $|G| = p^{2r+1}$, then $|G/G'| = p^{2r}$. Let $T = \{t_0 = 1, t_1, t_2, \ldots, t_{p^{2r}-1}\}$ be the set of all right transversal of G' in G. Then, the p^{2r} right cosets of G' in G are $G't_i = \{t_i, gt_i, g^2 t_i, \ldots, g^{p-1} t_i\}$ for $i = 0, 1, 2, \ldots, p^{2r} - 1$ and so $G = \bigcup_{i=0}^{p^{2r}-1} G't_i$.

Proposition 2.2 *Let G be an extra-special p-group of order p^{2r+1}. Then the following holds true:*

(a) *G has $p^{2r} + p - 1$ conjugacy classes; p of these has size 1 and the other $p^{2r} - 1$ conjugacy classes has size p, which is exactly the right cosets xG', $x \in G \backslash G'$.*
(b) *The number of linear characters of G is $|G/G'| = p^{2r}$.*
(c) *The total number of irreducible characters of G is equal to the number of conjugacy classes of G, that is, $|Irr(G)| = p^{2r} + p - 1$.*
(d) *The total number of non-principal characters of G is equal to $|Irr(G)| - |G/G'| = p - 1$.*

Proof Part (a) is a general result obtained from Sect. 2 in [1]. Part (b) follows from Corollary 2.23 in [2]. Part (c) follows from Corollary 2.7 in [2] together with part (a). Finally, part (d) is just a direct consequence from (b) and (c). Q.E.D.

Proposition 2.3 *Let G be an extra-special p-group of order p^{2r+1}. If χ is a linear character of G, then the following holds true:*

(a) *For all $g \in G$, $\chi(g) \neq 0$.*
(b) *For all $g \in G'$, $\chi(g) = 1$. In particular, $\chi(1) = 1$, that is, $deg(\chi) = 1$.*
(c) *For any $x, y \in gG'$, $\chi(x) = \chi(y)$.*

Proof Part (a) directly follows from the fact that χ is a homomorphism. For part (b), consider any $g \in G'$, $\chi(g) = \chi([x, y]) = \chi(x^{-1}y^{-1}xy) = 1$ for some $x, y \in G$. Finally, part (c) follows directly from (b). Q.E.D.

Proposition 2.4 *Let G be an extra-special p-group of order* p^{2r+1}. *If* χ *is a nonlinear character of G, then the following holds true:*

(a) $\chi(1) = p^r$.
(b) *For all* $g \in G'$, $\chi(g) \neq 0$.
(c) *For all* $g \notin G'$, $\chi(g) = 0$.
(d) *Fix a generator, g, of* G', *and let* ζ *be a primitive p-th root of unity in F. Then* $\chi_i(g) = p^r \zeta^i$ *for some* $i \in \{1, 2, \ldots, p-1\}$.

Proof To prove part (a), from Corollary 3.11 in [3], as χ is nonlinear, $G' \nsubseteq ker(\chi)$ and so $G' \cap ker(\chi) = \{1\}$. As $|G'| = p$, then $ker(\chi) = 1$ and together with part (e) of Lemma 2.27 in [2], we see that $Z(\chi) = Z(G)$. For all $z \in Z(G)$, $|\chi(z)| = \chi(1)$. Since χ is irreducible, from Corollary 2.17 in [2],

$$\begin{aligned} 1 &= [\chi, \chi] \\ &= \frac{1}{|G|} \sum_{z \in Z(G)} \chi(z)\overline{\chi(z)} \\ &= \frac{1}{p^{2r+1}} \sum_{z \in Z(G)} |\chi(z)|^2 \\ &= \frac{1}{p^{2r+1}} |Z(G)|\chi(1)^2. \end{aligned}$$

which implies $\chi(1) = p^r$.

For part (b), as $G' = Z(G) = Z(\chi)$, part (c) of Lemma 2.27 in [2] implies $\chi_{G'} = \chi(1)\lambda$, where λ is some linear character of G'. Therefore, $\chi_{G'}(g) = \chi(1)\lambda(g) \neq 0, \forall g \in G'$.

Next, since G has p^{2r} linear characters, which are, $\chi_1, \chi_2, \ldots, \chi_{p^{2r}}$, and let $\beta_1, \beta_2, \ldots, \beta_{p-1}$ be those nonlinear characters of G. Thus, for all $g \in G'$,

$$\begin{aligned} p^{2r} &= |C_G(g)| \\ &= \sum_{i=1}^{p^{2r}} |\chi(g)|^2 + \sum_{i=1}^{p-1} |\beta_i(g)|^2 \\ &= p^{2r} + \sum_{i=1}^{p-1} |\beta_i(g)|^2. \end{aligned}$$

implies $\sum_{i=1}^{p-1} |\beta_i(g)|^2 = 0$ and so $\beta_i(g) = 0$ for all $g \in G'$ and $i = 1, 2, \ldots, p-1$.

To prove part (d), we note from (b) that $\chi_{G'} = \chi(1)\lambda$. For all $g \in G'$, $\chi_{G'} = p^r\lambda(g)$. Furthermore, $\chi_{G'}(g) = \chi(g)$ for $\chi \in G^*$ and for all $g \in G'$. Since $g^p = 1$, then $\chi(g^p) = \chi(1) = p^r$, and so

$$p^r \lambda(g^p) = \chi(g^p) = p^r$$

implies $\lambda(g)^p = 1$ and $\lambda(g) = \zeta$, where ζ is a primitive p-th root of unity. Hence, (d) will follow directly. Q.E.D.

3 Idempotents

Since I_μ depends on those idemptents in μ, we start by deriving a formula for all idempotents in $\mathcal{F}_q G$ where G is the extra special p-group of order p^{2r+1}. From Theorem 2.12 in [2], any idempotent e_i in $\mathcal{F}_q G$ can be written uniquely as

$$e_i = \frac{\chi_i(1)}{|G|} \sum_{g \in G} \chi_i(g^{-1}) g.$$

As $G = \cup_{i=0}^{p^{2r}-1} G' t_0$ and $G' = \langle g | g^p = 1 \rangle$, then

$$e_i = \frac{\chi_i(1)}{p^{2r+1}} \sum_{j=0}^{p^{2r}-1} \sum_{i=0}^{p-1} \chi_i(g^{-i} t_j^{-1}) t_j g^i.$$

If χ_i is the principal character of G, that is, $\chi_i(g) = 1$ for all $g \in G$, then χ_i will define e_i, the principal idempotent which is usually denoted by $e_{principal}$ and

$$\begin{aligned} e_{principal} &= \frac{1}{p^{2r+1}} \sum_{j=0}^{p^{2r}-1} \sum_{i=0}^{p-1} t_j g^i \\ &= \frac{G}{|G|}. \end{aligned}$$

Note that for all $g \in G$,

$$e_{principal} g = \frac{G}{|G|} g = \frac{G}{|G|} = e_{principal}.$$

Next, let M_L be the set consisting of all linear idempotents of $\mathcal{F}_q G$, that is, all these idempotents correspond to the p^{2r} linear characters. Thus, $|M_L| = p^{2r}$. We know that from Proposition 2.3 for all $\chi_i \in G^*$ which is linear, then $\chi_i(g) = 1$ for all $g \in G'$. Therefore,

$$
\begin{aligned}
e_i &= \frac{1}{p^{2r+1}}\left(\sum_{i=0}^{p-1} g^i\right)\left(\sum_{j=0}^{p^{2r}-1} \chi_i(t_j^{-1})t_j\right)\\
&= \frac{G'}{p^{2r+1}}\left(\sum_{j=0}^{p^{2r}-1} \chi_i(t_j^{-1})t_j\right).
\end{aligned}
$$

It can be shown that for all $e \in M_L$, $ge = e$ for all $g \in G'$.

Finally, let M_N be the set consisting of all nonlinear idempotents of $\mathcal{F}_q G$, that is, all those idempotents corresponded to the nonlinear characters of G. Note that $|M_N| = p-1$. For all $e_i \in M_N$, since $G = G't_0 \bigcup \left(\bigcup_{i=1}^{p^{2r}-1} G't_i\right)$ and from Proposition 2.4, we have $\chi(G) = \chi(G') + \sum_{i=0}^{p^{2r}-1} \chi(G't_i) = \chi(G')$. Thus,

$$
\begin{aligned}
e_i &= \frac{p^r}{p^{2r+1}} \sum_{g\in G} \chi_i(g^{-1})g\\
&= \frac{1}{p^{r+1}}\left[\sum_{g\in G'} \chi_i(g^{-1})g + \sum_{g\in G\backslash G'} \chi_i(g^{-1})g\right]\\
&= \frac{1}{p^{r+1}} \sum_{g\in G'} \chi_i(g^{-1})g\\
&= \frac{1}{p^{r+1}} \sum_{j=0}^{p-1} p^r \zeta^{-ij} g^j\\
&= \frac{1}{p} \sum_{j=0}^{p-1} \zeta^{-ij} g^j.
\end{aligned}
$$

Clearly, for all $g \in G'$, $ge_i = e_i'$ where $e_i, e_i' \in M_N$. We collect all the facts above in the following Proposition.

Proposition 3.1 *Let G be an extra-special p-group of order p^{2r+1}.*

(a) *The principal idempotent of $\mathcal{F}_q G$ is $e_{principal} = \frac{G}{|G|}$.*

(b) $M_L\backslash\{e_{principal}\} = \left\{e_i = \frac{G'}{p^{2r+1}}\left(\sum_{j=0}^{p^{2r}-1} \chi_i(t_j^{-1})t_j\right) \mid i = 1, 2, \ldots, p^{2r}-1\right\}$.

(c) $M_N = \left\{e_i = \frac{1}{p}\sum_{j=0}^{p-1} \zeta^{-ij} g^j \mid i = 1, 2, \ldots, p-1\right\}$.

4 Dimension

In this section, we will determine the dimension for I_μ. We start by deriving two simple results.

Lemma 4.1 *For all* $e \in M_L, dim(\mathcal{F}_q Ge) = 1$.

Proof For all $h \in G$,

$$\begin{aligned} he &= \frac{1}{|G|}\sum_{g\in G}\chi(g^{-1})hg \\ &= \frac{1}{|G|}\sum_{g\in G}\chi(g^{-1}h^{-1})\chi(h)hg \\ &= \frac{\chi(h)}{|G|}\sum_{g\in G}\chi((hg)^{-1})hg \\ &= \frac{\chi(h)}{|G|}\sum_{w\in G}\chi(w^{-1})w \\ &= \chi(h)e. \end{aligned}$$

Next, for all $u = \sum_{h\in G} a_h h \in \mathcal{F}_q G$,

$$\begin{aligned} ue &= \left(\sum_{h\in G} a_h h\right) e \\ &= \sum_{h\in G} a_h(he) \\ &= \sum_{h\in G} a_h(\chi(h)e) \\ &= \left(\sum_{h\in G} a_h\chi(h)\right) e \in \langle e\rangle. \end{aligned}$$

Therefore, for all $u \in \mathcal{F}_q G$, $\mathcal{F}_q Ge = \langle e\rangle$ and so $dim(\mathcal{F}_q Ge) = 1$ for all $e \in M_L$.

Q.E.D.

Corollary 4.2 $dim(I_{M_N}) = p^{2r}$.

Proof From Lemma 1.1, we know that $I_{M_N} = \bigoplus_{e_i\in M\setminus M_N} FGe_i = \bigoplus_{e_i\in M\setminus M_L} FGe_i$. Thus, $dim(I_{M_N}) = \sum_{e_i\in M_L} dim(FGe_i)$ and so the result will follow from Lemma 4.1.

Q.E.D.

Lemma 4.3 *For all* $e \in M_N, dim(\mathcal{F}_q Ge) = p^{2r}$.

Proof Since each $e \in M_N$ corresponds to χ such that $deg(\chi) = p^r$ and $\mathcal{F}_q Ge \cong Mat_{n_j}(F)$ and so $dim(\mathcal{F}_q Ge_j) = n_j^2$ where $n_j = p^r = deg(\chi)$. Thus, $dim(\mathcal{F}_q Ge) = p^{2r}$.

Q.E.D.

Corollary 4.4 $dim(I_{M_L}) = p^{2r}(p-1)$.

Proof Follow from Lemma 1.1 and Lemma 4.3. Q.E.D.

Our main result on dimension of I_μ.

Theorem 4.5 *Let $\mu = \mu_L \cup \mu_N$, where $\mu_L \subseteq M_L$ and $\mu_N \subseteq M_N$. Then $dim(I_\mu) = p^{2r+1} - |\mu_L| - |\mu_N|p^{2r}$.*

Proof By the property of semisimple, as $\mathcal{F}_q G = \bigoplus_i \mathcal{F}_q G e_i$, then $|G| = dim(\mathcal{F}_q G) = \sum_i dim(F G e_i)$. Thus,

$$\begin{aligned} dim(I_\mu) &= dim\left(\bigoplus_{e_i \in M \setminus \mu} \mathcal{F}_q G e_i\right) \\ &= (|M_L| - |\mu_L|)dim(\mathcal{F}_q Ge) + (|M_N| - |\mu_N|)dim(\mathcal{F}_q Gf) \\ &= (|M_L| - |\mu_L|) + (|M_N| - |\mu_N|)p^{2r} \\ &= (p^{2r} - |\mu_L|) + (p - 1 - |\mu_N|)p^{2r} \\ &= p^{2r+1} - |\mu_L| - |\mu_N|p^{2r}, \end{aligned}$$

where $e \in M_L$ and $f \in M_N$. Q.E.D.

5 Minimum Distance

Theorem 5.1 $d(I_{e_{principal}}) = 2$.

Proof We first note that for all $u \in \mathcal{F}_q G$, $u = \sum_{j=0}^{p^{2r}-1} \sum_{i=0}^{p-1} \lambda_{ji} g^i t_j$. Then, $ue_{principal} = \left(\sum_{j=0}^{p^{2r}-1} \sum_{i=0}^{p-1} \lambda_{ji}\right) e_{principal}$. For all $u \in \mathcal{F}_q G$ with $wt(u) = 1$, that is, $u = \lambda g$ for some $g \in G$ and $\lambda \neq 0$. Then, $ue_{principal} = \lambda e_{principal} \neq 0$ and so $u \notin I_{e_{principal}}$, that is, $d(I_{e_{principal}}) \geq 2$. Next, for some $g, h \in G$ such that $g \neq h$, clearly, $u = \lambda g + (-\lambda)h \in I_{e_{principal}}$ as $ue_{principal} = (\lambda - \lambda)e_{principal} = 0$. Therefore, $d(I_{e_{principal}}) = 2$. Q.E.D.

From Theorem 4.5 and Theorem 5.1, we see that $I_{e_{principal}}$ is an $[p^{2r+1}, p^{2r+1}-1, 2]$-group algebra code which is an MDS code. Recalling from Sect. 2, any extra-special p-group G of order p^{2r+1} has the form $G = N_1 N_2 \ldots N_r$, where N_i is a normal subgroup of G.

Lemma 5.2 *Let e be an idempotent in $\mathcal{F}_q N_i$. Then $e \in \mathcal{F}_q G$.*

Proof Let $f = e_{principal}$, the principal idempotent of $\mathcal{F}_q G$ and let $\chi \in N_i^*$ be the character defined e. Then,

$$ef = \left(\frac{\chi(1)}{|N_i|} \sum_{g \in N_i} \chi(g^{-1}) g \right) \frac{G}{|G|}$$
$$= \frac{\chi(1)}{|G||N_i|} G \sum_{g \in N_i} \chi(g^{-1}).$$

If e is the principal idempotent of $\mathcal{F}_q N_i$, then $ef = \frac{1}{|G||N_i|} G|N_i| = f \in \mathcal{F}_q G$ and so $e \in \mathcal{F}_q G$. If e is a linear idempotent of $\mathcal{F}_q N_i$, then $ef = \frac{1}{|G||N_i|} G\chi(N_i) = \frac{1}{|G||N_i|} G|N_i| = f \in \mathcal{F}_q G$ and so $e \in \mathcal{F}_q G$. Finally, if e is a non-linear idempotent of $\mathcal{F}_q N_i$, then

$$ef = \frac{\chi(1)}{|G||N_i|} G \sum_{g \in N_i} \chi(g^{-1}) \in FG$$

and so $e\mathcal{F}_q G \subseteq \mathcal{F}_q G$ implies $e \in \mathcal{F}_q G$. Q.E.D.

From Lemma 2.1, we see that those results obtained in [1], can be used to help us to obtain a bound for minimum distance together with Lemma 1.2. Let ρ_L be all linear idempotents in $\mathcal{F}_q N_i$. From Lemma 5.2, since $\rho_L \subseteq \mu_L, d(I_{\mu_L}) \geq d(I_{\rho_L}) = 2$ follows from Theorem 4 in [1].

Next, let ρ_N be the set of all nonlinear idempotent in $\mathcal{F}_q N_i$. In [1], from Theorem 13, we know that if q is a primitive root modulo p, then $d(I_{\rho_N}) = |\rho_N| + 1$. From Lemma 5.2, since $\rho_N \subseteq \mu_N$, then $d(I_{\mu_N}) \geq d(I_{\rho_N}) = |\rho_N| + 1$.

Theorem 5.3 *Let $\mu = \mu_L \cup \mu_N$, where $\mu_L \subseteq M_L$ and $\mu_N \subseteq M_N$. Then I_μ is a $[p^{2r+1}, p^{2r+1} - |\mu_L| - |\mu_N| p^{2r}, \geq max\{2, |\rho_N| + 1\}]$- group algebra code where ρ_N is the set of all nonlinear idempotent in $\mathcal{F}_q N_i$.*

References

1. How, G.A., Wong, D.C.K.: Group codes defined using extra special p-group of order p^3. Bull. Malays. Math. Sci. Sco. **27**(2), 185–205 (2004)
2. Isaacs, I.M.: Character Theory of Finite Groups. AMS Chelsea Publishing, Providence (2006)
3. Isaacs, I.M.: Algebra, A Graduate Course. Brooks/Cole Publishing, Pacific Grove (1992)
4. Karpilovsky, G.: Group representations. Vol 1. Part B: Introduction to Group Representations and Characters. Elsevier Scicence Publishers B.V., Amsterdam (1992)
5. Sloane, N.J.A., Macwilliam, F.J.: The Theory of Error Correcting Codes. North Holland Publishing Co., Amsterdam (1978)
6. Berman, S.D.: Parameter of abelian codes in the group algebra KG of $G = \langle a \rangle \times \langle b \rangle, a^p = b^p = 1$, p is a prime, over a finite field K with a primitive pth root of unity and related MDS-codes. Contempary Math. 93 (1989)

New Subclasses of Meromorphic Functions Related to Cho–Kwon–Srivastava Operator

Firas Ghanim and Maslina Darus

Abstract Making use of a linear operator, which is defined here by means of the Hadamard product (or convolution), we introduce some new subclasses of the meromorphically univalent function class Σ and investigate their inclusion relationships.

1 Introduction

Let Σ denote the class of meromorphic functions $f(z)$ normalized by

$$f(z) = \frac{1}{z} + \sum_{n=1}^{\infty} a_n z^n, \tag{1}$$

which are analytic in the punctured unit disk $U = \{z : 0 < |z| < 1\}$. For $0 \leq \beta$, $S^*(\beta)$ and $k(\beta)$ denote the subclasses of Σ consisting of all meromorphic functions which are, respectively, starlike of order β and convex of order β in U.

For functions $f_j(z)(j = 1; 2)$ defined by

$$f_j(z) = \frac{1}{z} + \sum_{n=1}^{\infty} a_{n,j} z^n, \tag{2}$$

we denote the Hadamard product (or convolution) of $f_1(z)$ and $f_2(z)$ by

$$(f_1 * f_2) = \frac{1}{z} + \sum_{n=1}^{\infty} a_{n,1} a_{n,2} z^n. \tag{3}$$

M. Darus (✉)
School of Mathematical Sciences, Faculty of Science and Technology,
Universiti Kebangsaan Malaysia, Selangor, Malaysia
e-mail: maslina@ukm.my

F. Ghanim
Department of Mathematics, College of Sciences, University of Sharjah,
Sharjah, United Arab Emirates
e-mail: fgahmed@sharjah.ac.ae

A. Kilicman et al. (eds.), *International Conference on Mathematical Sciences and Statistics 2013,* DOI 10.1007/978-981-4585-33-0_14,

Let us define the function $\tilde{\phi}(\alpha, \beta; z)$ by

$$\tilde{\phi}(\alpha, \beta; z) = \frac{1}{z} + \sum_{n=0}^{\infty} \left| \frac{(\alpha)_{n+1}}{(\beta)_{n+1}} \right| z^n, \tag{4}$$

for $\beta \neq 0, -1, -2, \ldots$, and $\alpha \in \mathbb{C}/\{0\}$, where $(\lambda)n = \lambda(\lambda+1)_{n+1}$ is the Pochhammer symbol. We note that

$$\tilde{\phi}(\alpha, \beta; z) = \frac{1}{z} {}_2F_1(1, \alpha, \beta; z)$$

where

$${}_2F_1(b, \alpha, \beta; z) = \sum_{n=0}^{\infty} \frac{(b)_n (\alpha)_n}{(\beta)_n} \frac{z^n}{n!}$$

is the well-known Gaussian hypergeometric function.

Let us write

$$q_{\lambda,\mu}(z) = \frac{1}{z} + \sum_{n=1}^{\infty} \left(\frac{\lambda}{n+1+\lambda} \right)^{\mu} z^n, \qquad (\lambda > 0,\ \mu \geq 0).$$

Corresponding to the functions $\tilde{\phi}(\alpha, \beta; z)$ and $q_{\lambda,\mu}(z)$, and using the Hadamard product for $f(z) \in \Sigma$, we define a new linear operator $L(\alpha, \beta, \lambda, \mu)$ on Σ by

$$\begin{aligned} L(\alpha,\ \beta,\ \lambda,\ \mu) f(z) &= \left(f(z) * \tilde{\phi}(\alpha,\ \beta;\ z) * q_{\lambda,\mu}(z) \right) \\ &= \tfrac{1}{z} + \sum_{n=1}^{\infty} \left| \tfrac{(\alpha)_{n+1}}{(\beta)_{n+1}} \right| \left(\tfrac{\lambda}{n+1+\lambda} \right)^{\mu} a_n z^n. \end{aligned} \tag{5}$$

The meromorphic functions with the generalized hypergeometric functions were considered recently by Dziok and Srivastava [4], [5], Liu [14], Liu and Srivastava [12], [15], [16], and Cho and Kim [2].

For a function $f \in L(\alpha,\ \beta,\ \lambda,\ \mu) f(z)$ we define

$$I_{\lambda}^{\mu,0} f(\alpha, \beta; z) = L(\alpha,\ \beta,\ \lambda,\ \mu) f(z)$$

and for $k = 1, 2, 3, \ldots$,

$$\begin{aligned} I_{\lambda}^{\mu,k} f(\alpha, \beta; z) &= z \left(I^{k-1} L(\alpha, \beta, \lambda, \mu) f(z) \right)' + \frac{2}{z} \\ &= \frac{1}{z} + \sum_{n=1}^{\infty} n^k \left| \frac{(\alpha)_{n+1}}{(\beta)_{n+1}} \right| \left(\frac{\lambda}{n+1+\lambda} \right)^{\mu} a_n z^n. \end{aligned} \tag{6}$$

Note that if $n = \beta, k = 0$ the operator $I_{\lambda}^{\mu,0} f(\alpha, n; z)$ has been introduced by N. E. Cho, O. S. Kwon and H. M. Srivastava [1] for $\mu \in \mathbb{N}_0 = \mathbb{N} \cup 0$. It was known that the definition of the operator $I_{\lambda}^{\mu,0} f(\alpha, n; z)$ was motivated essentially by the

Choi–Saigo–Srivastava operator [3] for analytic functions, which includes a simpler integral operator studied earlier by Noor [20] and others (cf. [11], [13], [21]). Also note that the operator $I^{0,k} f(\alpha, \beta; z)$ has been recently introduced and studied by Ghanim and Darus [6], [7], [8] respectively. To our best knowledge, the recent work regarding operator $I_{\lambda}^{\mu,0} f(\alpha, n; z)$ was charmingly studied by Piejko and Sokól [22]. Moreover, the operator $I_{\lambda}^{\mu,k} f(\alpha, \beta; z)$ was defined and studied by Ghanim and Darus [10]. In the same direction, we will study for the operator $I_{\lambda}^{\mu,k} f(\alpha, \beta; z)$ given in (6).

Now, it follows from (5) and (6) that

$$z\left(I_{\lambda}^{\mu,k} f(\alpha, \beta; z)\right)' = \alpha I_{\lambda}^{\mu,k} f(\alpha+1, \beta; z) - (\alpha+1) I_{\lambda}^{\mu,k} f(\alpha, \beta; z). \tag{7}$$

Let A be the class of analytic functions $h(z)$ with $h(0) = 1$, which are convex and univalent in the open unit disk $U = U^* \cup \{0\}$ and for which

$$\Re\{h(z)\} > 0 \qquad (z \in U). \tag{8}$$

For functions f and g analytic in U, we say that f is subordinate to g and write

$$f \prec g \quad in \quad U \quad or \quad f(z) \prec g(z) \quad (z \in U)$$

if there exists an analytic function $w(z)$ in U such that $|w(z)| \leq |z|$ and $f(z) = g(w(z))$ $(z \in U)$. Furthermore, if the function g is univalent in U, then

$$f(z) \prec g(z) \Leftrightarrow f(0) = g(0) \qquad and \qquad f(U) = g(U), \qquad (z \in U).$$

Throughout this paper, we assume that

$$m \in N,\ \beta \notin Z_0^-,\ \ \varepsilon_m = \exp\left(\frac{2\pi i}{m}\right) \tag{9}$$

and

$$f_{m,\lambda,\mu}(\alpha, \beta; z) = \frac{1}{m} \sum_{j=0}^{m-1} \varepsilon_m^j \left(L(\alpha, \beta, \lambda, \mu) f\right)\left(\varepsilon_m^j z\right), \qquad f \in \Sigma. \tag{10}$$

Also, we define

$$I_{\lambda,m}^{\mu,k} f(\alpha, \beta; z) = \frac{1}{m} \sum_{j=0}^{m-1} \varepsilon_m^{j(k+1)} \left(I^k L(\alpha, \beta, \lambda, \mu) f\right)\left(\varepsilon_m^j z\right), \qquad k = 1, 2, 3 \ldots. \tag{11}$$

Making use of the linear operator $L(\alpha, \beta, \lambda, \mu)$ and the principle of subordination between analytic functions, we introduce and investigate the following subclasses of the meromorphically analytic function class Σ:

$$\Sigma_{\lambda,m}^{\mu,k}(\alpha, \beta; h),\ \ \mathrm{M}_{m,k}(\alpha, \beta; h),\ \ M_{\lambda,m}^{\mu,k}(\gamma; \alpha, \beta; h) \quad (h \in A).$$

Definition 1 A function $f \in \Sigma$ is said to be in the class $\Sigma_{\lambda,m}^{\mu,k}(\alpha,\beta;h)$ if it satisfies the following subordination condition:

$$\frac{-z\left(I^k L(\alpha,\beta,\lambda,\mu)f\right)'(z)}{I_{\lambda,m}^{\mu,k} f(\alpha,\beta;z)} \prec h(z) \qquad (z \in U) \tag{12}$$

where $h \in A$ and $I_{\lambda,m}^{\mu,k} f(\alpha,\beta;z) \neq 0 \qquad (z \in U^*)$.

Definition 2 A function $f \in \Sigma$ is said to be in the class $M_{\lambda,m}^{\mu,k}(\alpha,\beta;h)$ if it satisfies the following subordination condition:

$$\frac{-z\left(I^k L(\alpha,\beta,\lambda,\mu)f\right)'(z)}{I_{\lambda,m}^{\mu,k} g(\alpha,\beta;z)} \prec h(z) \qquad (z \in U) \tag{13}$$

for some $g \in \Sigma_{\lambda,m}^{\mu,k}(\alpha,\beta;h)$ where $h \in A$ and $I_{\lambda,m}^{\mu,k} g(\alpha,\beta;z)$ is defined as in (11).

Definition 3 A function $f \in \Sigma$ is said to be in the class $M_{\lambda,m}^{\mu,k}(\gamma;\alpha,\beta;h)$ if it satisfies the following subordination condition:

$$-\gamma\frac{z\left(I^k L(\alpha+1,\beta,\lambda,\mu)f\right)'(z)}{I_{\lambda,m}^{\mu,k} g(\alpha+1,\beta;z)} - (1-\gamma)\frac{z\left(I^k L(\alpha,\beta,\lambda,\mu)f\right)'(z)}{I_{\lambda,m}^{\mu,k} g(\alpha,\beta;z)} \prec h(z) \tag{14}$$

$(z \in U)$ for some $\gamma\ (\gamma \geq 0)$ and $g \in \Sigma_{\lambda,m}^{\mu,k}(\alpha,\beta;h)$, where $h \in A$ and $I_{\lambda,m}^{\mu,k} g(\alpha+1,\beta;z) \neq 0$.

2 Preliminary Results

In order to prove our main results, we need the following lemmas.

Lemma 1 *Let a $(a \geq 0)$ and γ be complex numbers and let $h(z)$ be analytic and convex univalent in U with $\Re\{a\,h(z)+\gamma\} > 0$. If $q(z)$ is analytic in U* with $q(0) = h(0)$, *then the subordination* [18]:

$$q(z) + \frac{zq'(z)}{aq(z)+\gamma} \prec h(z) \quad (z \in U)$$

implies that $q(z) \prec h(z) \quad (z \in U)$.

Lemma 2 *Let $h(z)$ be analytic and convex univalent in U and let $w(z)$ be analytic in U with $\Re\{w(z)\} \geq 0 (z \in U)$. If $q(z)$ is analytic in U* and $q(0) = h(0)$, *then the subordination* [19]:

$$q(z) + w(z)zq'(z) \prec h(z) \qquad (z \in U)$$

implies that $q(z) \prec h(z) \quad (z \in U)$.

Lemma 3 *Let* $f \in \Sigma_{\lambda,m}^{\mu,k}(\alpha,\beta;h)$. *Then*

$$-\frac{z\left(I_{\lambda,m}^{\mu,k} f'(\alpha,\beta;z)\right)}{I_{\lambda,m}^{\mu,k} f(\alpha,\beta;z)} \prec h(z) \qquad (z \in U). \tag{15}$$

Proof Making use of (11), we have

$$\begin{aligned} I_{\lambda,m}^{\mu,k} f(\alpha,\beta;\varepsilon_m^j z) &= \frac{1}{m}\sum_{n=0}^{m-1} \varepsilon_m^{n(k+1)} \left(I^k L(\alpha,\beta,\lambda,\mu) f\right)\left(\varepsilon_m^{n+j} z\right) \\ &= \frac{\varepsilon_m^{-j}}{m}\sum_{n=0}^{m-1} \varepsilon_m^{n(k+1)+j} \left(I^k L(\alpha,\beta,\lambda,\mu) f\right)\left(\varepsilon_m^{n+j} z\right) \\ &= \varepsilon_m^{-j} I_{\lambda,m}^{\mu,k} f(\alpha,\beta;z), \end{aligned}$$

$(j \in \{0,1,\ldots,m-1\})$ and

$$I_{\lambda,m}^{\mu,k} f'(\alpha,\beta;z) = \frac{1}{m}\sum_{j=0}^{m-1} \varepsilon_m^{j(k+2)} \left(I^k L(\alpha,\beta,\lambda,\mu) f\right)'\left(\varepsilon_m^{j} z\right).$$

Thus,

$$\begin{aligned} -\frac{z\left(I_{\lambda,m}^{\mu,k} f'(\alpha,\beta;z)\right)}{I_{\lambda,m}^{\mu,k} f(\alpha,\beta;z)} &= -\frac{1}{m}\sum_{j=0}^{m-1} \frac{\varepsilon_m^{j(k+2)} z \left(I^k L(\alpha,\beta,\lambda,\mu) f\right)'\left(\varepsilon_m^{j} z\right)}{I_{\lambda,m}^{\mu,k} f(\alpha,\beta;z)} \\ &= -\frac{1}{m}\sum_{j=0}^{m-1} \frac{\varepsilon_m^{jk} z \left(I^k L(\alpha,\beta,\lambda,\mu) f\right)'\left(\varepsilon_m^{j} z\right)}{I_{\lambda,m}^{\mu,k} f(\alpha,\beta;\varepsilon_m^j z)} \qquad (z \in U). \end{aligned} \tag{16}$$

Since $f \in \Sigma_{\lambda,m}^{\mu,k}(\alpha,\beta;h)$, it follows that

$$-\frac{\varepsilon_m^j z \left(I^k L(\alpha,\beta,\lambda,\mu) f\right)'\left(\varepsilon_m^{j} z\right)}{I_{\lambda,m}^{\mu,k} f(\alpha,\beta;\varepsilon_m^j z))} \prec h(z) \qquad (z \in U, j \in \{0,1,2,\ldots,k-1\}) \tag{17}$$

Since $h(z)$ is convex univalent in U, from (16) and (17) we conclude that (15) holds true.

3 Main Results

Theorem 1 *Let* $h \in A$ *with*

$$\Re\{h(z)\} < 1+\alpha \qquad (z \in U, \alpha > 0). \tag{18}$$

If $f \in \Sigma_{\lambda,m}^{\mu,k}(\alpha+1,\beta;h)$, *then* $f \in \Sigma_{\lambda,m}^{\mu,k}(\alpha,\beta;h)$,
provided that $I_{\lambda,m}^{\mu,k} f(\alpha,\beta;z) \neq 0 \qquad (z \in U^*)$.

Proof By using (7) and (11), we have

$$\begin{aligned}
&(\alpha+1)I_{\lambda,m}^{\mu,k}f(\alpha,\beta;z)+zI_{\lambda,m}^{\mu,k}f'(\alpha,\beta;z)\\
&=\frac{\alpha}{m}\sum_{j=0}^{m-1}\varepsilon_m^{j(k+1)}\left(I^kL(\alpha+1,\beta,\lambda,\mu)f\right)'\left(\varepsilon_m^j z\right)\\
&=\alpha I_{\lambda,m}^{\mu,k}f(\alpha+1,\beta;z), \qquad (f\in\Sigma). \qquad (19)
\end{aligned}$$

Let $f\in\Sigma_{\lambda,m}^{\mu,k}(\alpha+1,\beta;h)$ and suppose that

$$w(z)=-\frac{z\left(I_{\lambda,m}^{\mu,k}f'(\alpha,\beta;z)\right)}{I_{\lambda,m}^{\mu,k}f(\alpha,\beta;z)}. \qquad (20)$$

Then, $w(z)$ is analytic in U, with $w(0)=1$, and it follows from (19) and (20) that

$$\alpha+1-w(z)=\alpha\frac{I_{\lambda,m}^{\mu,k}f(\alpha+1,\beta;z)}{I_{\lambda,m}^{\mu,k}f(\alpha,\beta;z)}. \qquad (21)$$

Differentiating both sides of (21) with respect to z logarithmically and using (20), we obtain

$$w(z)+\frac{z\,w'(z)}{\alpha+1-w(z)}=\frac{z\left(I_{\lambda,m}^{\mu,k}f'(\alpha+1,\beta;z)\right)}{I_{\lambda,m}^{\mu,k}f(\alpha,\beta;z)}. \qquad (22)$$

From (22) and Lemma 3 (with a replaced by $\alpha+1$) we find that

$$w(z)+\frac{z\,w'(z)}{\alpha+1-w(z)}\prec h \qquad (z\in U). \qquad (23)$$

Now, in view of (18) and (23), an application of Lemma 1 yields

$$w(z)\prec h(z) \qquad (z\in U). \qquad (24)$$

Set

$$q(z)=\frac{-z\left(I^kL(\alpha,\beta,\lambda,\mu)f\right)'(z)}{I_{\lambda,m}^{\mu,k}f(\alpha,\beta;z)}. \qquad (25)$$

Then, $q(z)$ is analytic in U, with $q(0)=1$, and it follows from (18) and (25) that

$$\begin{aligned}
&I_{\lambda,m}^{\mu,k}f(\alpha,\beta;z)q(z)\\
&=-\alpha I^kL(\alpha+1,\beta,\lambda,\mu)f(z)+(1+\alpha)I^kL(\alpha,\beta,\lambda,\mu)f(z). \qquad (26)
\end{aligned}$$

Differentiating both sides of (26) with respect to z and using (25), we get

$$zq'(z) + \left(\alpha + 1 + \frac{z\left(I_{\lambda,m}^{\mu,k} f'(\alpha,\beta;z)\right)}{I_{\lambda,m}^{\mu,k} f(\alpha,\beta;z)} \right) q(z) = -\frac{\alpha z \left(I^k L(\alpha+1,\beta,\lambda,\mu) f\right)'(z)}{I_{\lambda,m}^{\mu,k} f(\alpha,\beta;z)}. \tag{27}$$

Furthermore, we find from (19), (20) and (27) that

$$q(z) + \frac{zq'(z)}{\alpha + 1 - w(z)} = -\frac{z\left(I^k L(\alpha+1,\beta,\lambda,\mu) f\right)'(z)}{I_{\lambda,m}^{\mu,k} f(\alpha+1,\beta;z)} \prec h(z) \qquad (z \in U), \tag{28}$$

since $f \in \Sigma_{\lambda,m}^{\mu,k}(\alpha+1,\beta;h)$. By (18) and (24), we see that $\Re\{\alpha+1-w(z)\} > 0$. Therefore, we deduce from (28) and Lemma 2 that $q(z) \prec h(z) \quad (z \in U)$, which implies that $f \in \Sigma_{\lambda,m}^{\mu,k}(\alpha,\beta;h)$ and the proof of Theorem 1 is thus complete.

Theorem 2 *Let $h \in A$ with*

$$\Re\{h(z)\} < 1 + \alpha \qquad (z \in U, \alpha > 0). \tag{29}$$

If $f \in M_{\lambda,m}^{\mu,k}(\alpha+1,\beta;h)$, with respect to $g \in \Sigma_{\lambda,m}^{\mu,k}(\alpha+1,\beta;h)$, then $f \in M_{\lambda,m}^{\mu,k}(\alpha,\beta;h)$ provided that $I_{\lambda,m}^{\mu,k} g(\alpha,\beta;z) \neq 0, \qquad (z \in U^*)$.

Proof According to the hypotheses of Theorem 2, we have

$$\frac{-z\left(I^k L(\alpha+1,\beta,\lambda,\mu) f\right)'(z)}{I_{\lambda,m}^{\mu,k} g(\alpha+1,\beta;z)} \prec h(z) \qquad (z \in U) \tag{30}$$

with $g \in \Sigma_{\lambda,m}^{\mu,k}(\alpha+1,\beta;h)$. Furthermore, it follows from Theorem 1 that $g \in \Sigma_{\lambda,m}^{\mu,k}(\alpha,\beta;h)$ and Lemma 3 yields

$$\Omega(z) = -\frac{z\left(I_{\lambda,m}^{\mu,k} g'(\alpha,\beta;z)\right)}{I_{\lambda,m}^{\mu,k} g(\alpha,\beta;z)} \prec h(z), \qquad (z \in U). \tag{31}$$

Suppose that

$$q(z) = -\frac{z\left(I^k L(\alpha,\beta,\lambda,\mu) f\right)'(z)}{I_{\lambda,m}^{\mu,k} g(\alpha,\beta;z)}. \tag{32}$$

By using (7), (32) can be written as follows:

$$I_{\lambda,m}^{\mu,k} g(\alpha,\beta;z) q(z) = -\alpha I^k L(\alpha+1,\beta,\lambda,\mu) f(z) + (1+\alpha) I^k L(\alpha,\beta,\lambda,\mu) f(z). \tag{33}$$

Differentiating both sides of (33) with respect to z and using (19) (with f replaced by g), we find that

$$q(z)+\frac{zq'(z)}{\alpha+1-\Omega(z)}=-\frac{z\left(I^{k}L(\alpha+1,\beta,\lambda,\mu)f\right)'(z)}{I_{\lambda,m}^{\mu,k}g(\alpha+1,\beta;z)}\qquad(z\in U).\tag{34}$$

Combining (30) and (34), we obtain

$$q(z)+\frac{zq'(z)}{\alpha+1-\Omega(z)}\prec h(z)\qquad(z\in U).\tag{35}$$

Consequently, in view of (29), (31) and (35), we deduce from Lemma 2 that

$$q(z)\prec h(z)\qquad(z\in U),$$

which shows that $f\in M_{\lambda,m}^{\mu,k}(\alpha,\beta;h)$ with respect to $g\in\Sigma_{\lambda,m}^{\mu,k}(\alpha,\beta;h)$.

Theorem 3 *Let $h\in A$ with*

$$\Re\{h(z)\}<1+\alpha\qquad(z\in U,\alpha>0).\tag{36}$$

Then,

$$M_{\lambda,m}^{\mu,k}(\gamma_1;\alpha,\beta;h)\subset M_{\lambda,m}^{\mu,k}(\gamma_2;\alpha,\beta;h)\qquad(0\leq\gamma_1<\gamma_2).$$

Proof For $f\in M_{\lambda,m}^{\mu,k}(\gamma_2;\alpha,\beta;h)$, there exists a function $g\in\Sigma_{\lambda,m}^{\mu,k}(\alpha,\beta;h)$ satisfying the following condition:

$$I_{\lambda,m}^{\mu,k}g(\alpha+1,\beta;z)\neq0\qquad(z\in U^{*}).$$

such that

$$-\gamma_2\frac{z\left(I^{k}L(\alpha+1,\beta,\lambda,\mu)f\right)'(z)}{I_{\lambda,m}^{\mu,k}g(\alpha+1,\beta;z)}$$

$$-(1-\gamma_2)\frac{z\left(I^{k}L(\alpha,\beta,\lambda,\mu)f\right)'(z)}{I_{\lambda,m}^{\mu,k}g(\alpha,\beta;z)}\prec h(z)\ (z\in U).\tag{37}$$

Set

$$q(z)=\frac{z\left(I^{k}L(\alpha,\beta,\lambda,\mu)f\right)'(z)}{I_{\lambda,m}^{\mu,k}g(\alpha,\beta;z)}\qquad(z\in U).$$

Since $g\in\Sigma_{\lambda,m}^{\mu,k}(\alpha,\beta;h)$ it follows from (31) to (34) (used in the proof of Theorem 2) and (37) together with Lemma 2 that $f\in M_{\lambda,m}^{\mu,k}(\gamma_1;\alpha,\beta;h)$. The details of the proofs are tedious, therefore we omit it. Thus, we complete the Theorem 3.

Remarks There are many other results that we can prove such as the convolution and the subordination properties for the three subclasses aforementioned.

Acknowledgement The work here is supported by LRGS/TD/2011/UKM/ICT/03/02.

References

1. Cho, N.E., Kwon, O.S., Srivastava, H.M.: Inclusion and argument properties for certain subclasses of meromorphic functions associated with a family of multiplier transformations. J. Math. Anal. Appl. **300**, 505–520 (2004)
2. Cho, N.E., Kim, I.H.: Inclusion properties of certain classes of meromorphic functions associated with the generalized hypergeometric function Appl. Math. Compu. **187**(1), 115–121 (2007)
3. Choi, J.H., Saigo, M., Srivastava, H.M.: Some inclusion properties of a certain family of integral operators. J. Math. Anal. Appl. **276**, 432–435 (2002)
4. Dziok, J., Srivastava, H.M.: Some subclasses of analytic functions with fixed argument of coefficients associated with the generalized hypergeometric function. Adv. Stud. Contemp. Math. Kyungshang. **5**(2), 115–125 (2002)
5. Dziok, J., Srivastava, H.M.: Certain subclasses of analytic functions associated with the generalized hypergeometric function. Integral. Transforms. Spec. Funct. **14**(1), 7–18 (2003)
6. Ghanim, F., Darus, M.: A new class of meromorphically analytic functions with applications to generalized hypergeometric functions. Abstract and applied analysis online article. http://www.hindawi.com/journals/aaa/2011/159405/ (2011)
7. Ghanim, F., Darus, M.: Some results of p-alent meromorphic functions defined by a linear operator. Far East J. Math. Sci. **44**(2), 155–165 (2010)
8. Ghanim, F., Darus, M.: Some properties Of certain subclass Of meromorphically multivalent functions defined by liner operator. J. Math. Stat. **6**(1), 34–41 (2010)
9. Ghanim, F., Darus, M.: New subclass of multivalent hypergeometric meromorphic functions. Inter. J. Pure Appl. Math. **61**(3), 269–280 (2010)
10. Ghanim, F., Darus, M.: Certain subclasses of meromorphic functions related to Cho-Kwon-Srivastava operator. Far East J. Math. Sci. (FJMS) **48**(2), 159–173 (2011)
11. Liu, J.L.: The Noor integral operator and strongly starlike functions. J. Math. Anal. Appl. **261**, 441–447 (2001)
12. Liu, J.L., Srivastava, H.M.: A linear operator and associated families of meromorphically multivalent functions, J. Math. Anal. Appl. **259**(2), 566–581 (2001)
13. Liu, J.L., Noor, K.I.: Some properties of Noor integral operator. J. Natur. Geom. **21**, 81–90 (2002)
14. Liu, J.L.: A linear operator and its applications on meromorphic p-valent functions. Bull. Inst. Math. Acad. Sin. **31**(1), 23–32 (2003)
15. Liu, J.L., Srivastava, H.M.: Certain properties of the Dziok-Srivastava operator. Appl. Math. Comput. **159**, 485–493 (2004)
16. Liu, J.L., Srivastava, H.M.: Classes of meromorphically multivalent functions associated with the generalized hypergeometric function. Math. Comput. Modelling. **39**(1), 21–34 (2004)
17. Miller, S.S., Mocanu, P.T.: Differential subordinations and univalent functions. Michigan. Math. J. **28**, 157–171 (1981)
18. Miller, S.S., Mocanu, P.T.: On some classes of first order differential subordinations. Michigan Math. J. **32**, 185–195 (1985)
19. Miller, S.S., Mocanu, P.T.: Differential subordinations and inequalities in the complex plane. MJ. Differ. Equations **67**, 199–211 (1987)
20. Noor, K.I.: On new classes of integral operators. J. Natur. Geom. **16**, 71–80 (1999)

21. Noor, K.I., Noor, M.A.: On integral operators. J. Natur. Geom. **238**, 341–352 (1999)
22. Piejko, K., Sokól, J.: Subclasses of meromorphic functions associated with the Cho–Kwon–Srivastava operator. J. Math. Anal. Appl. **337**, 1261–1266 (2008)
23. Ruscheweyh, S.: Convolutions in geometric function theory, Seacuteminaire de Mathe acutematiques Supeacuterieures. 83, Les Presses de l'Université de Montrèal, Montrèal, (1982)
24. Yang, D.G.: Some criteria for multivalent starlikeness. South Asian Bull. Math. **24**, 491–497 (2000)
25. Yang, D.G., Liu, J.L.: Multivalent functions associated with a linear operator. Applied. Math. Compu. **204**, 862–871 (2008)

The Description of Orbits Under an Action of GL_9 ON SLb_9

I. S. Rakhimov, S. K. Said Husain and F. Deraman

Abstract The paper concerns with the classification problem of a subclass of nine-dimensional complex filiform Leibniz algebras. The concepts of adapted basis and adapted transformations are given to express the action ("transport of structure") of the adapted transformations group on SLb_9. We present SLb_9 as a disjoint union of its subsets and specify the orbits under the action of adapted transformations. The list of isomorphism classes of SLb_9 with their representatives and the table of multiplications are given.

1 Introduction

The concept of Leibniz algebra has been introduced by Loday [3] in the study of Leibniz (co)homology as a noncommutative analog of Lie algebras (co)homology. A Leibniz algebra over a field K is a vector space over K, equipped with a K-bilinear map $[\cdot,\cdot] : L \times L \longrightarrow L$ satisfying the Leibniz identity

$$[x,[y,z]] = [[x,y],z] - [[x,z],y], \text{ for all } x,y,z \in L.$$

Clearly, a Lie algebra is a Leibniz algebra, and conversely, a Leibniz algebra L with property $[x,y] = -[y,x]$, for all $x,y \in L$ is a Lie algebra. Any Leibniz algebra L gives rise to a Lie algebra L_{Lie}, which is obtained as the quotient of L by relation $[x,x] = 0$.

I. S. Rakhimov (✉)
Laboratory of Cryptography, Analysis and Structure, (INSPEM)
& Department of Mathematics, Faculty of Science, Universiti Putra Malaysia,
43400 Serdang, Malaysia
e-mail: risamiddin@gmail.com

S. K. Said Husain
Department of Mathematics, Faculty of Science, Universiti Putra Malaysia,
43400 Serdang, Malaysia
e-mail: skartini@science.upm.edu.my

F. Deraman
Institute for Mathematical Research, Universiti Putra Malaysia,
43400 Serdang, Malaysia
e-mail: fatanahderaman@gmail.com

A. Kilicman et al. (eds.), *International Conference on Mathematical Sciences and Statistics 2013*, DOI 10.1007/978-981-4585-33-0_15,

If set $d_z(\cdot) = [\cdot, z]$ then the Leibniz identity expresses the Leibniz rule

$$d_z([x, y]) = [d_z(x), y] + [x, d_z(y)].$$

Further, all algebras considered are supposed to be over the field complex numbers $\mathbb{C}$.

For Leibniz algebras the notions of solvability and nilpotency imitate the corresponding notions in Lie theory. They have been introduced in [1].

Let L be a Leibniz algebra. We put:

$$L^1 = L, \quad L^{k+1} = [L^k, L], \ \ k \in \mathbb{N}.$$

Definition 1 A Leibniz algebra L is said to be nilpotent if there exists an integer $s \in \mathbf{N}$, such that

$$L^1 \supset L^2 \supset \ldots \supset L^s = 0.$$

Definition 2 An n-dimensional Leibniz algebra L is said to be *filiform* if dim $L^i = n - i$, where $2 \leq i \leq n$.

The filiform Leibniz algebra is the simplest nilpotent Leibniz algebra.

The set of all filiform Leibniz algebra structures on an n-dimensional vector space is denoted by Lb_n. According to Ayupov–Gomez–Omirov results (see [2], [5]), the class of all filiform Leibniz algebras is split into three subclasses which are invariant with respect to the "transport of structure" action of the linear group GL_n on Lb_n :

$$Lb_n = FLb_n \bigcup SLb_n \bigcup TLb_n.$$

There is a basis, called adapted, with respect to that each of the classes is given by a table of multiplication. The paper deals with SLb_n. This class is represented as follows.

$$SLb_{n+1} = \begin{cases} [e_0, e_0] = e_2, & \\ [e_i, e_0] = e_{i+1}, & 2 \leq i \leq n-1 \\ [e_0, e_1] = \beta_3 e_3 + \beta_4 e_4 + \ldots + \beta_n e_n, & \\ [e_1, e_1] = \gamma e_n, & \\ [e_j, e_1] = \beta_3 e_{j+2} + \beta_4 e_{j+3} + \ldots + \beta_{n+1-j} e_n, & \\ & 2 \leq j \leq n-2. \end{cases}$$

Elements of SLb_{n+1} are denoted by $L(\beta_3, \beta_4, \ldots, \beta_n, \gamma)$.

In this paper we give complete list of isomorphism classes and appropriate invariants of GL_9 acting on SLb_9. As for other classes in low dimensional cases they have been studied in [4], [6]–[8]. The theoretical background of all these has been given in [5].

Definition 3 An action of algebraic group G on a variety X is a morphism $\sigma : G \times X \to X$ with

i) $\sigma(e,x) = x$ for $\forall x \in X$, where e is the identity element of G.
ii) $\sigma(g,\sigma(h,x)) = \sigma(gh,x)$, for all $g,h \in G$ and $x \in X$.

One writes gx for $\sigma(g,x)$ and calls X as a G-variety. The orbit of the action is the set $O(x) = \{y \in X | \exists g \in G \text{such that} y = gx\}$. A function $f : X \to K$ is said to be invariant if $f(gx) = f(x)$ for any $g \in G$ and $x \in X$.

The action of the reductive linear group GL_9 on SLb_9 can be reduced to the action of a subgroup, which consists of linear transformations of adapted basis.

Definition 4 A basis $\{e_0, e_1, \ldots, e_n\}$ of L is said to be adapted if its multiplication table has one of the form SLb_{n+1}.

Recall that the adapted basis always exists. Hence, the isomorphism problem can be considered with respect to adapted transformations.

Definition 5 The basis transformation $f \in GL(V)$ is said to be adapted for the structure of L, if the basis $\{f(e_0), f(e_1), \ldots, f(e_n)\}$ is adapted.

The subgroup of $GL(V)$ consisting of all linear transformations sending one adapted basis to another is said to be *adapted* for the structure of L. This subgroup is denoted by G_{ad}. Thus, consider $G = G_{ad}$ and $X = SLb_{n+1}$. Then, the orbits with respect to the "transport of structure" action of G_{ad} on SLb_{n+1} consist of all algebras isomorphic to other algebras.

In G_{ad}, we consider the following elementary base change.

$$\text{First type: } \sigma(b,n) = \begin{cases} f(e_0) = e_0, \\ f(e_1) = e_1 + be_n, \\ f(e_{i+1}) = [f(e_i), f(e_0)], \; 2 \le i \le n-1, \\ f(e_2) = [f(e_0), f(e_0)], \end{cases}$$

$$\text{Second type: } \eta(a,k) = \begin{cases} f(e_0) = e_0 + ae_k, \qquad 2 \le k \le n, \\ f(e_1) = e_1, \\ f(e_{i+1}) = [f(e_i), f(e_0)], \; 2 \le i \le n-1, \\ f(e_2) = [f(e_0), f(e_0)], \end{cases}$$

$$\text{Third type: } \delta(a,b,d) = \begin{cases} f(e_0) = ae_0 + be_1, \\ f(e_1) = de_1 - \frac{bd\gamma}{a} e_{n-1}, \quad ad \neq 0, \\ f(e_{i+1}) = [f(e_i), f(e_0)], \; 2 \le i \le n-1, \\ f(e_2) = [f(e_0), f(e_0)], \end{cases}$$

where $a, b, d \in \mathbb{C}$.

The proofs of the following propositions are straightforward.

Proposition 1 *Let f be an adapted transformation of SLb_{n+1}, then*

$$f = \sigma(B_n, n) \circ \eta(A_n, n) \circ \eta(A_{n-1}, n-2) \circ \ldots \circ \eta(A_2, 2) \circ \delta(A_0, A_1, B_1).$$

Proposition 2 *The transformation*

$$g = \sigma(B_n, n) \circ \eta(A_n, n) \circ \eta(A_{n-1}, n-2) \circ \ldots \circ \eta(A_2, 2)$$

preserves the structure constants of algebras from SLb_{n+1}.

As a result the adapted base change is expressed as follows:

$$e_0' = Ae_0 + Be_1,$$

$$e_1' = De_1 - \frac{BD}{A}\gamma e_3,$$

$$e_2' = A^2e_2 + AB\beta_3e_3 + AB\beta_4e_4 + AB\beta_5e_5 + AB\beta_6e_6 + AB\beta_7e_7 + B(A\beta_8 + B\gamma)e_8,$$

$$e_3' = A^3e_3 + 2A^2B\beta_3e_4 + \left(2A^2B\beta_4 + AB^2\beta_3^2\right)e_5 + (2A^2B\beta_5 + 2AB^2\beta_3\beta_4)e_6$$
$$+\left(2A^2B\beta_6 + 2AB^2\beta_3\beta_5 + AB^2\beta_4^2\right)e_7 + (2A^2B\beta_7 + 2AB^2\beta_3\beta_6 + 2AB^2\beta_4\beta_5)e_8,$$

$$e_4' = A^4e_4 + 3A^2B\beta_3e_5 + \left(3A^3B\beta_4 + 3A^2B^2\beta_3^2\right)e_6$$
$$+\left(3A^3B\beta_5 + 6A^2BB^2\beta_3\beta_4 + AB^3\beta_3^3\right)e_7$$
$$+\left(3A^3B\beta_6 + 6A^2B^2\beta_3\beta_5 + 3A^2B^2\beta_4^2 + 3AB^3\beta_3^2\beta_4\right)e_8$$

$$e_5' = A^5e_5 + 4A^4B\beta_3e_6 + \left(4A^4B\beta_4 + 6A^3B^2\beta_3^2\right)e_7$$
$$+\left(4A^4B\beta_5 + 12A^3B^2\beta_3\beta_4 + 4A^2B^3\beta_3^3\right)e_8,$$

$$e_6' = A^6e_6 + 5A^5B\beta_3e_7 + \left(5A^5B\beta_4 + 10A^4B^2\beta_3^2\right)e_8,$$

$$e_7' = A^7e_7 + 6A^6B\beta_3e_8,$$

$$e_8' = A^8e_8.$$

This base change effects to the structure constants $\beta_3, \beta_4, \ldots, \beta_8, \gamma$ of $L(\beta_3, \beta_4, \ldots, \beta_8, \gamma)$ as follows.

$$\beta_3' = \frac{1}{A}\frac{D}{A}\beta_3,$$

$$\beta_4' = \frac{1}{A^2}\frac{D}{A}\left(\beta_4 - 2\frac{B}{A}\beta_3^2\right),$$

$$\beta_5' = \frac{1}{A^3}\frac{D}{A}\left(\beta_5 - 5\frac{B}{A}\beta_3\beta_4 + 5\left(\frac{B}{A}\right)^2\beta_3^3\right),$$

$$\beta_6' = \frac{1}{A^4}\frac{D}{A}\left(\beta_6 - 6\frac{B}{A}\beta_3\beta_5 + 21\left(\frac{B}{A}\right)^2\beta_3^2\beta_4 - 3\frac{B}{A}\beta_4^2 - 14\left(\frac{B}{A}\right)^3\beta_3^4\right), \quad (1)$$

$$\beta_7' = \frac{1}{A^5}\frac{D}{A}\left(\beta_7 - 7\frac{B}{A}\beta_3\beta_6 + 28\left(\frac{B}{A}\right)^2\beta_3^2\beta_5 + 28\left(\frac{B}{A}\right)^2\beta_3\beta_4^2 - 7\frac{B}{A}\beta_4\beta_5\right.$$
$$\left. -84\left(\frac{B}{A}\right)^3\beta_3^3\beta_4 + 42\left(\frac{B}{A}\right)^4\beta_3^5\right),$$

$$\beta_8' = \frac{1}{A^6}\frac{D}{A}\left(\frac{B}{A}\gamma + \beta_8 - 8\frac{B}{A}\beta_3\beta_7 + 36\left(\frac{B}{A}\right)^2\beta_3^2\beta_6 + 72\left(\frac{B}{A}\right)^2\beta_3\beta_4\beta_5\right.$$
$$-8\frac{B}{A}\beta_4\beta_6 + 12\left(\frac{B}{A}\right)^2\beta_4^3 - 180\left(\frac{B}{A}\right)^3\beta_3^2\beta_4^2 - 120\left(\frac{B}{A}\right)^3\beta_3^3\beta_5$$
$$\left. +330\left(\frac{B}{A}\right)^4\beta_3^4\beta_4 - 4\frac{B}{A}\beta_5^2 - 132\left(\frac{B}{A}\right)^5\beta_3^6\right),$$

$$\gamma' = \frac{1}{A^6}\left(\frac{D}{A}\right)^2\gamma.$$

This is the isomorphism criterion, i.e. two algebras $L(\beta_3', \beta_4', \ldots, \beta_8', \gamma')$ and $L(\beta_3, \beta_4, \ldots, \beta_8, \gamma)$ are isomorphic if and only if there exist complex numbers A, B and D with $AD \neq 0$ such that the equalities (15.1) hold. Thus, the isomorphism problem led to finding such complex numbers A, B and D. To do this first we change the basis such a way that the resulting effect to the structure constant $\beta_3, \beta_4, \ldots, \beta_8, \gamma$ is given as below.

$$\Lambda_3 = \beta_3, \quad \Lambda_4 = \beta_4, \quad \Lambda_5 = 4\beta_3\beta_5 - 5\beta_4^2, \quad \Lambda_6 = 4\beta_3^2\beta_6 - 7\beta_4^3,$$
$$\Lambda_7 = 8\beta_3^3\beta_7 - 21\beta_4^4, \quad \Lambda_8 = 8\beta_3^4\beta_8 - 33\beta_4^5.$$

Due to the notations the isomorphism criterion above for two algebras $L(\Lambda')$ and $L(\Lambda)$ from SLb_9, where $\Lambda' = (\Lambda_3', \Lambda_4', \Lambda_5', \Lambda_6', \Lambda_7', \Lambda_8', \gamma')$ and $\Lambda = (\Lambda_3, \Lambda_4, \Lambda_5, \Lambda_6, \Lambda_7, \Lambda_8, \gamma)$ is written as follows. Two algebras $L(\Lambda')$ and $L(\Lambda)$ from SLb_9 are isomorphic if and only if there exist complex numbers A, B and D such that $AD \neq 0$ and the following system of equalities holds:

$$\Lambda_3' = \frac{1}{A}\frac{D}{A}\Lambda_3,$$
$$\Lambda_4' = \frac{1}{A^2}\frac{D}{A}\left(\Lambda_4 - 2\frac{B}{A}\Lambda_3^2\right),$$
$$\Lambda_5' = \frac{1}{A^4}\left(\frac{D}{A}\right)^2\Lambda_5,$$
$$\Lambda_6' = \frac{1}{A^6}\left(\frac{D}{A}\right)^3\left(\Lambda_6 - 6\frac{B}{A}\Lambda_4^2\Lambda_5\right), \tag{2}$$
$$\Lambda_7' = \frac{1}{A^8}\left(\frac{D}{A}\right)^4\left(\Lambda_7 - 14\frac{B}{A}\Lambda_3^2\Lambda_6 + 56\left(\frac{B}{A}\right)^2\Lambda_3^4\Lambda_5\right.$$

$$-14\frac{B}{A}\Lambda_3^2\Lambda_4\Lambda_5\Big),$$

$$\Lambda_8^{'} = \frac{1}{A^{10}}\left(\frac{D}{A}\right)^5\Bigg(8\frac{B}{A}\Lambda_3^4\gamma + \Lambda_8 - 8\frac{B}{A}\Lambda_3^2\Lambda_7 + 72\left(\frac{B}{A}\right)^2$$

$$\Lambda_3^4\Lambda_6 + 144\left(\frac{B}{A}\right)^2\Lambda_3^4\Lambda_4\Lambda_5 - 16\frac{B}{A}\Lambda_3^2\Lambda_4\Lambda_6 - 2\frac{B}{A}\Lambda_3^2$$

$$\Lambda_5^2 - 240\left(\frac{B}{A}\right)^3\Lambda_3^6\Lambda_5 - 20\frac{B}{A}\Lambda_3^2\Lambda_4^2\Lambda_5\Bigg),$$

$$\gamma^{'} = \frac{1}{A^6}\left(\frac{D}{A}\right)^2\gamma.$$

We introduce the following functions:

$$t_3(X) = X_5,$$
$$t_4(X) = X_6 - 3X_4X_5,$$
$$t_5(X) = X_7 - 7X_4X_6 + 7X_4^2X_5.$$
$$t_6(X,Y) = 4X_3^2X_4Y + X_8 - 4X_3^3X_5 - X_4X_5^2 + 10X_4^2X_6 - 4X_4X_7,$$

where $X = (X_3, X_4, \ldots, X_8)$.

2 Isomorphism Classes

To find the isomorphism classes we split SLb_9 into its two disjoint G_{ad}-invariant subsets as follows.

$$SLb_9 = SLb_9^a \bigcup SLb_9^b, \text{ where } SLb_9^a = \{L \in SLb_9 : \beta_3 \neq 0\} \text{ and}$$
$$SLb_9^b = \{L \in SLb_9 : \beta_3 = 0\}.$$

2.1 *Isomorphism Classes in* SLb_9^a

By using the set of functions the isomorphism criterion for SLb_9^a can be stated as follows. Two algebras $L(\Lambda')$ and $L(\Lambda)$ from SLb_9^a are isomorphic if and only if there exist complex numbers A, B and D such that $AD \neq 0$ and the following system of equalities holds:

$$\Lambda_3' = \frac{1}{A}\frac{D}{A}\Lambda_3,$$

$$\Lambda_4' = \frac{1}{A^2}\frac{D}{A}\left(\Lambda_4 - 2\frac{B}{A}\Lambda_3^2\right),$$
$$t_3(\Lambda') = \frac{1}{A^4}\left(\frac{D}{A}\right)^2 t_3(\Lambda),$$
$$t_4(\Lambda') = \frac{1}{A^6}\left(\frac{D}{A}\right)^3 t_4(\Lambda), \tag{3}$$
$$t_5(\Lambda') = \frac{1}{A^8}\left(\frac{D}{A}\right)^4 t_5(\Lambda).$$
$$t_6(\Lambda', \gamma') = \frac{1}{A^{10}}\left(\frac{D}{A}\right)^5 t_6(\Lambda, \gamma),$$
$$t_7(\gamma') = \frac{1}{A^6}\left(\frac{D}{A}\right)^2 t_7(\gamma).$$

We split the class SLb_9^a into its disjoint subsets as follows:

$$SLb_9^a = \bigcup_{i=1}^{9} U_i \bigcup_{j=21}^{24} U_j,$$

where

$$U_1 = \{L(\Lambda) \in SLb_9^a : t_3(\Lambda) \neq 0,\ t_4(\Lambda) \neq 0\},$$
$$U_2 = \{L(\Lambda) \in SLb_9^a : t_3(\Lambda) \neq 0,\ t_4(\Lambda) = 0,\ t_6(\Lambda, \gamma) \neq 0\},$$
$$U_3 = \{L(\Lambda) \in SLb_9^a : t_3(\Lambda) \neq 0,\ t_4(\Lambda) = 0,\ t_6(\Lambda, \gamma) = 0\},$$
$$U_4 = \{L(\Lambda) \in SLb_9^a : t_3(\Lambda) = 0,\ t_4(\Lambda) \neq 0,\ t_5(\Lambda) \neq 0\},$$
$$U_5 = \{L(\Lambda) \in SLb_9^a : t_3(\Lambda) = 0,\ t_4(\Lambda) \neq 0,\ t_5(\Lambda) = 0,\ t_6(\Lambda, \gamma) \neq 0\},$$
$$U_6 = \{L(\Lambda) \in SLb_9^a : t_3(\Lambda) = 0,\ t_4(\Lambda) \neq 0,\ t_5(\Lambda) = 0,\ t_6(\Lambda, \gamma) = 0,\ t_7(\gamma) \neq 0\},$$
$$U_7 = \{L(\Lambda) \in SLb_9^a : t_3(\Lambda) = 0,\ t_4(\Lambda) = 0,\ t_5(\Lambda) \neq 0,\ t_6(\Lambda, \gamma) \neq 0\},$$
$$U_8 = \{L(\Lambda) \in SLb_9^a : t_3(\Lambda) = 0,\ t_4(\Lambda) = 0,\ t_5(\Lambda) \neq 0,\ t_6(\Lambda, \gamma) = 0,\ t_7(\gamma) \neq 0\},$$
$$U_9 = \{L(\Lambda) \in SLb_9^a : t_3(\Lambda) = 0,\ t_4(\Lambda) = 0,\ t_5(\Lambda) = 0,\ t_7(\gamma) \neq 0,\ t_6(\Lambda, \gamma) \neq 0\},$$
$$U_{21} = \{L(\Lambda) \in SLb_9^a : t_3(\Lambda) = 0,\ t_4(\Lambda) \neq 0,\ t_5(\Lambda) = 0,\ t_6(\Lambda, \gamma) = 0,\ t_7(\gamma) = 0\},$$
$$U_{22} = \{L(\Lambda) \in SLb_9^a : t_3(\Lambda) = 0,\ t_4(\Lambda) = 0,\ t_5(\Lambda) \neq 0,\ t_6(\Lambda, \gamma) = 0,\ t_7(\gamma) = 0\},$$
$$U_{23} = \{L(\Lambda) \in SLb_9^a : t_3(\Lambda) = 0,\ t_4(\Lambda) = 0,\ t_5(\Lambda) = 0,\ t_7(\gamma) = 0,\ t_6(\Lambda, \gamma) \neq 0\},$$
$$U_{24} = \{L(\Lambda) \in SLb_9^a : t_3(\Lambda) = 0,\ t_4(\Lambda) = 0,\ t_5(\Lambda) = 0,\ t_7(\gamma) \neq 0,\ t_6(\Lambda, \gamma) = 0\}.$$

We show that each of these subsets either is a single orbit or is a union of infinitely many orbits under the action of G_{ad}.

2.2 Isomorphism Classes in SLb_9^b

The algebras in SLb_9^b is represented by $L(0, \beta_4, \ldots, \beta_8, \gamma)$. The isomorphism criterion for SLb_9^b is given as follows. Two algebras $L(0, \beta_4', \ldots, \beta_8', \gamma')$ and $L(0, \beta_4, \ldots, \beta_8, \gamma)$ are isomorphic if and only if there exi st complex numbers A, B and D with $AD \neq 0$ such that the following system of equalities holds:

$$
\begin{aligned}
\beta_4' &= \frac{1}{A^2}\frac{D}{A}\beta_4,\\
\beta_5' &= \frac{1}{A^3}\frac{D}{A}\beta_5,\\
\beta_6' &= \frac{1}{A^4}\frac{D}{A}\left(\beta_6 - 3\frac{B}{A}\beta_4^2\right),\\
\beta_7' &= \frac{1}{A^5}\frac{D}{A}\left(\beta_7 - 7\frac{B}{A}\beta_4\beta_5\right),\\
\beta_8' &= \frac{1}{A^6}\frac{D}{A}\left(\frac{B}{A}\gamma + \beta_8 - 8\frac{B}{A}\beta_4\beta_6 + 12\left(\frac{B}{A}\right)^2\beta_4^3 - 4\frac{B}{A}\beta_5^2\right),\\
\gamma' &= \frac{1}{A^6}\left(\frac{D}{A}\right)^2\gamma.
\end{aligned}
\tag{4}
$$

The class SLb_9^b is written as a union of its disjoint G_{ad}-invariant subsets as follows:

$$SLb_9^b = \bigcup_{i=10}^{20} U_i \bigcup_{j=25}^{36} U_j,$$

$$
\begin{aligned}
U_{10} &= \{L(\beta) \in SLb_9^b : \beta_4 \neq 0,\ \beta_5 \neq 0\},\\
U_{11} &= \{L(\beta) \in SLb_9^b : \beta_4 \neq 0,\ \beta_5 = 0,\ \beta_7 \neq 0,\ \gamma\beta_6 + 3\beta_4^2\beta_8 - 4\beta_4\beta_6^2 \neq 0\},\\
U_{12} &= \{L(\beta) \in SLb_9^b : \beta_4 \neq 0,\ \beta_5 = 0,\ \beta_7 \neq 0,\ \gamma\beta_6 + 3\beta_4^2\beta_8 - 4\beta_4\beta_6^2 = 0, \gamma \neq 0\},\\
U_{13} &= \{L(\beta) \in SLb_9^b : \beta_4 \neq 0, \beta_5 = 0,\ \beta_7 = 0,\ \gamma\beta_6 + 3\beta_4^2\beta_8 - 4\beta_4\beta_6^2 \neq 0,\ \gamma \neq 0\},\\
U_{14} &= \{L(\beta) \in SLb_9^b : \beta_4 = 0,\ \beta_5 \neq 0,\ 4\beta_5^2 - \gamma \neq 0,\ \beta_6 \neq 0\},\\
U_{15} &= \{L(\beta) \in SLb_9^b : \beta_4 = 0,\ \beta_5 \neq 0,\ 4\beta_5^2 - \gamma \neq 0,\ \beta_6 = 0, \beta_7 \neq 0\},\\
U_{16} &= \{L(\beta) \in SLb_9^b : \beta_4 = 0,\ \beta_5 \neq 0,\ 4\beta_5^2 - \gamma \neq 0,\ \beta_6 = 0, \beta_7 = 0\},\\
U_{17} &= \{L(\beta) \in SLb_9^b : \beta_4 = 0,\ \beta_5 \neq 0,\ 4\beta_5^2 - \gamma = 0,\ \beta_6 \neq 0\},\\
U_{18} &= \{L(\beta) \in SLb_9^b : \beta_4 = 0,\ \beta_5 \neq 0,\ 4\beta_5^2 - \gamma = 0,\ \beta_6 \neq 0, \beta_7 \neq 0, \beta_8 \neq 0\},
\end{aligned}
$$

$$U_{19} = \{L(\beta) \in SLb_9^b : \beta_4 = 0,\ \beta_5 = 0,\ \beta_6 \neq 0,\ \gamma \neq 0,\ \beta_7 \neq 0\},$$
$$U_{20} = \{L(\beta) \in SLb_9^b : \beta_4 = 0,\ \beta_5 = 0,\ \beta_6 \neq 0,\ \gamma = 0,\ \beta_7 \neq 0\},$$
$$U_{25} = \{L(\beta) \in SLb_9^b : \beta_4 \neq 0,\ \beta_5 = 0,\ \beta_7 \neq 0,\ \gamma\beta_6 + 3\beta_4^2\beta_8 - 4\beta_4\beta_6^2 = 0,\ \gamma = 0\},$$
$$U_{26} = \{L(\beta) \in SLb_9^b : \beta_4 \neq 0,\ \beta_5 = 0,\ \beta_7 = 0,\ 3\beta_4\beta_8 - 4\beta_6^2 \neq 0,\ \gamma = 0\},$$
$$U_{27} = \{L(\beta) \in SLb_9^b : \beta_4 = 0,\ \beta_5 \neq 0,\ 4\beta_5^2 - \gamma = 0,\ \beta_6 \neq 0,\ \beta_7 \neq 0,\ \beta_8 = 0\},$$
$$U_{28} = \{L(\beta) \in SLb_9^b : \beta_4 = 0,\ \beta_5 = 0,\ \beta_6 \neq 0,\ \gamma \neq 0,\ \beta_7 = 0\},$$
$$U_{29} = \{L(\beta) \in SLb_9^b : \beta_4 = 0,\ \beta_5 = 0,\ \beta_6 \neq 0,\ \gamma = 0,\ \beta_7 = 0,\ \beta_8 \neq 0\},$$
$$U_{30} = \{L(\beta) \in SLb_9^b : \beta_4 = 0,\ \beta_5 = 0,\ \beta_6 \neq 0,\ \gamma = 0,\ \beta_7 = 0,\ \beta_8 = 0\},$$
$$U_{31} = \{L(\beta) \in SLb_9^b : \beta_4 = 0,\ \beta_5 = 0,\ \beta_6 = 0,\ \gamma \neq 0,\ \beta_7 \neq 0\},$$
$$U_{32} = \{L(\beta) \in SLb_9^b : \beta_4 = 0,\ \beta_5 = 0,\ \beta_6 = 0,\ \gamma \neq 0,\ \beta_7 = 0\},$$
$$U_{33} = \{L(\beta) \in SLb_9^b : \beta_4 = 0,\ \beta_5 = 0,\ \beta_6 = 0,\ \gamma = 0,\ \beta_7 \neq 0,\ \beta_8 \neq 0\},$$
$$U_{34} = \{L(\beta) \in SLb_9^b : \beta_4 = 0,\ \beta_5 = 0,\ \beta_6 = 0,\ \gamma = 0,\ \beta_7 \neq 0,\ \beta_8 = 0\},$$
$$U_{35} = \{L(\beta) \in SLb_9^b : \beta_4 = 0,\ \beta_5 = 0,\ \beta_6 = 0,\ \gamma = 0,\ \beta_7 = 0,\ \beta_8 \neq 0\},$$
$$U_{36} = \{L(\beta) \in SLb_9^b : \beta_4 = 0,\ \beta_5 = 0,\ \beta_6 = 0,\ \gamma = 0, \beta_7 = 0,\ \beta_8 = 0\}.$$

We show that each of these subsets is also either a single orbit or is a union of infinitely many orbits under the action of G_{ad}.

3 Description of Orbits and Invariant Functions

In the given collection of disjoint subsets, the isomorphism problem in each of these subsets will be studied separately. Propositions 3 gives that the subsets $U_1 - U_{20}$ are a union of infinitely many orbits and Proposition 4 shows that the subsets $U_1 - U_{20}$ are single orbits.

Proposition 3 i) *Two algebras $L(t(\Lambda))$ and $L(t(\Lambda'))$ from U_1 are isomorphic if and only if*

$$\frac{t_3^3(\Lambda')}{t_4^2(\Lambda')} = \frac{t_3^3(\Lambda)}{t_4^2(\Lambda)}, \quad \frac{t_3(\Lambda')\cdot t_5(\Lambda')}{t_4^2(\Lambda')} = \frac{t_3(\Lambda)\cdot t_5(\Lambda)}{t_4^2(\Lambda)},$$
$$\frac{t_3^2(\Lambda')\cdot t_6(\Lambda',\gamma')}{t_4^3(\Lambda')} = \frac{t_3^2(\Lambda)\cdot t_6(\Lambda,\gamma)}{t_4^3(\Lambda)},$$

and
$$\frac{t_7(\gamma')\cdot t_1^2(\Lambda')\cdot t_5^4(\Lambda')}{t_4^2(\Lambda')} = \frac{t_7(\gamma)\cdot t_1^2(\Lambda)\cdot t_5^4(\Lambda)}{t_4^4(\Lambda)}.$$

ii) *For any* $\lambda_1 \in c^*$ $\lambda_2, \lambda_3, \lambda_4 \in C$ *there exists* $L(t(\Lambda)) \in U_1$ *such that,*

$$\lambda_1 = \frac{t_3^3(\Lambda)}{t_4^2(\Lambda)}, \quad \lambda_2 = \frac{t_3(\Lambda)\cdot t_5(\Lambda)}{t_4^2(\Lambda)},$$

$$\lambda_3 = \frac{t_3^2(\Lambda)\cdot t_6(\Lambda,\gamma)}{t_4^3(\Lambda)} \quad \textit{and} \quad \lambda_4 = \frac{t_7(\gamma)\cdot t_1^2(\Lambda)\cdot t_5^4(\Lambda)}{t_4(\Lambda)}.$$

Therefore, the orbits in U_1 *can be represented as* $L(1, 0, \lambda_1, \lambda_1, \lambda_2, \lambda_3, \lambda_4)$.

Proof (i) **Necessity** Suppose, $L(\Lambda)$ and $L(\Lambda^{'})$ are isomorphic. This means that the isomorphism criterion (1) holds true. The system of Eq. (2) is also to be true since the system of Eq. (2) is equivalent to (1). Then, substitute the values of $\Lambda_i^{'}$ where $i = 3, 4, .., 8$ and $\gamma^{'}$ into the following functions to obtain the required equations.

$$\frac{t_3^3(\Lambda^{'})}{t_4^2(\Lambda^{'})} = \frac{\left[\frac{1}{A^4}\left(\frac{D}{A}\right)^2 t_3(\Lambda)\right]^3}{\left[\frac{1}{A^6}\left(\frac{D}{A}\right)^3 t_3(\Lambda)\right]^2} = \frac{t_3^3(\Lambda)}{t_4^2(\Lambda)}$$

$$\frac{t_3(\Lambda^{'})\cdot t_5(\Lambda^{'})}{t_4^2(\Lambda^{'})} = \frac{\left[\frac{1}{A^4}\left(\frac{D}{A}\right)^2 t_3(\Lambda)\right]\left[\frac{1}{A^8}\left(\frac{D}{A}\right)^4 t_5(\Lambda)\right]}{\left[\frac{1}{A^6}\left(\frac{D}{A}\right)^3 t_3(\Lambda)\right]^2} = \frac{t_3(\Lambda)\cdot t_5(\Lambda)}{t_4^2(\Lambda)}$$

$$\frac{t_3^2(\Lambda^{'})\cdot t_6(\Lambda^{'},\gamma^{'}))}{t_4^3(\Lambda^{'})} = \frac{\left[\frac{1}{A^4}\left(\frac{D}{A}\right)^2 t_3(\Lambda)\right]^2\left[\frac{1}{A^{10}}\left(\frac{D}{A}\right)^5 t_6(\Lambda,\gamma)\right]}{\left[\frac{1}{A^6}\left(\frac{D}{A}\right)^3 t_3(\Lambda)\right]^3} = \frac{t_3^2(\Lambda)\cdot t_6(\Lambda,\gamma)}{t_4^3(\Lambda)}$$

$$\frac{t_7(\gamma^{'})\cdot t_1^2(\Lambda^{'})\cdot t_5^4(\Lambda^{'})}{t_4^4(\Lambda^{'})} = \frac{\left[\frac{1}{A^6}\left(\frac{D}{A}\right)^2 t_7(\gamma)\right]\left[\frac{1}{A}\left(\frac{D}{A}\right) t_1(\Lambda)\right]^2\left[\frac{1}{A^4}\left(\frac{D}{A}\right)^2 t_3(\Lambda)\right]^4}{\left[\frac{1}{A^6}\left(\frac{D}{A}\right)^3 t_3(\Lambda)\right]^4} = \frac{t_7(\gamma)\cdot t_1^2(\Lambda)\cdot t_5^4(\Lambda)}{t_4^4(\Lambda)}.$$

(ii) **Sufficiency** Let the equations in Proposition 3 hold. Suppose,

$$A_0 = \frac{t_4(\Lambda)}{t_1(\Lambda)\cdot t_3(\Lambda)}, \quad B_0 = \frac{2\cdot t_1(\Lambda)\cdot t_4(\Lambda)}{t_2(\Lambda)\cdot t_3(\Lambda)}, \quad D_0 = \frac{t_4^2(\Lambda)}{t_1^3(\Lambda)\cdot t_3^2(\Lambda)}$$

$$A_0^{'} = \frac{t_4(\Lambda^{'})}{t_1(\Lambda^{'})\cdot t_3(\Lambda^{'})}, \quad B_0^{'} = \frac{2\cdot t_1(\Lambda^{'})\cdot t_4(\Lambda^{'})}{t_2(\Lambda^{'})\cdot t_3(\Lambda^{'})}, \quad D_0^{'} = \frac{t_4^2(\Lambda^{'})}{t_1^3(\Lambda^{'})\cdot t_3^2(\Lambda^{'})}.$$

Let $A = \frac{A_0}{A_0^{'}}$, $B = \frac{B_0 D_0^{'} - B_0^{'} D_0}{A_0^{'} D_0^{'}}$ and $D = \frac{D_0}{D_0^{'}}$, that is

$$A = \frac{t_1(\Lambda^{'}) \cdot t_3(\Lambda^{'}) \cdot t_4(\Lambda)}{t_1(\Lambda) \cdot t_3(\Lambda) \cdot t_4(\Lambda^{'})}, \quad D = \frac{t_1^3(\Lambda^{'}) \cdot t_3^2(\Lambda^{'}) \cdot t_4^2(\Lambda)}{t_1^3(\Lambda) \cdot t_3^2(\Lambda) \cdot t_4^2(\Lambda^{'})}$$

$$B = \frac{t_4(\Lambda^{'}) \cdot t_3^2(\Lambda^{'})}{t_4^2(\Lambda^{'})} \left[\frac{\Lambda_4 \cdot t_4(\Lambda) \cdot t_4(\Lambda^{'})}{2 \cdot t_1^3(\Lambda) \cdot t_3(\Lambda) \cdot t_3(\Lambda^{'})} - \frac{t_2(\Lambda^{'}) \cdot t_4^2(\Lambda)}{2 \cdot t_1^3(\Lambda) \cdot t_3^2(\Lambda)} \right].$$

Substitute the values as the following;

a)

$$\begin{aligned}
\frac{1}{A} \frac{D}{A} t_1(\Lambda) &= \left[\frac{t_6(\Lambda^{'}, \gamma^{'}) \cdot t_1(\Lambda) \cdot t_3^2(\Lambda)}{t_1(\Lambda^{'}) \cdot t_3^2(\Lambda^{'}) \cdot t_6(\Lambda, \gamma)} \right] \left[\frac{t_6(\Lambda, \gamma) \cdot t_1^2(\Lambda^{'}) \cdot t_3^2(\Lambda^{'})}{t_6(\Lambda^{'}, \gamma^{'}) \cdot t_1^2(\Lambda) \cdot t_3^2(\Lambda)} \right] t_1(\Lambda) \\
&= t_1(\Lambda^{'}).
\end{aligned}$$

b)

$$\begin{aligned}
\frac{1}{A^2} \frac{D}{A} t_2(\Lambda) &= \left[\frac{t_6(\Lambda^{'}, \gamma^{'}) \cdot t_1(\Lambda) \cdot t_3^2(\Lambda)}{t_1(\Lambda^{'}) \cdot t_3^2(\Lambda^{'}) \cdot t_6(\Lambda, \gamma)} \right]^2 \left[\frac{t_6(\Lambda, \gamma) \cdot t_1^2(\Lambda^{'}) \cdot t_3^2(\Lambda^{'})}{t_6(\Lambda^{'}, \gamma^{'}) \cdot t_1^2(\Lambda) \cdot t_3^2(\Lambda)} \right] t_2(\Lambda) \\
&= \left[\frac{t_6(\Lambda^{'}, \gamma^{'}) \cdot t_3^2(\Lambda)}{t_3^2(\Lambda^{'}) \cdot t_6(\Lambda, \gamma)} \right] \left[\frac{t_3(\Lambda^{'}) \cdot t_2(\Lambda^{'}) \cdot t_4(\Lambda)}{t_3(\Lambda) \cdot t_4(\Lambda^{'})} \right] \\
&= \frac{t_6(\Lambda^{'}, \gamma^{'}) \cdot t_3(\Lambda) \cdot t_4(\Lambda) \cdot t_2(\Lambda^{'})}{t_3(\Lambda^{'}) \cdot t_3(\Lambda^{'}) \cdot t_6(\Lambda, \gamma)} \quad \text{from } \lambda_3 \\
&= \frac{t_4^2(\Lambda^{'}) \cdot t_3^3(\Lambda) \cdot t_2(\Lambda^{'})}{t_3^3(\Lambda^{'}) \cdot t_4^2(\Lambda)} \quad \text{from } \lambda_1 \\
&= t_2(\Lambda^{'}) \qquad \text{Hint}: t_2(\Lambda) = \frac{t_3(\Lambda^{'}) \cdot t_2(\Lambda^{'}) \cdot t_4(\Lambda)}{t_3(\Lambda) \cdot t_4(\Lambda^{'})}.
\end{aligned}$$

c)

$$\begin{aligned}
\frac{1}{A^4} \left(\frac{D}{A} \right)^2 t_3(\Lambda) &= \left[\frac{t_6(\Lambda^{'}, \gamma^{'}) \cdot t_1(\Lambda) \cdot t_3^2(\Lambda)}{t_1(\Lambda^{'}) \cdot t_3^2(\Lambda^{'}) \cdot t_6(\Lambda, \gamma)} \right]^4 \left[\frac{t_6(\Lambda, \gamma) \cdot t_1^2(\Lambda^{'}) \cdot t_3^2(\Lambda^{'})}{t_6(\Lambda^{'}, \gamma^{'}) \cdot t_1^2(\Lambda) \cdot t_3^2(\Lambda)} \right]^2 t_3(\Lambda) \\
&= \left[\frac{t_6^4(\Lambda^{'}, \gamma^{'}) \cdot t_1^4(\Lambda) \cdot t_3^8(\Lambda)}{t_1^4(\Lambda^{'}) \cdot t_3^8(\Lambda^{'}) \cdot t_6^4(\Lambda, \gamma)} \right] \left[\frac{t_6^2(\Lambda, \gamma) \cdot t_1^4(\Lambda^{'}) \cdot t_3^4(\Lambda^{'})}{t_6^2(\Lambda^{'}, \gamma^{'}) \cdot t_1^4(\Lambda) \cdot t_3^4(\Lambda)} \right] t_3(\Lambda) \\
&= \frac{t_6^2(\Lambda^{'}, \gamma^{'}) \cdot t_3^5(\Lambda)}{t_3^4(\Lambda^{'}) \cdot t_6^2(\Lambda, \gamma)}
\end{aligned}$$

$$= \left[\frac{t_6^2(\Lambda^{'},\gamma^{'})}{t_3^4(\Lambda^{'})}\right]\left[\frac{t_5(\Lambda^{'})\cdot t_3^3(\Lambda^{'})}{t_6^2(\Lambda^{'},\gamma^{'})}\right]\left[\frac{t_3^2(\Lambda^{'})}{t_5(\Lambda^{'})}\right] \text{from } \lambda_3$$

$$= t_3(\Lambda^{'})$$

d)

$$\frac{1}{A^6}\left(\frac{D}{A}\right)^3 t_4(\Lambda) = \left[\frac{t_6(\Lambda^{'},\gamma^{'})\cdot t_1(\Lambda)\cdot t_3^2(\Lambda)}{t_1(\Lambda^{'})\cdot t_3^2(\Lambda^{'})\cdot t_6(\Lambda,\gamma)}\right]^6\left[\frac{t_6(\Lambda,\gamma)\cdot t_1^2(\Lambda^{'})\cdot t_3^2(\Lambda^{'})}{t_6(\Lambda^{'},\gamma^{'})\cdot t_1^2(\Lambda)\cdot t_3^2(\Lambda)}\right]^3 t_4(\Lambda)$$

$$= \frac{t_3^3(\Lambda)\cdot t_4^3(\Lambda^{'})}{t_3^3(\Lambda^{'})\cdot t_4^2(\Lambda)}$$

$$= \frac{t_3^3(\Lambda^{'})\cdot t_4^3(\Lambda^{'})}{t_3^3(\Lambda^{'})\cdot t_4^2(\Lambda^{'})} \text{from } \lambda_1$$

$$= t_4(\Lambda)$$

e)

$$\frac{1}{A^8}\left(\frac{D}{A}\right)^4 t_5(\Lambda) = \left[\frac{t_6(\Lambda^{'},\gamma^{'})\cdot t_1(\Lambda)\cdot t_3^2(\Lambda)}{t_1(\Lambda^{'})\cdot t_3^2(\Lambda^{'})\cdot t_6(\Lambda,\gamma)}\right]^8\left[\frac{t_6(\Lambda,\gamma)\cdot t_1^2(\Lambda^{'})\cdot t_3^2(\Lambda^{'})}{t_6(\Lambda^{'},\gamma^{'})\cdot t_1^2(\Lambda)\cdot t_5^2(\Lambda)}\right]^4 t_5(\Lambda)$$

$$= \frac{t_6^4(\Lambda^{'},\gamma^{'})\cdot t_3^8(\Lambda)\cdot t_5(\Lambda)}{t_3^8(\Lambda^{'})\cdot t_6^4(\Lambda,\gamma)}$$

$$= \frac{t_6^4(\Lambda^{'},\gamma^{'})}{t_3^8(\Lambda^{'})}\cdot\frac{t_3(\Lambda^{'})\cdot t_5(\Lambda^{'})}{t_4^2(\Lambda^{'})}\cdot\frac{t_3^7(\Lambda)\cdot t_4^2(\Lambda)}{t_6^4(\Lambda,\gamma)} \text{from } \lambda_2$$

$$= \frac{t_6^4(\Lambda^{'},\gamma^{'})t_5(\Lambda^{'})}{t_3^7(\Lambda^{'})\cdot t_4^2(\Lambda^{'})}\frac{t_4^{12}(\Lambda^{'})}{t_3^8(\Lambda^{'})\cdot t_6^4(\Lambda^{'},\gamma^{'})}\cdot\frac{t_3^{15}(\Lambda)}{t_4^{10}(\Lambda)} \text{from } \lambda_3$$

$$= \frac{t_5(\Lambda^{'})}{t_3^7(\Lambda^{'})}\cdot\frac{t_4^{10}(\Lambda^{'})}{t_3^8(\Lambda^{'})}\cdot\frac{t_3^{15}(\Lambda^{'})}{t_4^{10}(\Lambda^{'})} \text{from } \lambda_1$$

$$= t_5(\Lambda^{'})$$

f)

$$\frac{1}{A^{10}}\left(\frac{D}{A}\right)^5 t_6(\Lambda,\gamma) = \left[\frac{t_6(\Lambda^{'},\gamma^{'})\cdot t_1(\Lambda)\cdot t_3^2(\Lambda)}{t_1(\Lambda^{'})\cdot t_3^2(\Lambda^{'})\cdot t_6(\Lambda,\gamma)}\right]^{10}\left[\frac{t_6(\Lambda,\gamma)\cdot t_1^2(\Lambda^{'})\cdot t_3^2(\Lambda^{'})}{t_6(\Lambda^{'},\gamma^{'})\cdot t_1^2(\Lambda)\cdot t_5^2(\Lambda)}\right]^5 t_6(\Lambda,\gamma)$$

$$= \frac{t_3^5(\Lambda)\cdot t_4^5(\Lambda^{'})\cdot t_6(\Lambda,\gamma)}{t_3^5(\Lambda^{'})\cdot t_4^5(\Lambda)}$$

$$= \frac{t_4^2(\Lambda^{'})\cdot t_3^3(\Lambda)\cdot t_6(\Lambda^{'},\gamma^{'})}{t_3^3(\Lambda^{'})\cdot t_4^2(\Lambda)} \text{from } \lambda_1$$

$$= t_6(\Lambda^{'},\gamma^{'}) \text{ from } \lambda_2.$$

Union of orbits	Representative
U_2	$L(1,0,\lambda_1,0,\lambda_2,\lambda_1,\lambda_3)$ where $\lambda_1,\lambda_2,\lambda_3 \in \mathbb{C}$ $\lambda_1 = \frac{t_3^2(\Lambda)}{t_5(\Lambda)}, \lambda_2 = \frac{t_3^3(\Lambda)\cdot t_5(\Lambda)}{t_6^2(\Lambda,\gamma)}, \lambda_3 = \frac{t_7(\gamma)}{t_3(\Lambda)}$
U_3	$L(1,0,1,0,\lambda_1,0,\lambda_2)$ where $\lambda_1,\lambda_2 \in \mathbb{C}$, $\lambda_1 = \frac{t_5(\Lambda)}{t_3^2(\Lambda)}, \lambda_2 = \frac{t_7(\gamma)}{t_3(\Lambda)}$
U_4	$L(1,0,0,\lambda_1,\lambda_1,\lambda_2,\lambda_3)$ where $\lambda_1 \in \mathbb{C}^*, \lambda_2,\lambda_3 \in \mathbb{C}$ $\lambda_1 = \frac{t_4^3(\Lambda)}{t_1(\Lambda)\cdot t_5^2(\Lambda)}, \lambda_3 = \frac{t_1^2(\Lambda)\cdot t_4^4(\Lambda)\cdot t_7(\gamma)}{t_5^2(\Lambda)}$, $\lambda_2 = \frac{t_4(\Lambda)\cdot t_6(\Lambda,\gamma)}{t_5^2(\Lambda)}$
U_5	$L(1,0,0,\lambda_1,0,\lambda_1,\lambda_2)$ where $\lambda_1 \in \mathbb{C}^*, \lambda_2 \in \mathbb{C}$ $\lambda_1 = \frac{t_6^3(\Lambda,\gamma)}{t_4^3(\Lambda)}$ $\lambda_2 = \frac{t_1^2(\Lambda)\cdot t_6^4(\Lambda,\gamma)\cdot t_7(\gamma)}{t_4^8(\Lambda)}$

g)

$$\frac{1}{A^6}\left(\frac{D}{A}\right)^2 t_7(\gamma) = \left[\frac{t_6(\Lambda',\gamma')\cdot t_1(\Lambda)\cdot t_3^2(\Lambda)}{t_1(\Lambda')\cdot t_3^2(\Lambda')\cdot t_6(\Lambda,\gamma)}\right]^6 \left[\frac{t_6(\Lambda,\gamma)\cdot t_1^2(\Lambda')\cdot t_3^2(\Lambda')}{t_6(\Lambda',\gamma')\cdot t_1^2(\Lambda)\cdot t_5^2(\Lambda)}\right]^2 t_7(\gamma)$$

$$= \frac{t_1^2(\Lambda)\cdot t_3^4(\Lambda)\cdot t_4^4(\Lambda')\cdot t_7(\gamma)}{t_1^2(\Lambda')\cdot t_3^4(\Lambda')\cdot t_4^4(\Lambda)}$$

$$= t_7(\gamma') \text{ from } \lambda_3.$$

This change transforms $L(\Lambda)$ into

$$L\left(1,0,\frac{t_3^3(\Lambda)}{t_4^2(\Lambda)},\frac{t_3^3(\Lambda)}{t_4^2(\Lambda)},\frac{t_3(\Lambda)\cdot t_5(\Lambda)}{t_4^2(\Lambda)},\frac{t_3^2(\Lambda)\cdot t_6(\Lambda,\gamma)}{t_4^3(\Lambda)},\frac{t_7(\gamma)\cdot t_1^2(\Lambda)\cdot t_5^4(\Lambda)}{t_4(\Lambda)}\right).$$

An analogous base change with "prime" transforms $L(\Lambda')$ into $L(1,0,\frac{t_3^3(\Lambda')}{t_4^2(\Lambda')},\frac{t_3^3(\Lambda')}{t_4^2(\Lambda')},\frac{t_3(\Lambda')\cdot t_5(\Lambda')}{t_4^2(\Lambda')},\frac{t_3^2(\Lambda')\cdot t_6(\Lambda',\gamma')}{t_4^3(\Lambda')},\frac{t_7(\gamma')\cdot t_1^2(\Lambda')\cdot t_5^4(\Lambda')}{t_4(\Lambda')})$. Since, $\frac{t_3^3(\Lambda')}{t_4^2(\Lambda')} = \frac{t_3^3(\Lambda)}{t_4^2(\Lambda)}$, $\frac{t_3(\Lambda')\cdot t_5(\Lambda')}{t_4^2(\Lambda')} = \frac{t_3(\Lambda)\cdot t_5(\Lambda)}{t_4^2(\Lambda)}$, $\frac{t_3^2(\Lambda')\cdot t_6(\Lambda',\gamma')}{t_4^3(\Lambda')} = \frac{t_3^2(\Lambda)\cdot t_6(\Lambda,\gamma)}{t_4^3(\Lambda)}$, and $\frac{t_7(\gamma')\cdot t_1^2(\Lambda')\cdot t_5^4(\Lambda')}{t_4^2(\Lambda')} = \frac{t_7(\gamma)\cdot t_1^2(\Lambda)\cdot t_5^4(\Lambda)}{t_4^4(\Lambda)}$.
The algebras are isomorphic.

(ii) Evidently, for any $\lambda_1 \in C^*$ $\lambda_2,\lambda_3,\lambda_4 \in C$ there exists $L(\Lambda) \in U_1$ such that,

$$\lambda_1 = \frac{t_3^3(\Lambda)}{t_4^2(\Lambda)}, \qquad \lambda_2 = \frac{t_3(\Lambda)\cdot t_5(\Lambda))}{t_4^2(\Lambda)},$$
$$\lambda_3 = \frac{t_3^2(\Lambda)\cdot t_6(\Lambda,\gamma)}{t_4^3(\Lambda)} \text{ and } \lambda_4 = \frac{t_7(\gamma)\cdot t_1^2(\Lambda)\cdot t_5^4(\Lambda)}{t_4^4(\Lambda)}.$$

It can be found by solving the system of Eq. (2) with respect to Λ_3, Λ_4, Λ_5, Λ_6, Λ_7, Λ_8 and γ.

The proofs for $U_2 - U_{20}$ are the similar to that of Proposition 1.

Proposition 4 *The subsets* $U_{21}, U_{22}, U_{23}, U_{24}, U_{25}, U_{26}, U_{27}, U_{28}, U_{29}, U_{30}, U_{31}, U_{32}, U_{33}, U_{34}, U_{35}$, and U_{36} *are single orbits with representatives* $L(1,0,0,1,0,0,0)$,

Orbit	Representative
U_{11}	$L(0,1,0,0,\lambda_1,\lambda_1,\lambda_2)$ where $\lambda_1 \in \mathbb{C}^*, \lambda_2 \in \mathbb{C}$ $\lambda_1 = \frac{\beta_3^5\beta_7^4}{\left(\gamma\beta_6+3\beta_4^2\beta_8-4\beta_4\beta_6^2\right)^3}$, $\lambda_2 = \frac{\gamma\beta_4^2\beta_7^2}{\left(\gamma\beta_6+3\beta_4^2\beta_8-4\beta_4\beta_6^2\right)^2}$
U_{12}	$L(0,1,0,0,\lambda,0,\lambda)$ where $\lambda \in \mathbb{C}^*$ $\lambda = \frac{\gamma^3}{\beta_4^4\beta_7^2}$
U_{13}	$L(0,1,0,0,0,\lambda,1)$ where $\lambda \in \mathbb{C}$ $\lambda = \frac{\left(\gamma\beta_3+3\beta_4^2\beta_8-4\beta_4\beta_6^2\right)^4\beta_4}{\gamma^2}$
U_{14}	$L(0,0,1,1,\lambda_1,0,\lambda_2)$ where $\lambda_1 \in \mathbb{C},\ \lambda_2 \neq 4$ $\lambda_1 = \frac{\beta_5\beta_7}{\beta_6^2}$ $\lambda_2 = \frac{\gamma}{\beta_5^2}$
U_{15}	$L(0,0,1,0,1,0,\lambda)$ where $\lambda \in \mathbb{C}$ $\lambda = \frac{\gamma}{\beta_5^2}$

Orbit	Representative
U_{16}	$L(0,0,1,0,0,0,\lambda)$ where $\lambda \in \mathbb{C}$ $\lambda = \frac{\gamma}{\beta_5^2}$
U_{17}	$L(0,0,1,1,\lambda_1,\lambda_2,4)$ where $\lambda_1,\lambda_2 \in \mathbb{C}$ $\lambda_1 = \frac{\beta_5\beta_7}{\beta_6^2}$ $\lambda_2 = \frac{\beta_5^2\beta_8}{\beta_6^2}$
U_{18}	$L(0,0,1,0,\lambda,\lambda,4)$ where $\lambda \in \mathbb{C}^*$ $\lambda = \frac{\beta_7^3}{\beta_5\beta_8^2}$
U_{19}	$L(0,0,0,1,1,0,\lambda)$ where $\lambda \in \mathbb{C}^*$ $\lambda = \frac{\gamma\beta_7^2}{\beta_6^4}$
U_{20}	$L(0,0,0,1,1,\lambda,0)$ where $\lambda \in \mathbb{C}$ $\lambda = \frac{\beta_6\beta_8}{\beta_7^2}$

$L(1,0,0,0,1,0,0)$, $L(1,0,0,0,0,1,0)$, $L(1,0,0,0,0,0,1)$, $L(0,1,0,0,1,0,0)$, $L(0,1,0,0,0,1,0)$, $L(0,0,1,0,1,0,4)$, $L(0,0,0,1,0,0,1)$, $L(0,0,0,1,0,1,0)$, $L(0,0,0,1,0,0,0)$, $L(0,0,0,0,1,0,1)$, $L(0,0,0,0,0,0,1)$, $L(0,0,0,0,1,1,0)$, $L(0,0,0,0,1,0,0)$, $L(0,0,0,0,0,1,0)$, and $L(0,0,0,0,0,0,0)$ respectively.

Proof The proof can be carried out by giving the base change explicitly leading to these representatives in each subsets $U_{21},\ldots,U_{36}$ case. For example, for U_{21}, we consider $\Lambda_5 = 0$, $\Lambda_7 = 7\Lambda_4\Lambda_6$, $\Lambda_8 = 18\Lambda_4^2\Lambda_6\ \gamma = 0$.

Put $A = \left(\frac{\Lambda_6}{\Lambda_3^3}\right)^{\frac{1}{3}}$, $D = \frac{A^2}{\Lambda_3}$, and $B = \frac{\Lambda_4}{2\Lambda_3^2} \cdot A$.

Acknowledgement This research was supported by the Universiti Putra Malaysia (UPM) through the research grant 05-02-12-2188RU.

References

1. Ayupov, S.A., Omirov, B.A.: On some classes of nilpotent Leibniz algebras. Siberian Math. J. **42**(1), 18–29 (2001)
2. Gómez, J.R., Omirov, B.A.: On classification of complex filiform Leibniz algebras http://front.math.ucdavis.edu/,ArXiv:0612735v1 [math.RA] (2006)
3. Loday, J.-L.: Une version non commutative des algèbres de Lie: les algèbres de Leibniz. Ens. Math. **39**, 269–293 (1993)
4. Munther, A.H.: Isomorphism classes and invariants of a subclass of complex filiform Leibniz algebras. Ph.D Thesis, Universiti Putra Malaysia (2010)
5. Rakhimov, I.S., Bekbaev, U.D.: On isomorphism classes and invariants of finite dimensional complex filiform Leibniz algebras. Commun. Algebra. **38**(12), 4705–4738 (2010)
6. Rakhimov, I.S., Said Husain, S.K.: Classification of a subclass of low-dimensional complex filiform Leibniz algebras. Linear and Multilinear Algebra. **59**(3), 339–354 (2011)
7. Rakhimov, I.S., Said Husain, S.K.: On isomorphism classes and invariants of a subclass of low-dimensional complex filiform Leibniz algebras. Linear and Multilinear Algebra. **59**(2), 205–220 (2011)
8. Sozan, J.O., Rakhimov, I.S., Kamel, A.M.A.: Classification of First Class of Complex Filiform Leibniz Algebras. Lampert Academic Publishing, Germany (2010)

Effects of Magnetohydrodynamic on the Stagnation Point Flow past a Stretching Sheet in the Presence of Thermal Radiation with Newtonian Heating

Muhammad Khairul Anuar Mohamed, Muhammad Imran Anwar, Sharidan Shafie, Mohd Zuki Salleh and Anuar Ishak

Abstract In this study, the effects of magnetohydrodynamic on the stagnation point flow past a stretching surface in the presence of thermal radiation generated by Newtonian heating are studied, where the heat transfer rate from the bounding surface with a finite heat capacity is proportional to the local surface temperature. The transformed boundary layer equations are solved numerically using the Keller-box method. Numerical solutions are obtained for the local heat transfer coefficient, the surface temperature as well as the velocity and temperature profiles. The features of the flow and heat transfer characteristics for various values of the magnetic parameter and thermal radiation parameter are analyzed and discussed.

M. K. A. Mohamed (✉)
Faculty of Information and Interactive Technology,
University College Shahputra, 25200 Kuantan, Pahang, Malaysia
e-mail: baa_khy@yahoo.com

M. I. Anwar · S. Shafie
Department of Mathematical Sciences, Faculty of Science,
Universiti Teknologi Malaysia, 81310 Skudai, Johor, Malaysia
e-mail: jeotarar@yahoo.com

S. Shafie
Department of Mathematical Sciences, Faculty of
Science, Universiti Teknologi Malaysia, 81310 Skudai, Johor, Malaysia
e-mail: sharidan@mel.fs.utm.my

M. Z. Salleh
Faculty of Industrial Science and Technology, Universiti
Malaysia Pahang, 26300 UMP Kuantan, Pahang, Malaysia
e-mail: zukikuj@yahoo.com

A. Ishak
School of Mathematical Sciences, Faculty of Science and Technology,
Universiti Kebangsaan Malaysia, 43600 UKM Bangi, Selangor, Malaysia
e-mail: anuarishak001@yahoo.com

A. Kilicman et al. (eds.), *International Conference on Mathematical Sciences and Statistics 2013*, DOI 10.1007/978-981-4585-33-0_16,

1 Introduction

Problems related to convection boundary layer flow are important in engineering and industrial activities. Such flows are applied to manage thermal effects in many industrial outputs for example in electronic devices, computer power supply and also in engine cooling system such as a heat sink in a car radiator. Sakiadis [1] was the first to study the boundary layer flow on a continuous solid surface moving at constant speed. Due to entrainment of the ambient fluid, this boundary layer flow is quite different from Blasius flow past a flat plate. Sakiadis's theoretical predictions for Newtonian fluids were later corroborated experimentally by Tsou et al. [2].

Flow of a viscous fluid past a stretching sheet is a classical problem in fluid dynamics. Crane [3] was the first to study the convection boundary layer flow over a stretching sheet. The heat and mass transfer on a stretching sheet with suction or blowing was investigated by Gupta and Gupta [4]. They considered an isothermal moving plate and obtained the temperature and concentration distributions. Chen and Char [5] studied the laminar boundary layer flow and heat transfer from a linearly stretching, continuous sheet subjected to suction or blowing with prescribed wall temperature and heat flux.

Ishak et al. [6], [7] studied the post stagnation-point and mixed convection towards vertical, linearly and continuosly stretching sheet, respectively. This type of problem was then extended to viscous fluids, viscoelastic fluids or micropolar fluids by many investigators by considering the usually applied boundary conditions, either prescribed wall temperature or prescribed wall heat flux.

The effects of thermal radiation and magnetohydrodynamic (MHD) on the flow and heat transfer characteristics have been done by many investigators. Bataller [8] studied the effects of radiation on the Blasius and Sakiadis flows. Ishak et al. [9], [10] investigated the MHD stagnation point flow towards a stretching sheet with prescribed surface temperature, then extended it to the prescribed surface heat flux and taking into consideration the radiation effects on the thermal boundary layer flow over a moving plate (sakiadis flow). Makinde and Aziz [11] considered the MHD mixed convection flow in a porous medium while the MHD flow with energy dissipation, internal heat source and thermal radiation has been studied by Chen [12]. Recently, Salleh et al. [13] considered the free convection flow over a permeable horizontal flat plate in a porous medium with radiation effect taken into consideration.

Generally, in modeling the convection boundary layer flow, the boundary conditions that were usually applied are (1) constant or prescribed wall temperature and (2) constant or prescribed surface heat flux. However, the Newtonian heating condition, (3) has been used only quite recently by Merkin [14], Pop et al. [15] and Lesnic et al. [16] to study the free convection boundary layer flow over vertical and horizontal surfaces embedded in a porous medium. The asymptotic solution near the leading edge and the full numerical solution along the whole plate domain have been obtained numerically, whilst the asymptotic solution for downstream along the plate has been obtained analytically. Recently, Salleh et al. [17], [18] and Salleh and Nazar [19] studied the forced convection boundary layer flow at a forward stagnation point

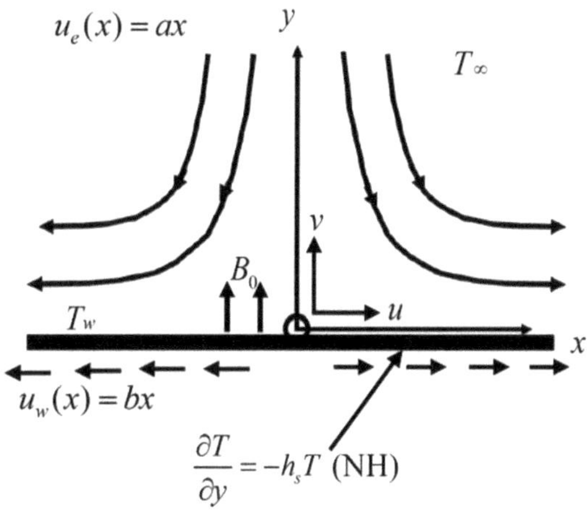

Fig. 1 Physical model and the coordinate system

with Newtonian heating. Very recently, Mohamed et al. [20] studied the stagnation point flow over a stretching sheet and Hayat et al. [21] investigated the flow of a second grade fluid over a stretching surface with Newtonian heating.

The aim of this study is to extend the work of Mohamed et al. [20], by introducing the effects of magnetohydrodynamic on the stagnation point flow over a stretching surface in the presence of thermal radiation, with Newtonian heating conditions. To the best of our knowledge, this problem has not been considered before, so that the reported results are new.

2 Mathematical Formulation

A steady two-dimensional stagnation-point flow over a stretching/shrinking plate immersed in an incompressible viscous fluid of ambient temperature T_∞ is considered. It is assumed that the external velocity $u_e(x)$ and the stretching velocity $u_w(x)$ are of the forms $u_e(x) = ax$ and $u_w(x) = \mathrm{b}x$ where a and b are constants. Further, a uniform magnetic field of strength B_0 is assumed to be applied in the positive y-direction normal to the stretching sheet. The magnetic Reynolds number is assumed to be small, and thus the induced magnetic field is negligible. The physical model and coordinate system of this problem is shown in Fig. 1. It is further assumed that the plate is subjected to a Newtonian heating [14]. The boundary layer equations are

$$\frac{\partial u}{\partial x} + \frac{\partial v}{\partial y} = 0 \tag{1}$$

$$u\frac{\partial u}{\partial x} + v\frac{\partial u}{\partial y} = u_e\frac{du_e}{dx} + \nu\frac{\partial^2 u}{\partial y^2} + \frac{\sigma B_0^2}{\rho}(u_e(x) - u) \tag{2}$$

$$u\frac{\partial T}{\partial x} + v\frac{\partial T}{\partial y} = \alpha\frac{\partial^2 T}{\partial y^2} - \frac{\partial q_r}{\partial y}. \tag{3}$$

They are subject to the boundary conditions

$$u = u_w(x), \quad v = 0, \quad \frac{\partial T}{\partial y} = -h_s T \text{ (NH) at } y = 0$$

$$u \to u_e(x), \quad T \to T_\infty \text{ as } y \to \infty \tag{4}$$

where u and v are the velocity components along the x and y, the directions, respectively. Further, T is the fluid temperature in the boundary layer, ν is the kinematic viscosity, ρ is the fluid density, σ is the electric conductivity, k is the thermal conductivity, α is the thermal diffusivity and h_s is the heat transfer coefficient.

By using the Rosseland approximation for radiation [8], the radiative heat flux is simplified as

$$q_r = -\frac{4\sigma^*}{3k^*}\frac{\partial T^4}{\partial y} \tag{5}$$

where σ^* and k^* are the Stefan-Boltzmann constant and the mean absorption coefficient, respectively. We assume that the temperature differences within the flow through the fluid such that the term T^4 may be expressed as a linear function of temperature. Hence, expanding T^4 in a Taylor series about T_∞ and neglecting higher-order terms, we get

$$T^4 \cong 4T_\infty^3 T - 3T_\infty^4. \tag{6}$$

In view of (5) and (6), (3) reduces to

$$u\frac{\partial T}{\partial x} + v\frac{\partial T}{\partial y} = \left(\alpha + \frac{16\sigma^* T_\infty^3}{3k^*}\right)\frac{\partial^2 T}{\partial y^2}. \tag{7}$$

From the above equation it is seen that the effect of radiation is to enhance the thermal diffusivity. If we take $N_R = \frac{\alpha k^*}{4\sigma^* T_\infty^3}$ as the radiation parameter, (7) becomes

$$u\frac{\partial T}{\partial x} + v\frac{\partial T}{\partial y} = \alpha\left(1 + \frac{4}{3N_R}\right)\frac{\partial^2 T}{\partial y^2}. \tag{8}$$

We now introduce the following similarity variables (see [18] and [22]):

$$\eta = \left(\frac{u_E}{\nu x}\right)^{1/2} y, \quad \psi = (\nu x u_E)^{1/2} f(\eta), \quad \theta(\eta) = \frac{T - T_\infty}{T_\infty} (NH) \tag{9}$$

where ψ is the stream function defined as $u = \frac{\partial \psi}{\partial y}$ and $u = -\frac{\partial \psi}{\partial x}$, which identically satisfies (1). Thus, we have

$$u = axf'(\eta), \quad v = -(a\nu)^{1/2} f(\eta) \tag{10}$$

where prime denotes differentiation with respect to η.

Table 1 Comparison between the present solution of the equations, (11) and (12), with previously published results when $\gamma = 1, M = 0, N_R = \infty$ and $\varepsilon = 0$

Pr	Salleh [25]		Mohamed et al. [20]		Present	
	$\theta(0)$	$-\theta'(0)$	$\theta(0)$	$-\theta'(0)$	$\theta(0)$	$-\theta'(0)$
5	23.0042	24.0042	23.0239	23.0239	23.0042	23.0042
7	5.6872	6.6872	5.6062	6.6062	5.6873	6.6873
10	2.9226	3.9226	2.9516	3.9516	2.9227	3.9227

Substituting (9) and (10) into (2) and (8), we obtain the following nonlinear ordinary differential equations:

$$f''' + ff'' + 1 - f'^2 - M(f' - 1) = 0 \tag{11}$$

$$\frac{1}{\Pr}\left(1 + \frac{4}{3N_R}\right)\theta'' + f\theta' = 0 \tag{12}$$

where $\Pr = \frac{\nu}{\alpha}$ is the Prandtl number and $M = \frac{\sigma B_0^2}{\rho a}$ is the magnetic parameter.

The boundary conditions corresponding to conditions (4) become

$$f(0) = 0, \quad f'(0) = \varepsilon, \quad \theta'(0) = -\gamma(1 + \theta(0))(NH) \tag{13}$$

$$f'(\eta) \to 1, \quad \theta(\eta) \to 0, \text{ as } \eta \to \infty \tag{14}$$

where $\varepsilon = \frac{b}{a}$ is the stretching parameter. Further, $\gamma = h_s\left(\frac{\nu}{a}\right)^{1/2}$ is the conjugate parameter for the Newtonian heating. It is noticed that $\gamma = 0$ is for the insulated plate and $\gamma \to \infty$ is when the surface temperature is prescribed.

3 Results and Discussion

The equations, (11) and (12) subject to the boundary conditions (13) and (14), were solved numerically using the Keller-box method [23]. In order to validate the efficiency of the method used, the comparison of the values of the wall temperature $\theta(0)$ and heat transfer coefficient $-\theta'(0)$ has been made. Due to the decoupled boundary layer equations, (11) and (12), for $\varepsilon = 0$ and $M = 0$, it is found that the value of the skin friction coefficient, $f''(0) = 1.23258766$ is in very good agreement with the classical value $f''(0) = 1.232588$ obtained by [24]. Table 1 presents the comparison between the present results with the previously reported results by [25] and [20] for various values of the Prandtl number Pr when $\gamma = 1, M = 0, N_R = \infty$ and $\varepsilon = 0$. It is found that they are in a good agreement. We can conclude that this method works efficiently for the present problem and we are also confident that results presented here are accurate.

Table 2 presents the values of $\theta(0)$ and $-\theta'(0)$ for various values of N_R when $\Pr = 7$, $\varepsilon = 3$, $\gamma = 1$ and $M = 0{,}5{,}10$. It is observed that, when the magnetohydrodynamic

Table 2 Values of $\theta(0)$ and $-\theta'(0)$ from the equations, (11) and (12), for various values of N_R when Pr = 7, $\varepsilon = 3$, $\gamma = 1$ and $M = 0,5,10$

N_R	$M=0$		$M=5$		$M=10$	
	$\theta(0)$	$-\theta'(0)$	$\theta(0)$	$-\theta'(0)$	$\theta(0)$	$-\theta'(0)$
1	0.8955	1.8955	0.9744	1.9744	1.0323	2.0323
2	0.6454	1.6454	0.6884	1.6884	0.7208	1.7208
3	0.5684	1.5684	0.6031	1.6031	0.6282	1.6282
5	0.5094	1.5094	0.5382	1.5382	0.5588	1.5588
7	0.4846	1.4846	0.5102	1.5102	0.5291	1.5291
10	0.4657	1.4657	0.4894	1.4894	0.5076	1.5076
100	0.4267	1.4267	0.4472	1.4472	0.4625	1.4625
1000	0.4227	1.4227	0.4430	1.443	0.4582	1.4582
∞	0.4223	1.4223	0.4426	1.4426	0.4578	1.4578

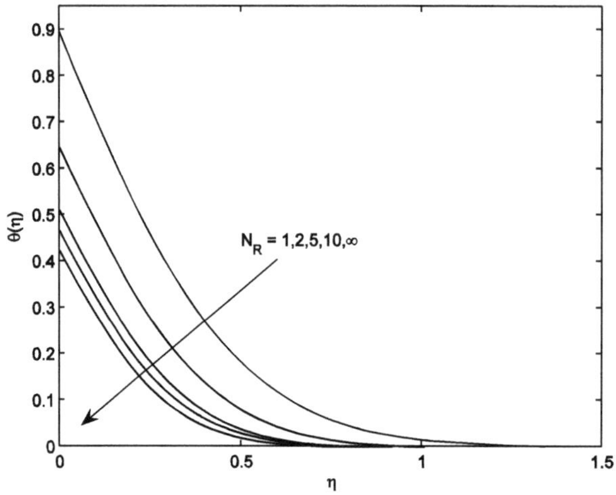

Fig. 2 Temperature profiles $\theta(\eta)$ for various values of N_R when $M = 0$

(MHD) effect is not considered ($M = 0$), an increasing of N_R, results the values of $\theta(0)$ and $-\theta'(0)$ decrease. Also, the same trends occur for case $M = 5$ and $M = 10$. When N_R is fixed, an increasing of M results the increase of $\theta(0)$ and $-\theta'(0)$.

Figure 2 shows the temperature profiles for various values of N_R. It is found that as N_R increases, the temperature in the boundary layer decreases, and the thermal boundary layer thickness also decreases. The temperature profiles presented in Fig. 3 show that when the value of M increases, the temperature also increases, contrary to the temperature profiles with various values of N_R presented in Fig. 2.

Figure 4 presents the velocity profile $f'(\eta)$ for various values of M when $\varepsilon = 3$. When the stretching velocity exceeds the external velocity ($\varepsilon > 1$), the flow has a boundary layer structure. From this figure, it can be seen that when M increases, the thickness of the boundary layer decreases which implies increasing manner of the magnitude of the velocity gradient at the surface which implies an increase of the skin friction coefficient $f''(0)$. Besides, N_R, Pr and γ do not influence the velocity profile which is clear from (11) and (12). Lastly, Fig. 5 shows the velocity profiles for different values of ε which produces $f'(0) = \varepsilon$ and $f'(\eta) = 1$ as $\eta \to \infty$ when $M = 1$. When the stretching velocity exceeds the external velocity ($\varepsilon > 1$), it is found that

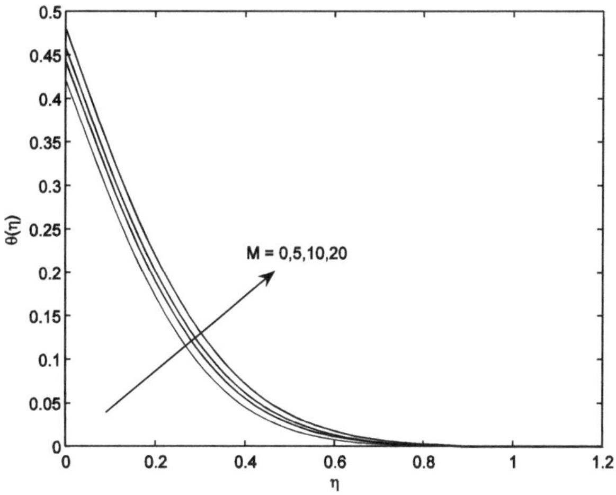

Fig. 3 Temperature profiles $\theta(\eta)$ for various values of M when $N_R = \infty$

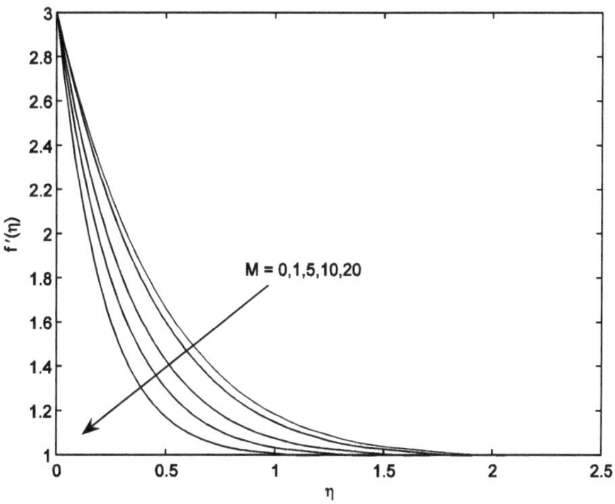

Fig. 4 Velocity profiles $f'(\eta)$ for various values of M when $\varepsilon = 3$

the flow has an inverted boundary layer structure. The thickness of the boundary layer increases with an increase of ε. For the case of stretching velocity is equal to the external velocity ($\varepsilon = 1$), the velocity is constant which implies zero skin friction coefficient ($f''(\eta) = 0$) since the surface and the fluid move with the same velocity. On the other hand, when $\varepsilon < 1$, the flow has a boundary layer structure. It is found that the thickness of the boundary layer increases with a decrease of ε.

4 Conclusion

In this paper, we have numerically studied the problem of magnetohydrodynamic effects on the stagnation point flow past a stretching sheet in the presence of thermal radiation generated by Newtonian heating conditions. It is shown how the magnetic

Fig. 5 Velocity profiles $f'(\eta)$ for various values of ε when $M = 1$

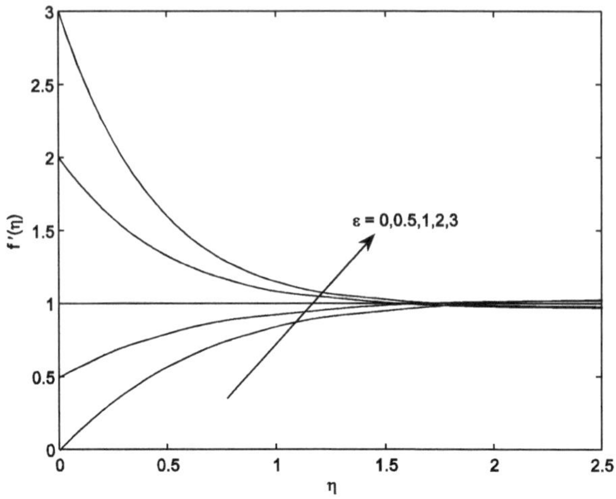

parameter M and thermal radiation parameter N_R affects the values of the surface temperature $\theta(0)$ and the heat transfer coefficient $-\theta'(0)$. We can conclude that the thermal boundary layer thickness depends strongly on these parameters. Further, it is seen that an increase in thermal radiation parameter N_R results in a decrease of the temperature. However, the increase of magnetic parameter M lead to an increase of the surface temperature $\theta(0)$.

Acknowledgments The authors gratefully acknowledge the financial support received in the form of research grants from the Universiti Malaysia Pahang (RDU110108 and RDU110390).

References

1. Sakiadis, B.C.: Boundary-layer behavior on continuous solid surfaces: I. Boundary-layer equations for two-dimensional and axisymmetric flow. AIChE J. **7**(1), 26–28 (1961)
2. Tsou, F.K., et al.: Flow and heat transfer in the boundary layer on a continuous moving surface. Int. J. Heat Mass Trans. **10**(2), 219–235 (1967)
3. Crane, L.J.: Flow past a stretching plate. ZAMP. **21**, 645–647 (1970)
4. Gupta, P.S., Gupta, A.S.: Heat and mass transfer on a stretching sheet with suction or blowing. Can. J. Chem. Eng. **55**(6), 744–746 (1977)
5. Chen, C.K., Char, M.: Heat transfer on a continuous, stretching surface with suction and blowing. J. Math. Anal. Appl. **135**, 568–580 (1988)
6. Ishak, A., et al.: Mixed convection on the stagnation point flow toward a vertical, continuously stretching sheet. J. Heat Trans. **129**(8), 1087–1090 (2007)
7. Ishak, A., et al.: Post stagnation point boundary layer flow and mixed convection heat transfer over a vertical, linearly stretching sheet. Arch. Mech. **60**, 303–322 (2008)
8. Bataller, R.C.: Radiation effects for the Blasius and Sakiadis flows with a convective surface boundary condition. Appl. Math. Comput. **206**(2), 832–840 (2008)
9. Ishak, A., et al.: MHD stagnation point flow towards a stretching sheet. Phys. A: Stat. Mech. Appl. **388**(17), 3377–3383 (2009)

10. Ishak, A., et al.: Radiation effects on the thermal boundary layer flow over a moving plate with convective boundary condition. Meccanica. **46**(4), 795–801 (2011)
11. Makinde, O.D., Aziz, A.: MHD mixed convection from a vertical plate embedded in a porous medium with a convective boundary condition. Int. J. Therm. Sci. **49**(9), 1813–1820 (2010)
12. Chen, C.H.: On the analytic solution of MHD flow and heat transfer for two types of viscoelastic fluid over a stretching sheet with energy dissipation, internal heat source and thermal radiation. Int. J. Heat Mass Trans. **53**(19–20), 4264–4273 (2010)
13. Salleh, M.Z., et al.: Free convection over a permeable horizontal flat plate embedded in a porous medium with radiation effects and mixed thermal boundary conditions. J. Math. Stat. **8**(1), 122–128 (2012)
14. Merkin, J.H.: Natural-convection boundary-layer flow on a vertical surface with Newtonian heating. Int. J. Heat Fluid Flow. **15**(5), 392–398 (1994)
15. Pop, I., et al.: Asymptotic solutions for the free convection boundary-layer flow along a vertical surface in a porous medium with Newtonian heating. Hybrid Methods Eng. **2**, 31–40 (2000)
16. Lesnic, D., et al.: Free convection boundary-layer flow above a nearly horizontal surface in a porous medium with Newtonian heating. Heat Mass Trans. **40**(9), 665–672 (2004)
17. Salleh, M.Z., et al.: Forced convection boundary layer flow at a forward stagnation point with Newtonian heating. Chem. Eng. Commun. **196**, 987–996 (2009)
18. Salleh, M.Z., et al.: Boundary layer flow and heat transfer over a stretching sheet with Newtonian heating. J. Taiwan Inst. Chem. Eng. **41**(6), 651–655 (2010)
19. Salleh, M.Z., Nazar, R.: Pemodelan Matematik Bagi Aliran Lapisan Sempadan Olakan Bebas Terhadap Silinder Bulat Mengufuk Dengan Pemanasan Newtonan. Sains Malays. **39**(4), 671–676 (2010)
20. Mohamed, M.K.A., et al.: Stagnation point flow over a stretching sheet with Newtonian heating. Sains Malays. **41**(11), 1467–1473 (2012)
21. Hayat, T., et al.: Flow of a second grade fluid over a stretching surface with Newtonian heating. J. Mech. **28**(1), 209–216 (2012)
22. Aziz, A.: A similarity solution for laminar thermal boundary layer over a flat plate with a convective surface boundary condition. Commun. Nonlinear Sci. Numer. Simulat. **14**(4), 1064–1068 (2009)
23. Na, T.Y.: Computational methods in engineering boundary value problems. Academic Press, New York (1979)
24. Hiemenz, K.: DIe Grenzschicht an einem in den gleichformigen Flussigkeitsstrom eingetauchten ger-aden Kreiszylinder. DIngl. Polytech. J. **32**, 321–410 (1911)
25. Salleh, M.Z.: Mathematical modelling for convection boundary layer flows with Newtonian heating. Ph.D Thesis, Universiti Kebangsaan Malaysia (2011)

Description of Three Dimensional Solvable Evaluation Algebras

Farrukh Mukhamedov, Bakhrom Omirov and Izzat Qaralleh

Abstract The concept of evolution algebras lies between algebras and dynamical system. In this paper we are going to describe three dimensional nilpotent, solvable evolution algebras. It is seen that three dimensional nilpotent evolution algebras are classified into three classes, and solvable ones are classified into five classes.

1 Introduction

In the present paper we are going to describe certain class of evolution algebras. The foundation of evolution algebra has been developed in [4]. Generally speaking, evolution algebras have many connections with Graph theory, Group theory, and dynamical system, etc. (see [4]). In [3] there has been an established equivalence between nil, right nilpotent evolution algebras. Moreover, all 2-D complex evolution algebras are classified. Some other properties of the evolution algebras were investigated in [2]. These investigations show that classification of evolution algebras is one of the important problems in this field. Therefore, in the present paper we are going to describe 3-D nilpotent and solvable evolution algebras.

2 Preliminaries

Let us define the main object of this work—evolution algebra.

F. Mukhamedov (✉) · I. Qaralleh
Department of Computational & Theoretical Sciences, Faculty of Science,
International Islamic University Malaysia, Kuantan, Pahang 25200, Malaysia
e-mail: farrukh_m@iium.edu.my

I. Qaralleh
e-mail: izzat_math@yahoo.com

B. Omirov
Institute of Mathematics, National University of Uzbekistan, 29,
Do'rmon Y'oli Str., Tashkent 100125, Uzbekistan
e-mail: omirovb@mail.ru

A. Kilicman et al. (eds.), *International Conference on Mathematical Sciences and Statistics 2013,* DOI 10.1007/978-981-4585-33-0_17,

Definition 1 Let $(E, \cdot)$ be an algebra over a field F. If it admits a basis $\{e_1, e_2, \dots\}$ such that [4]

$$e_i \cdot e_j = 0, \quad for\ i \neq j, \qquad e_i \cdot e_i = \sum_k a_{i,k} e_k, \quad for\ any\ i,$$

then this algebra is called *evolution algebra.*

It is remarkable that this type of algebra depends on evolution basis $\{e_1, e_2, \dots\}$. The basis $\{e_1, e_2, \dots\}$ is said to be *natural basis.*

By $\mathbf{A}= (a_{ij})$ w denote the matrix of the structural constants of the evolution algebra E. The following properties are known [4]:

1. Evolution algebras are not associative in general.
2. Evolution algebras are commutative and flexible.
3. Evolution algebras are not power-associative.
4. The direct sum of evolution algebras is also an evolution algebra.
5. The Kronecker product of evolution algebras is an evolution algebra.

Definition 2 If a is element of an evolution algebra E then a is called *nill* if there exists $n \in N$ such that $a^n = 0$. An evolution algebra E is called *nill* if every element of the algebra is nill.

For an evolution algebra E we introduce the following sequences

$$E^{[1]} = E^{\langle 1 \rangle} = E, \quad E^{[k+1]} = E^{[k]} E^{[k]}, \quad E^{\langle k+1 \rangle} = E^{\langle k \rangle} E,$$

$$E^1 = E, \quad E^k = \sum_{i=1}^{k-1} E^i E^{k-i}, \ k \geq 1.$$

The following inclusions hold:

$$E^{\langle k \rangle} \subseteq E^k, \quad E^{[k+1]} \subseteq E^{2^k}.$$

Since, E is a commutative algebra, then one gets

$$E^k = \sum_{i=1}^{\lfloor \frac{k}{2} \rfloor} E^i E^{k-i},$$

where $\lfloor x \rfloor$ denotes the integer part of x.

Definition 3 An evolution algebra E is called

(i) *right nilpotent* if there exists some $s \in N$ such that $E^{\langle s \rangle} = 0$. The smallest s such that $E^{\langle s \rangle} = 0$ is called the index of right nilpotency;
(ii) *nilpotent* if there exists some $n \in N$ such that $E^n = 0$. The smallest n such that $E^n = 0$ is called the index of nilpotency;
(iii) *solvable* if there exists some $t \in N$ such that $E^{[t]} = 0$. The smallest t such that $E^{[t]} = 0$ is called the index of solvability.

Theorem 1 *Let E be an n-dimensional evolution algebra. Then the following statements are equivalent [3].*

(i) *The matrix corresponding to E can be written as*

$$\widehat{A} = \begin{pmatrix} 0 & a_{12} & a_{13} & \dots & a_{1n} \\ 0 & 0 & a_{23} & \dots & a_{2n} \\ 0 & 0 & 0 & \dots & a_{3n} \\ \vdots & \vdots & \vdots & \dots & \vdots \\ 0 & 0 & 0 & \dots & 0 \end{pmatrix}$$

(ii) *E is right nilpotent algebra.*
(iii) *E is a nill algebra.*

Proposition 1 *Let E be a nilpotent evolution algebra with index of nilpotency not equal to $2^{n-1}+1$. Then it is not greater than $2^{n-2}+1$ [2].*

3 Three Dimensional Nilpotent Evolution Algebras

In this section we are going to describe all three dimensional nilpotent evolution algebras.

Theorem 2 *Let E be a three dimensional nilpotent evolution algebra then it is isomorphic to one of the following algebras,*

(1) $A_1 : e_1^2 = e_2, \quad e_2^2 = 0, \quad e_3^2 = 0$
(2) $A_2 : e_1^2 = e_3, \quad e_2^2 = e_3, \quad e_3^2 = 0$
(3) $A_3 : e_1^2 = e_2, \quad e_2^2 = e_3, \quad e_3^2 = 0$

Proof In the considered case we have $DimE = n = 3$. Since, E is nilpotent, then according to Proposition 1 one concludes that the index of nilpotency $N \leq 5$. In [2] it is shown that if $N = 5$ then one gets the following algebra:

$$e_1^2 = ae_2 + be_3, \quad e_2^2 = ce_3, \quad e_3^2 = 0.$$

Now, consider several cases:
Case 1 Let $a = 0$, Then

$$e_1^2 = be_3, \quad e_2^2 = ce_3, \quad e_3^2 = 0.$$

Case 1.1 Assume that $c = 0$. Then taking into account that evolution algebra is not abelian (algebra with zero products) we have that $b \neq 0$. Then, by scaling of basis and renumerating them we can assume

$$A_1 : \; e_1^2 = e_2, \quad e_2^2 = 0, \quad e_3^2 = 0.$$

Case 1.2 Let $c \neq 0$. Then we need to consider two cases:

Case 1.2.1 Let $b \neq 0$. Then by simple scaling of bases we get

$$A_2: \ e_1^2 = e_3, \quad e_2^2 = e_3, \quad e_3^2 = 0.$$

Case 1.2.2 Let $b = 0$. Then we have the following algebra

$$e_1^2 = 0, \ e_2^2 = ce_3, \ e_3^2 = 0$$

which is isomorphic to algebra A_1.
Case 2 Let $a \neq 0$. Then, since $N \leq 3$ we conclude that $c = 0$ and we have the family of algebras

$$e_1^2 = ae_2 + be_3, \quad e_2^2 = ce_3, \quad e_3^2 = 0.$$

If we take the following change of bases

$$e_1^{'} = e_1, \quad e_2^{'} = ae_2 + be_3, \quad e_3^{'} = e_3$$

we obtain the algebra A_1. Finally we have the following nilpotent algebras:

$$A_1: \ e_1^2 = e_2, \quad e_2^2 = 0, \quad e_3^2 = 0$$
$$A_2: \ e_1^2 = e_3, \quad e_2^2 = e_3, \quad e_3^2 = 0$$
$$A_3: \ e_1^2 = e_2, \quad e_2^2 = e_3, \quad e_3^2 = 0.$$

□

Using the proved theorem, we can establish the following

Theorem 3 *Let E be a three dimensional nilpotent evolution algebra. Then, the following assertions hold true:*

(i) One dimensional subalgebra of A_1 is generated by $\langle be_2 + ce_3 \rangle$ for any values of b, c.
(ii) One dimensional subalgebra of A_2 is generated by $\langle ce_3 \rangle$ for any value of c.
(iii) One dimensional subalgebra of A_3 is generated by $\langle ce_3 \rangle$ for any value of c.

4 On Three Dimensional Solvable Evolution Algebra

In this section we are going to describe some three dimensional solvable evolution algebra.

Theorem 4 *Let E be a three-dimensional solvable evolution algebra with the table*:

$$e_1^2 = a_1e_1 + a_2e_2 + a_3e_3, \ e_2^2 = b_1e_1 + b_2e_2 + b_3e_3,$$
$$e_3^2 = c_1e_1 + c_2e_2 + c_3e_3,$$

then the following statements hold true:

(i) *If* $Dim E^2 = 0$, *then* E *is nilpotent.*
(ii) *If* $Dim E^2 = 1$, *then* E *is isomorphic to one of the following algebras*:

(1) E_1: $e_1^2 = e_2 + e_3$,
$e_2^2 = e_2 + e_3, \ e_3^2 = -(e_2 + e_3)$;
(2) E_2: $e_1^2 = e_2, \ e_3^2 = e_2$;
(3) E_3: $e_1^2 = e_1 + e_2, \ e_2^2 = (e_1 + e_2)$,
$e_3^2 = e_1 + e_2$;
(4) E_4: $e_1^2 = e_1 + e_2 + e_3, \ e_2^2 = A(e_1 + e_2 + e_3)$,
$e_3^2 = (1 - A)(e_1 + e_2 + e_3)$.

(iii) *If* $Dim E^2 = 2$ *and* $E^2E^2 = 0$ *then* E *is isomorphic to the following algebra*

$$E_5(a): e_1^2 = e_1 + a_2e_2$$
$$e_2^2 = e_1 + a_2e_2 - \left(\frac{1}{a_2^2}\right) e_3, \ (a_2 \neq 0), \ e_3^2 = 0.$$

Proof In order to prove this theorem we have to consider three distinct cases:
Case 1 If $Dim E^2 = 0$. Then, we get algebra with zero multiplication, i.e. E is nilpotent.
Case 2 If $Dim E^2 = 1$. Then, without loss of generality we can assume that $E^2 = \{e_1^2\}$ and $e_2^2 = Ae_1^2, \ e_3^2 = Be_1^2$. Since E is solvable, then $(E^2)^2 = 0$, and $e_1^2e_1^2 = (a_1e_1 + a_2e_2 + a_3e_3)(a_1e_1 + a_2e_2 + a_3e_3) = a_1^2(a_1e_1 + a_2e_2 + a_3e_3) + Aa_2^2(a_1e_1 + a_2e_2 + a_3e_3) + Ba_3^2(a_1e_1 + a_2e_2 + a_3e_3) = \left(a_1^3 + Aa_2^2a_1 + Ba_3^2a_1\right) e_1 + \left(a_1^2a_2 + Aa_2^3 + Ba_3^2a_1\right) e_2 + \left(a_1^2a_3 + Aa_2^2a_3 + Ba_3^2\right) e_3 = 0$ compering the coefficients we get the following system:

$$\begin{aligned} a_1^3 + Aa_2^2a_1 + Ba_3^2a_1 &= 0, \\ a_1^2a_2 + Aa_2^3 + Ba_3^2a_1 &= 0, \\ a_1^2a_3 + Aa_2^2a_3 + Ba_3^2 &= 0. \end{aligned}$$

Case 2.1 If $a_1 = 0$. Then $(a_2, a_3) \neq (0, 0)$, and since a_2 and a_3 are not zero simultaneously, therefore, one can assume that $a_2 \neq 0$. Putting $e_2^{'} = a_2e_2$ one gets the following table of multiplications:

$$e_1^2 = e_2 + a_3e_3, \ e_2^2 = A(e_2 + a_3e_3), \ e_3^2 = B(e_2 + a_3e_3),$$

with condition:

$$A + Ba_3^2 = 0, \quad a_3\left(A + Ba_3^2\right) = 0.$$

It is easy to see that $B \neq 0$.
Case 2.1.1 If $a_3 \neq 0$. Then, by setting $e_3^{'} = a_3e_3$ one gets

$$e_1^2 = e_2 + e_3, \ e_2^2 = A(e_2 + e_3), \ e_3^2 = B(e_2 + e_3),$$

with condition:

$$A + B = 0.$$

Thus, we obtain the algebra:

$$e_1^2 = e_2 + e_3, \ e_2^2 = A(e_2 + e_3), \ e_3^2 = -A(e_2 + e_3).$$

Now if we take the change of basis: $e_1' = \frac{e_1}{\sqrt{A}}$, $e_2' = \frac{e_2}{A}$, $e_3' = \frac{e_3}{A}$; then one finds the solvable (not nilpotent) evolution algebra:

$$E_1 : e_1^2 = e_2 + e_3, \;\; e_2^2 = e_2 + e_3, \;\; e_3^2 = -e_2 - e_3.$$

Case 2.1.2 If $a_3 = 0$. Then, we have

$$e_1^2 = e_2, \;\; e_2^2 = Ae_2, \;\; e_3^2 = Be_2,$$

with condition $A = 0$, and by scaling $e_3' = \frac{e_3}{\sqrt{B}}$ we get

$$E_2 : \;\; e_1^2 = e_2, \;\; e_3^2 = e_2.$$

Case 2.2 If $a_1 \neq 0$. Then, by scaling of basis element $e_1' = \frac{e_1}{\sqrt{a_1}}$ we have

$$e_1^2 = e_1 + a_2e_2 + a_3e_3, \;\; e_2^2 = A(e_1 + a_2e_2 + a_3e_3),$$
$$e_3^2 = B(e_1 + e_2 + a_3e_3),$$

with conditions:

$$e_1 + Aa_2^2 + Ba_3^2 = 0, \;\; a_2 + Aa_2^3 + Ba_3^2a_2 = 0,$$
$$a_3 + Aa_2^2a_3 + Ba_3^3 = 0.$$

From the above conditions we see that $(a_2, a_3) \neq (0, 0)$ and without loss of generality we can assume that $a_2 \neq 0$. Then, by scaling $(e_2' = a_2e_2)$ we can suppose that algebra has the form:

$$e_1^2 = e_1 + e_2 + a_3e_3, \;\; e_2^2 = A(e_1 + e_2 + a_3e_3),$$
$$e_3^2 = B(e_1 + e_2 + a_3e_3),$$

with condition:

$$e_1 + A + Ba_3^2 = 0.$$

Case 2.2.1 If $a_3 = 0$. Then, we have the following algebra:

$$e_1^2 = e_1 + e_2, \;\; e_2^2 = -(e_1 + e_2), \;\; e_3^2 = B(e_1 + e_2).$$

Note that $B \neq 0$ (otherwise we get split solvable algebra). Setting $e_3' = \frac{e_3}{\sqrt{B}}$ we derive the evolution algebra:

$$E_3 : e_1^2 = e_1 + e_2, \;\; e_2^2 = -(e_1 + e_2), \;\; e_3^2 = e_1 + e_2.$$

Case 2.2.2 If $a_3 \neq 0$. Then, by putting $e_3' = a_3e_3$ we have

$$E_4(A) : e_1^2 = e_1 + e_2 + e_3, \;\; e_2^2 = A(e_1 + e_2 + e_3),$$

$$e_3^2 = (-1 - A)(e_1 + e_2 + e_3).$$

It is evident that $A = 0$ or $-1 - A = 0$, then algebra is split by taking appropriate change of basis.
Case 3 $Dim E^2 = 2$. Then, we can assume that $E^2 = \{e_1^2, e_2^2\}$. So, we have

$$\begin{aligned} e_1^2 &= a_1e_1 + a_2e_2 + a_3e_3, \ \ e_2^2 = b_1e_1 + b_2e_2 + b_3e_3, \\ e_3^2 &= Ae_1^2 + Be_2^2. \end{aligned}$$

where e_1^2, e_2^2 are linearly independent. Hence, dim $E^2E^2 = 1$ or dim $E^2E^2 = 0$.
Case 3.1 Let dim $E^2E^2 = 0$. Then, we have that $e_1^2e_1^2 = e_2^2e_2^2 = e_1^2e_2^2 = 0$, and one can get an evolution algebra:

$$e_1^2 = a_1e_1 + a_2e_2 + a_3e_3, \ \ e_2^2 = b_1e_1 + b_2e_2 + b_3e_3, \ \ e_3^2 = Ae_1^2 + Be_2^2$$

and the condition $E^2E^2 = 0$ implies the restrictions:

$$\begin{aligned} &a_1^2a_1 + a_2^2b_1 + a_3^2(Aa_1 + Bb_1) = 0, \ \ a_1^2a_2 + a_2^2b_2 + a_3^2(Aa_2 + Bb_2) = 0, \\ &a_1^2a_3 + a_2^2b_3 + a_3^2(Aa_3 + Bb_3) = 0, \ \ b_1^2a_1 + b_2^2b_1 + b_3^2(Aa_1 + Bb_1) = 0, \\ &b_1^2a_2 + b_2^2b_2 + b_3^2(Aa_2 + Bb_2) = 0, \ \ b_1^2a_3 + b_2^2b_3 + b_3^2(Aa_3 + Bb_3) = 0, \\ &a_1b_1a_1 + a_2b_2b_1 + a_3b_3(Aa_1 + Bb_1) = 0, \ \ a_1b_1a_2 + a_2b_2b_2 + a_3b_3(Aa_2 + Bb_2) = 0, \\ &a_1b_1a_3 + a_2b_2b_3 + a_3b_3(Aa_3 + Bb_3) = 0. \end{aligned}$$

Case 3.1.1 Let $a_1 = 0$. Then, we have

$$\begin{aligned} e_1^2 &= a_2e_2 + a_3e_3, \ \ e_2^2 = b_1e_1 + b_2e_2 + b_3e_3, \\ e_3^2 &= A(a_2e_2 + a_3e_3) + B(b_1e_1 + b_2e_2 + b_3e_3). \end{aligned}$$

If we suppose that $a_2 = 0$, then we have $a_3 \neq 0$ (hence $b_1 \neq 0$ or $b_2 \neq 0$) and without loss of generality we can assume that $a_3 = 1$. Then from the above restrictions one finds

$$\begin{aligned} &Bb_1 = 0, \ \ Bb_2 = 0, \ \ A + Bb_3 = 0, \\ &b_2^2b_1 + b_3^2Bb_1 = 0, \ \ b_2^2b_2 + b_3^2Bb_2 = 0, \\ &b_1^2b_2^2b_3 + b_3^2(A + Bb_3) = 0, \ \ b_3Bb_1 = 0, \\ &b_3Bb_2 = 0, \ \ b_3(A + Bb_3) = 0. \end{aligned}$$

Thus, we conclude that $a_2 \neq 0$. Without loss of generality we can assume that $a_2 = 1$. Suppose that $b_1 \neq 0$, then we have evolution algebra

$$e_1^2 = e_2 + a_3e_3, \ \ e_2^2 = b_1e_1 + b_2e_2 + b_3e_3,$$

$$e_3^2 = A(e_2 + a_3e_3) + B(b_1e_1 + b_2e_2 + b_3e_3)$$

with restrictions:

$$a_3^2B = -1, \ \ Bb_3^2 = -b_2^2, \ \ a_3b_3B = -b_2, \ \ a_3A = 0.$$

From here one can find $b_1 = 0$ which is contradiction with our assumption. Thus, we conclude that $b_1 = 0$ and we get the following evolution algebra:

$$e_1^2 = e_2 + a_3e_3, \ \ e_2^2 = b_2e_2 + b_3e_3,$$
$$e_3^2 = A(e_2 + a_3e_3) + B(b_2e_2 + b_3e_3)$$

with restrictions:

$$b_2 + a_3^2(A + Bb_2) = 0, \ \ b_3 + a_3^2(Aa_3 + Bb_3) = 0,$$
$$b_2^2b_2 + b_3^2(A + Bb_2) = 0, \ \ b_2^2b_3 + b_3^2(Aa_3 + Bb_3) = 0,$$
$$b_2^2 + a_3b_3(A + Bb_2) = 0, \ \ b_2b_3 + a_3b_3(Aa_3 + Bb_3) = 0.$$

The linear independence of e_1^2, e_2^2 implies that $b_2 \neq 0$. So, we may assume that $b_2 = 1$. It is evident that $a_3 \neq 0$ and $b_3 \neq 0$. We can assume that $a_3 = 1$, and we derive the algebra

$$e_1^2 = e_2 + e_3, \ \ e_2^2 = e_2 + b_3e_3,$$
$$e_3^2 = A(e_2 + e_3) + B(e_2 + b_3e_3)$$

with restrictions: $A + B = -1$ and $b_3 = 1$; but we get that e_1^2 and e_2^2 are linearly dependent which is a contradiction. Hence, in this case there is no solvable algebra.
Case 3.1.2 Let $a_1 \neq 0$. Then, we can conclude that $a_1 = 1$ and we have

$$e_1^2 = e_1 + a_2e_2 + a_3e_3, \ \ e_2^2 = b_1e_1 + b_2e_2 + b_3e_3,$$
$$e_3^2 = A(e_1 + a_2e_2 + a_3e_3) + B(b_1e_1 + b_2e_2 + b_3e_3)$$

and the condition $E^2E^2 = 0$ implies the restrictions:

$$1 + a_2^2b_1 + a_3^2(A + Bb_1) = 0, \ \ a_2 + a_2^2b_2 + a_3^2(Aa_2 + Bb_2) = 0,$$
$$a_3 + a_2^2b_3 + a_3^2(Aa_3 + Bb_3) = 0, \ \ b_1^2 + b_2^2b_1 + b_3^2(A + Bb_1) = 0,$$
$$b_1^2a_2 + b_2^2b_2 + b_3^2(Aa_2 + Bb_2) = 0,$$
$$b_1^2a_3 + b_2^2b_3 + b_3^2(Aa_3 + Bb_3) = 0, \ \ b_1 + a_2b_2b_1 + a_3b_3(A + Bb_1) = 0,$$
$$b_1a_2 + a_2b_2b_2 + a_3b_3(Aa_2 + Bb_2) = 0, \ \ b_1a_3 + a_2b_2b_3 + a_3b_3(Aa_3 + Bb_3) = 0.$$

It is easy to see that $b_1 \neq 0$. Indeed if $b_1 = 0$ then from above restriction we derive $b_2 = b_3 = 0$. Thus, $b_1 \neq 0$ and we can assume (by appropriate change of basis) that $b_1 = 1$, and we get the evolution algebra:

$$e_1^2 = e_1 + a_2e_2 + a_3e_3, \ \ e_2^2 = e_1 + b_2e_2 + b_3e_3,$$

$$e_3^2 = A(e_1 + a_2e_2 + a_3e_3) + B(e_1 + b_2e_2 + b_3e_3)$$

and the condition $E^2E^2 = 0$ implies the restrictions:

$$\begin{aligned}
&1 + a_2^2 + a_3^2(A + B) = 0, \;\; a_2 + a_2^2b_2 + a_3^2(Aa_2 + Bb_2) = 0,\\
&a_3 + a_2^2b_3 + a_3^2(Aa_3 + Bb_3) = 0, \;\; 1 + b_2^2 + b_3^2(A + B) = 0,\\
&a_2 + b_2^2b_2 + b_3^2(Aa_2 + Bb_2) = 0, \;\; a_3 + b_2^2b_3 + b_3^2(Aa_3 + Bb_3) = 0,\\
&1 + a_2b_2 + a_3b_3(A + B) = 0, \;\; a_2 + a_2b_2b_2 + a_3b_3(Aa_2 + Bb_2) = 0,\\
&a_3 + a_2b_2b_3 + a_3b_3(Aa_3 + Bb_3) = 0.
\end{aligned}$$

Note that $a_2 \neq 0$. Indeed, if $a_2 = 0$, then from above restrictions we have $a_3 \neq 0, b_2 = 0, a_3 = b_3$ and $1 + a_3^2(A + B) = 0$, from which one concludes that e_1^2, e_2^2 are linearly dependent. Therefore, $a_2 \neq 0$. Similarly, we find that $b_2 \neq 0$ and $a_3 \neq 0$ (since the restrictions lead to contradiction). One can assume that $a_3 = 1$ and one has the evolution algebra:

$$\begin{aligned}
&e_1^2 = e_1 + a_2e_2 + e_3, \;\; e_2^2 = e_1 + b_2e_2 + b_3e_3,\\
&e_3^2 = A(e_1 + a_2e_2 + e_3) + B(e_1 + b_2e_2 + b_3e_3)
\end{aligned}$$

and the condition $E^2E^2 = 0$ implies the restrictions:

$$\begin{aligned}
&1 + a_2^2 + (A + B) = 0, \;\; a_2 + a_2^2b_2 + (Aa_2 + Bb_2) = 0,\\
&1 + a_2^2b_3 + (A + Bb_3) = 0, \;\; 1 + b_2^2 + b_3^2(A + B) = 0,\\
&a_2 + b_2^2b_2 + b_3^2(Aa_2 + Bb_2) = 0, \;\; 1 + b_2^2b_3 + b_3^2(A + Bb_3) = 0,\\
&1 + a_2b_2 + a_3b_3(A + B) = 0, \;\; a_2 + a_2b_2b_2 + a_3b_3(Aa_2 + Bb_2) = 0,\\
&1 + a_2b_2b_3 + b_3(A + Bb_3) = 0.
\end{aligned}$$

If we suppose $A + B \neq 0$, then from the above restrictions we conclude that $a_2 = b_2$ and $b_3 = 1$, which implies the linearly dependent of e_1^2 and e_2^2, but this is impossible. Hence, $A + B = 0$ and we have

$$\begin{aligned}
&e_1^2 = e_1 + a_2e_2 + e_3, \;\; e_2^2 = e_1 + b_2e_2 + b_3e_3,\\
&e_3^2 = A(e_1 + a_2e_2 + e_3) - A(e_1 + b_2e_2 + b_3e_3)
\end{aligned}$$

and the condition $E^2E^2 = 0$ implies the restrictions:

$$\begin{aligned}
&a_2^2 = -1, \;\; a_2 + a_2^2b_2 + (Aa_2 - Ab_2) = 0,\\
&1 + a_2^2b_3 + (A - Ab_3) = 0,\\
&b_2^2 = -1, \;\; a_2 + b_2^2b_2 + b_3^2(Aa_2 - Ab_2) = 0,\\
&1 + b_2^2b_3 + b_3^2(A - Ab_3) = 0, \;\; a_2b_2 = -1,\\
&a_2 + a_2b_2b_2 + a_3b_3(Aa_2 - Ab_2) = 0,
\end{aligned}$$

$$1 + a_2b_2b_3 + b_3(A - Ab_3) = 0.$$

From the above restrictions one can find $a_2 = b_2$, then $b_3 \neq 1$,

$$\begin{aligned} &a_2^2 = -1, \ \ a_2 = b_2, \\ &1 + a_2^2 b_3 + (A - Ab_3) = 0 \\ &1 + a_2^2 b_3 + b_3^2(A - Ab_3) = 0 \\ &1 + a_2^2 b_3 + b_3(A - Ab_3) = 0 \end{aligned}$$

Also from the above restrictions one can find $A(b_3 + 1) = 0$. If $b_3 = -1$, then we get contradiction. Therefore, $A = 0$, $1 + a_2^2 b_3 = 0$. Hence, $b_3 = \frac{-1}{a_2^2}$. Thus, in the this case there is a solvable evolution algebra

$$\begin{aligned} E_5(a_2) : e_1^2 &= e_1 + a_2e_2 + e_3, \ \ e_2^2 = e_1 + b_2e_2 + b_3e_3, \\ e_3^2 &= 0, \ \ (a_2 \neq 0). \end{aligned}$$

□

Remark 1 We have to state that the case $Dim E^2E^2 = 1$ is tricky. Therefore, it will be considered elsewhere.

Acknowledgements The authors acknowledge the MOHE grants FRGS11-022-0170, ERGS13-024-0057 and the Junior Associate scheme of the Abdus Salam International Centre for Theoretical Physics, Trieste, Italy.

References

1. Camacho, L., Gómez, J., Omirov, B., Turdibaev, R.: Bull. Korean Math. Soc. (to appear) arXiv:1004.1987
2. Casas, J., Ladra, M., Omirov, B., Rozikov, U.: Alg. Colloq (to appear) arXiv:1004.1050
3. Rozikov, U., Tian, J.P.: Lobachevskii J. Math. **32**, 270–277 (2011)
4. Tian, J.P.: Evolution Algebras and Their Applications, Springer, 2007

On Quasi Quantum Quadratic Operators of $M_2(C)$

Farrukh Mukhamedov and Abduaziz Abduganiev

Abstract In the present paper we study quasi quantum quadratic operators (q.q.o) acting on the algebra of 2×2 matrices $M_2(C)$. We describe quasi q.q.o. with Haar state, and prove that if a symmetric quasi q.q.o. with Haar state is q-pure, then it cannot be positive.

1 Introduction

It is known that there are many systems which are described by nonlinear operators. One of the simplest nonlinear case is quadratic one. Quadratic dynamical systems have been proved to be a rich source of analysis for the investigation of dynamical properties and modeling in different domains, such as population dynamics [1], [6], [8], physics [20], [23], economy [3], mathematics [9], [24], [25]. The problem of studying the behavior of trajectories of quadratic stochastic operators was stated in [24]. The limit behavior and ergodic properties of trajectories of such operators were studied in [10], [11], [13], [25]. However, such kind of operators do not cover the case of quantum systems. Therefore, in [4], [5] quantum quadratic operators acting on a von Neumann algebra were defined and studied. Certain ergodic properties of such operators were studied in [14], [15]. In those papers basically, dynamics of quadratic operators were defined according to some recurrent rule (an analog of Kolmogorov–Chapman equation) which creates a possibility to study asymptotic behaviors of such operators. However, with a given quadratic operator one can also define a non-linear operator whose dynamics (in non-commutative setting) is not well studied till date. Some class of such kind of operators defined on $M_2(C)$ has been studied in [18]. Note that in [12] another construction of nonlinear quantum maps were suggested and some physical explanations of such nonlinear quantum dynamics were discussed. Very recently, in [7] convergence of ergodic averages associated

F. Mukhamedov (✉) · A. Abduganiev
Department of Computational & Theoretical Sciences, Faculty of Science,
International Islamic University Malaysia, Selayang, Selangor, Malaysia
e-mail: farrukh_m@iium.edu.my

A. Abduganiev
e-mail: azizi85@yandex.ru

A. Kilicman et al. (eds.), *International Conference on Mathematical Sciences and Statistics 2013,* DOI 10.1007/978-981-4585-33-0_18,

with mentioned non-linear operator are studied by means of absolute contractions of von Neumann algebras. In all these investigations, the said quantum quadratic operators by definition are positive. In general, to study the nonlinear dynamics the positivity is strong condition. Therefore, in the present paper we are going to introduce a weaker than positivity condition, and corresponding operators are called quasi quantum quadratic. In the paper we concentrate ourselves to such operators with Haar state acting on $M_2(C)$. Each such operator defines a quadratic operator acting on state space of $M_2(C)$. We call such an operator pure, if it maps pure state to pure ones. In this paper, we first describe all such kind of operators. Moreover, we prove that any pure symmetric quasi quantum quadratic operator is not positive. Here the purity means that quadratic operator sends pure state to pure.

2 Preliminaries

Let $M_2(C)$ be an algebra of 2×2 matrices over complex field C. By $M_2(C) \otimes M_2(C)$ we mean tensor product of $M_2(C)$ into itself. We note that such a product can be considered as an algebra of 4×4 matrices $M_4(C)$ over C. In the sequel $\mathbb{I}$ means an identity matrix, i.e. $\mathbb{I} = \begin{pmatrix} 1 & 0 \\ 0 & 1 \end{pmatrix}$. By $M_2(C)^*$ it is usually denoted as the conjugate space of $M_2(C)$. We recall that a linear functional $\varphi \in M_2(C)^*$ is called *positive* if $\varphi(x) \geq 0$ whenever $x \geq 0$. The set of all positive linear functionals is denoted by $M_2(C)^*_+$. A positive functional φ is called *state* if $\varphi(\mathbb{I}) = 1$. By $S(M_2(C))$ we denote the set of all states defined on $M_2(C)$.

Let $\Delta : M_2(C) \to M_2(C) \otimes M_2(C)$ be a linear operator. Then Δ defines a conjugate operator $\Delta^* : (M_2(C) \otimes M_2(C))^* \to M_2(C)^*$ by

$$\Delta^*(f)(x) = f(\Delta x), \ f \in (M_2(C) \otimes M_2(C))^*, \ x \in M_2(C).$$

One can define an operator V_Δ by

$$V_\Delta(\varphi) = \Delta^*(\varphi \otimes \varphi), \ \varphi \in M_2(C)^*.$$

Let $U : M_2(C) \otimes M_2(C) \to M_2(C) \otimes M_2(C)$ be a linear operator such that $U(x \otimes y) = y \otimes x$ for all $x, y \in M_2(C)$.

Definition 1 *A linear operator $\Delta : M_2(C) \to M_2(C) \otimes M_2(C)$ is said to be*

(a) – a *quasi quantum quadratic operator (quasi q.q.o) if it is unital (i.e. $\Delta\mathbb{I} = \mathbb{I} \otimes \mathbb{I}$), *-preserving (i.e. $\Delta(x^*) = \Delta(x)^*$, $\forall x \in M_2(C)$) and*

$$V_\Delta(\varphi) \in M_2(C)^*_+ \ \text{ whenever } \varphi \in M_2(C)^*_+;$$

(b) – a *quantum quadratic operator (q.q.o.) if it is unital (i.e. $\Delta\mathbb{I} = \mathbb{I} \otimes \mathbb{I}$) and positive (i.e. $\Delta x \geq 0$ whenever $x \geq 0$);*

(c) – *a quantum convolution if it is a q.q.o. and satisfies coassociativity condition:*

$$(\Delta \otimes id) \circ \Delta = (id \otimes \Delta) \circ \Delta,$$

where id *is the identity operator of* $M_2(C)$.

(d) – *a symmetric if one has* $U\Delta = \Delta$.

One can see that if Δ is q.q.o. then it is a quasi q.q.o. A state $h \in S(M_2(C))$ is called *a Haar state* for a quasi q.q.o. Δ if for every $x \in M_2(C)$ one has

$$(h \otimes id) \circ \Delta(x) = (id \otimes h) \circ \Delta(x) = h(x)\mathbb{1}. \quad (1)$$

Remark 2 *Note that if a quantum convolution* Δ on $M_2(C)$ *becomes a* $*$ - *homomorphic map with a condition*

$$\overline{\text{Lin}}((\mathbb{1} \otimes M_2(C))\Delta(M_2(C))) = \overline{\text{Lin}}((M_2(C) \otimes \mathbb{1})\Delta(M_2(C))) = M_2(C) \otimes M_2(C)$$

then a pair $(M_2(C), \Delta)$ *is called a compact quantum group* [22], [26]. *It is known* [26] *that for given any compact quantum group there exists a unique Haar state w.r.t.* Δ.

Remark 3 *Let* $U : M_2(C) \otimes M_2(C) \to M_2(C) \otimes M_2(C)$ *be a linear operator such that* $U(x \otimes y) = y \otimes x$ *for all* $x, y \in M_2(C)$. *If a q.q.o.* Δ *satisfies* $U\Delta = \Delta$, *then* Δ *is called a quantum quadratic stochastic operator. Such kind of operators were studied and investigated in* [14], [16].

Remark 4 *We note that there is another approach to nonlinear quantum operators on* C^**-algebras* (see [12]).

Note that from unitality of Δ we conclude that for any quasi q.q.o. V_Δ maps $S(M_2(C))$ into itself. In some literature operator V_Δ is called *quadratic convolution* (see for example [7]). In [18] certain dynamical properties of V_Δ associated with q.q.o. are investigated. In [17] Kadison–Schwarz property of q.q.o. has been studied.

Definition 5 *A quasi q.q.o.* Δ *is called q-pure if for any pure state* φ *the state* $V_\Delta(\varphi)$ *is also pure.*

3 Quasi Quantum Quadratic Operators on $M_2(C)$

In this section we are going to describe quantum quadratic operators on $M_2(C)$.

Recall [2] that the identity and Pauli matrices $\{\mathbb{1}, \sigma_1, \sigma_2, \sigma_3\}$ form a basis for $M_2(C)$, where

$$\sigma_1 = \begin{pmatrix} 0 & 1 \\ 1 & 0 \end{pmatrix} \quad \sigma_2 = \begin{pmatrix} 0 & -i \\ i & 0 \end{pmatrix} \quad \sigma_3 = \begin{pmatrix} 1 & 0 \\ 0 & -1 \end{pmatrix}.$$

In this basis every matrix $x \in M_2(C)$ can be written as $x = w_0\mathbb{1} + \mathbf{w}\sigma$ with $w_0 \in C$, $\mathbf{w} = (w_1, w_2, w_3) \in C^3$, here $\mathbf{w}\sigma = w_1\sigma_1 + w_2\sigma_2 + w_3\sigma_3$. In what follows, we frequently use notation $\overline{\mathbf{w}} = (\overline{w_1}, \overline{w_2}, \overline{w_3})$.

Lemma 1 [21] *The following assertions hold true:*

(a) *x is self-adjoint iff $w_0, \mathbf{w}$ are reals;*
(b) $\mathrm{Tr}(x) = 1$ *iff* $w_0 = 0.5$, *here* Tr *is the trace of a matrix* x;
(c) $x > 0$ *iff* $\|\mathbf{w}\| \leq w_0$, *where* $\|\mathbf{w}\| = \sqrt{|w_1|^2 + |w_2|^2 + |w_3|^2}$;
(d) *A linear functional φ on $M_2(C)$ is a state iff it can be represented by*

$$\varphi(w_0 \mathbb{1} + \mathbf{w}\sigma) = w_0 + \langle \mathbf{w}, \mathbf{f} \rangle, \tag{2}$$

where $\mathbf{f} = (f_1, f_2, f_3) \in R^3$ such that $\|\mathbf{f}\| \leq 1$. Here, as before $\langle \cdot, \cdot \rangle$ stands for the scalar product in C^3.
(e) *A state φ is a pure if and only if $\|\mathbf{f}\| = 1$. Hence, pure states can be seen as the elements of unit sphere in R^3.*

In the sequel we shall identify a state with a vector $\mathbf{f} \in R^3$. By τ we denote a normalized trace, i.e. $\tau(x) = \frac{1}{2}\mathrm{Tr}(x)$, $x \in M_2(C)$.

Let $\Delta : M_2(C) \to M_2(C) \otimes M_2(C)$ be a quasi q.q.o. Then we write the operator Δ in terms of a basis in $M_2(C) \otimes M_2(C)$ formed by the Pauli matrices. Namely,

$$\Delta \mathbb{1} = \mathbb{1} \otimes \mathbb{1};$$

$$\Delta(\sigma_i) = b_i(\mathbb{1} \otimes \mathbb{1}) + \sum_{j=1}^{3} b_{ji}^{(1)}(\mathbb{1} \otimes \sigma_j)$$

$$+ \sum_{j=1}^{3} b_{ji}^{(2)}(\sigma_j \otimes \mathbb{1}) + \sum_{m,l=1}^{3} b_{ml,i}(\sigma_m \otimes \sigma_l), \tag{3}$$

where $i = 1, 2, 3$.

In general, a description of positive operators is one of the main problems of quantum information. In the literature most tractable maps are positive and trace-preserving ones, since such maps arise naturally in quantum information theory (see [19]). Therefore, in the sequel we shall restrict ourselves to q.q.o. which has a Haar state τ. Hence, we would like to describe all such kind of maps.

Proposition 2 *Let $\Delta : M_2(C) \to M_2(C) \otimes M_2(C)$ be a quasi q.q.o. with a Haar state τ, then in (3) one has $b_j = 0$, $b_{ij}^{(1)} = b_{ij}^{(2)} = 0$ and $b_{ij,k}$ are real for every $i, j, k \in \{1, 2, 3\}$. Moreover, Δ has the following form:*

$$\Delta(x) = w_0 \mathbb{1} \otimes \mathbb{1} + \sum_{m,l=1}^{3} \langle \mathbf{b}_{ml}, \overline{\mathbf{w}} \rangle \sigma_m \otimes \sigma_l, \tag{4}$$

where $x = w_0 \mathbb{1} + \mathbf{w}\sigma$, $\mathbf{b}_{ml} = (b_{ml,1}, b_{ml,2}, b_{ml,3})$. Here, as before $\langle \cdot, \cdot \rangle$ stands for the standard scalar product in C^3.

Proof From the *-preserving condition we get

$$\Delta(\sigma_i^*) = \overline{b_i}(\mathbb{I}\otimes\mathbb{I}) + \sum_{j=1}^{3}\overline{b_{ji}^{(1)}}(\mathbb{I}\otimes\sigma_j) + \sum_{j=1}^{3}\overline{b_{ji}^{(2)}}(\sigma_j\otimes\mathbb{I}) + \sum_{m,l=1}^{3}\overline{b_{ml,i}}(\sigma_m\otimes\sigma_l).$$

This yields that $b_i = \overline{b_i}$, $b_{ji}^{(k)} = \overline{b_{ji}^{(k)}}$ $(k = 1, 2)$ and $b_{ml,i} = \overline{b_{ml,i}}$, i.e. all coefficients are real numbers.

From (1) one finds

$$\tau\otimes\tau(\Delta x) = \tau(\tau\otimes id)(\Delta(x)) = \tau(x)\tau(\mathbb{I}) = \tau(x), \quad x\in M_2(C),$$

which means that τ is an invariant state for Δ. Hence, we have $\tau\otimes\tau(\Delta(\sigma_i)) = \tau(\sigma_i) = 0$ which yields $b_j = 0$, $j = 1, 2, 3$.

Again using the equality (1) with $h = \tau$, one gets

$$(id\otimes\tau)\Delta(\sigma_i) = (id\otimes\tau)\left[\sum_{j=1}^{3}(b_{ji}^{(1)}(\mathbb{I}\otimes\sigma_j) + b_{ji}^{(2)}(\sigma_j\otimes\mathbf{I})) + \sum_{m,l=1}^{3}b_{ml,i}(\sigma_m\otimes\sigma_l)\right]$$

$$= \sum_{j=1}^{3}b_{ji}^{(2)}\sigma_j = \tau(\sigma_j)\mathbb{I} = 0.$$

Therefore, $b_{ji}^{(2)} = 0$, for all $i, j = 1, 2, 3$. Similarly, one finds

$$(\tau\otimes id)\Delta(\sigma_j) = \sum_{j=1}^{3}b_{ji}^{(1)}\sigma_j = \tau(\sigma_j)\mathbb{I},$$

which means $b_{ji}^{(1)} = 0$. Hence, Δ has the following form

$$\Delta(\sigma_i) = \sum_{m,l=1}^{3}b_{ml,i}(\sigma_m\otimes\sigma_l), \quad i = 1, 2, 3. \tag{5}$$

Denoting

$$\mathbf{b}_{ml} = (b_{ml,1}, b_{ml,2}, b_{ml,3}). \tag{6}$$

and taking any $x = w_0\mathbb{I} + \mathbf{w}\sigma \in M_2(C)$, from (5) we immediately find (4). This completes the proof.

4 Symmetric Q-Pure Quasi Quantum Quadratic Operators on $M_2(C)$

In this section we are going to describe symmetric q-pure quasi q.q.o. Denote

$$\mathbf{B} = \left\{\mathbf{p} = (p_1, p_2, p_3) \in R : p_1^2 + p_2^2 + p_3^2 \leq 1\right\},$$
$$\mathbf{S} = \left\{\mathbf{p} = (p_1, p_2, p_3) \in R : p_1^2 + p_2^2 + p_3^2 = 1\right\}.$$

Let Δ be a symmetric quasi q.q.o. on $M_2(C)$ with a Haar state τ. Then due to Lemma 1 (d) and Proposition 2 the functional $\Delta^*(\varphi \otimes \psi)$ is a state if and only if the vector

$$\mathbf{f}_{\Delta^*(\varphi,\psi)} = \left(\sum_{i,j=1}^{3} b_{ij,1} f_i p_j, \sum_{i,j=1}^{3} b_{ij,2} f_i p_j, \sum_{i,j=1}^{3} b_{ij,3} f_i p_j\right).$$

satisfies $\|\mathbf{f}_{\Delta^*(\varphi,\psi)}\| \leq 1$.

Let us consider the quadratic operator, which is defined by $V_\Delta(\varphi) = \Delta^*(\varphi\otimes\varphi)$, $\varphi \in S(M_2(C))$. In the last expression we found that

$$V_\Delta(\varphi)(\sigma_k) = \sum_{i,j=1}^{3} b_{ij,k} f_i f_j, \quad \mathbf{f} \in \mathbf{B}.$$

This suggests us the consideration of a nonlinear operator $V : \mathbf{B} \to \mathbf{B}$ defined by

$$V(\mathbf{f})_k = \sum_{i,j=1}^{3} b_{ij,k} f_i f_j, \quad k = 1, 2, 3. \tag{7}$$

where $\mathbf{f} = (f_1, f_2, f_3) \in S$.

From the definition and Lemma 1 (e) we conclude that the Δ is q-pure if and only if $V(\mathbf{S}) \subset \mathbf{S}$.

Example Let us consider an example of pure symmetric quasi q.q.o. Let

$$\begin{aligned}\Delta(x) = w_0 \mathbb{I}\otimes\mathbb{I} \ &+ \ w_1\sigma_1\otimes\sigma_2 + w_1\sigma_2\otimes\sigma_1 \\ &+ \ w_2\sigma_1\otimes\sigma_1 - w_2\sigma_2\otimes\sigma_2 - w_2\sigma_3\otimes\sigma_3 \\ &+ \ w_3\sigma_1\otimes\sigma_3 + w_3\sigma_3\otimes\sigma_1,\end{aligned}$$

Then the corresponding quadratic operator has the following form

$$V(\mathbf{f}) = \begin{cases} 2f_1 f_2 \\ f_1^2 - f_2^2 - f_3^2 \\ 2f_1 f_3 \end{cases}$$

Let us show V maps S to S. Indeed, let $\mathbf{f} = (f_1, f_2, f_3) \in \mathbf{S}$, i.e. $f_1^2 + f_2^2 + f_3^2 = 1$. Then we have

$$(2f_1 f_2)^2 + \left(f_1^2 - f_2^2 - f_3^2\right)^2 + (2f_1 f_3)^2 = 4f_1^2 f_2^2 + \left(2f_1^2 - 1\right)^2 + 4f_1^2 f_3^2$$

$$= 4f_1^2\left(1 - f_1^2 - f_3^2\right) + 4f_1^4 - 4f_1^2 + 1 + 4f_1^2 f_3^2 = 1$$

This shows that Δ is pure.

Now, let us rewrite the quadratic operator V (see (7)) as follows

$$V(\mathbf{f}) = \begin{cases} a_1 f_1^2 + b_1 f_2^2 + c_1 f_3^2 + \alpha_1 f_1 f_2 + \beta_1 f_2 f_3 + \gamma_1 f_1 f_3 \\ a_2 f_1^2 + b_2 f_2^2 + c_2 f_3^2 + \alpha_2 f_1 f_2 + \beta_2 f_2 f_3 + \gamma_2 f_1 f_3 \\ a_3 f_1^2 + b_3 f_2^2 + c_3 f_3^2 + \alpha_3 f_1 f_2 + \beta_3 f_2 f_3 + \gamma_3 f_1 f_3 \end{cases} \tag{8}$$

where $\mathbf{f} \in S$.

Theorem 1 *The operator V given by (8) maps $\mathbf{S}$ into itself if and only if the followings hold true*

(i) $\|\mathbf{a}\| = 1,\ \|\mathbf{b}\| = 1,\ \|\mathbf{c}\| = 1;$
(ii) $\|A\| = \|\mathbf{a} - \mathbf{b}\|,\ \|\Gamma\| = \|\mathbf{a} - \mathbf{c}\|,\ \|B\| = \|\mathbf{b} - \mathbf{c}\|;$
(iii) $\langle \mathbf{a}, B\rangle + \langle A, \Gamma\rangle = 0,\ \langle \mathbf{b}, \Gamma\rangle + \langle A, B\rangle = 0, \langle \mathbf{c}, A\rangle + \langle B, \Gamma\rangle = 0;$
(iv) $\langle \mathbf{a}, A\rangle = 0,\ \langle \mathbf{a}, \Gamma\rangle = 0,\ \langle \mathbf{b}, A\rangle = 0,\ \langle \mathbf{b}, B\rangle = 0, \langle \mathbf{c}, \Gamma\rangle = 0,\ \langle \mathbf{c}, B\rangle = 0,$

where $\mathbf{a} = (a_1, a_2, a_3),\ \mathbf{b} = (b_1, b_2, b_3),\ \mathbf{c} = (c_1, c_2, c_3),\ \Gamma = (\gamma_1, \gamma_2, \gamma_3),\ A = (\alpha_1, \alpha_2, \alpha_3),\ B = (\beta_1, \beta_2, \beta_3)$.

Proof "only if" part. It is enough to show

$$(V(\mathbf{f})_1)^2 + (V(\mathbf{f})_2)^2 + (V(\mathbf{f})_3)^2 = 1 \tag{9}$$

for any $\mathbf{f} = (f_1, f_2, f_3)$ with $f_1^2 + f_2^2 + f_3^2 = 1$.

Let us rewrite (8) as follows

$$V(\mathbf{f}) = \begin{cases} (a_1 - c_1) f_1^2 + (b_1 - c_1) f_2^2 + c_1 + \alpha_1 f_1 f_2 + \beta_1 f_2 f_3 + \gamma_1 f_1 f_3 \\ (a_2 - c_2) f_1^2 + (b_2 - c_2) f_2^2 + c_2 + \alpha_2 f_1 f_2 + \beta_2 f_2 f_3 + \gamma_2 f_1 f_3 \\ (a_3 - c_3) f_1^2 + (b_3 - c_3) f_2^2 + c_3 + \alpha_3 f_1 f_2 + \beta_3 f_2 f_3 + \gamma_3 f_1 f_3 \end{cases} \tag{10}$$

From (9), (10) we derive

$$\begin{aligned} & \left((a_1 - c_1) f_1^2 + (b_1 - c_1) f_2^2 + c_1 + \alpha_1 f_1 f_2 + \beta_1 f_2 f_3 + \gamma_1 f_1 f_3\right)^2 \\ + & \left((a_2 - c_2) f_1^2 + (b_2 - c_2) f_2^2 + c_2 + \alpha_2 f_1 f_2 + \beta_2 f_2 f_3 + \gamma_2 f_1 f_3\right)^2 \\ + & \left((a_3 - c_3) f_1^2 + (b_3 - c_3) f_2^2 + c_3 + \alpha_3 f_1 f_2 + \beta_3 f_2 f_3 + \gamma_3 f_1 f_3\right)^2 = 1 \end{aligned}$$

After some calculations we obtain the following

$$\begin{aligned} & \left(\|\mathbf{a}\|^2 + \|\mathbf{c}\|^2 - \|\Gamma\|^2 - 2\langle \mathbf{a}, \mathbf{c}\rangle\right) f_1^4 + \left(\|\mathbf{b}\|^2 + \|\mathbf{c}\|^2 - \|B\|^2 - 2\langle \mathbf{b}, \mathbf{c}\rangle\right) f_2^4 \\ + & (2\langle \mathbf{a}, A\rangle - 2\langle B, \Gamma\rangle - 2\langle \mathbf{c}, A\rangle) f_1^3 f_2 + (2\langle \mathbf{a}, \Gamma\rangle - 2\langle \mathbf{c}, \Gamma\rangle) f_1^3 f_3 \\ + & (2\langle \mathbf{b}, A\rangle - 2\langle B, \Gamma\rangle - 2\langle \mathbf{c}, A\rangle) f_1 f_2^3 + (2\langle \mathbf{b}, B\rangle - 2\langle \mathbf{c}, B\rangle) f_2^3 f_3 \\ + & \left(2\|\mathbf{c}\|^2 + \|A\|^2 - \|B\|^2 - \|\Gamma\|^2 + 2\langle \mathbf{a}, \mathbf{b}\rangle - 2\langle \mathbf{b}, \mathbf{c}\rangle - 2\langle \mathbf{a}, \mathbf{c}\rangle\right) f_1^2 f_2^2 \end{aligned}$$

$$
\begin{aligned}
&+ \quad (2\langle \mathbf{a}, B\rangle + 2\langle A, \Gamma\rangle - 2\langle \mathbf{c}, B\rangle) f_1^2 f_2 f_3 + \left(\|\Gamma\|^2 + 2\langle \mathbf{a}, \mathbf{c}\rangle - 2\|\mathbf{c}\|^2\right) f_1^2 \\
&+ \quad (2\langle \mathbf{b}, \Gamma\rangle + 2\langle A, B\rangle - 2\langle \mathbf{c}, \Gamma\rangle) f_1 f_2^2 f_3 + \left(\|B\|^2 + 2\langle \mathbf{b}, \mathbf{c}\rangle - 2\|\mathbf{c}\|^2\right) f_2^2 \\
&+ \quad (2\langle \mathbf{c}, A\rangle + 2\langle B, \Gamma\rangle) f_1 f_2 + 2\langle \mathbf{c}, \Gamma\rangle f_1 f_3 + 2\langle \mathbf{c}, B\rangle f_2 f_3 + \|\mathbf{c}\|^2 - 1 = 0
\end{aligned}
$$

which is satisfied if one has

$$\|\mathbf{a}\| = 1, \ \|\mathbf{b}\| = 1, \ \|\mathbf{c}\| = 1, \tag{11}$$

$$\|A\|^2 + 2\langle \mathbf{a}, \mathbf{b}\rangle = 2, \ \|\Gamma\|^2 + 2\langle \mathbf{a}, \mathbf{c}\rangle = 2, \ \|B\|^2 + 2\langle \mathbf{b}, \mathbf{c}\rangle = 2, \tag{12}$$

$$\langle \mathbf{a}, B\rangle + \langle A, \Gamma\rangle = 0, \ \langle \mathbf{b}, \Gamma\rangle + \langle A, B\rangle = 0, \ \langle \mathbf{c}, A\rangle + \langle B, \Gamma\rangle = 0, \tag{13}$$

$$\langle \mathbf{a}, A\rangle = 0, \ \langle \mathbf{a}, \Gamma\rangle = 0, \ \langle \mathbf{b}, A\rangle = 0, \ \langle \mathbf{b}, B\rangle = 0, \ \langle \mathbf{c}, \Gamma\rangle = 0, \ \langle \mathbf{c}, B\rangle = 0 \tag{14}$$

Then, from (11), (12) one gets

$$\|A\| = \|\mathbf{a} - \mathbf{b}\|, \ \|\Gamma\| = \|\mathbf{a} - \mathbf{c}\|, \ \|B\| = \|\mathbf{b} - \mathbf{c}\|.$$

"if" part. Let (i), (ii), (iii) and (iv) be satisfied. Then (9) yields. This completes the proof.

Now, let us investigate a linear operator Δ corresponding to (8). According to Proposition 2 Δ has the following form

$$
\begin{aligned}
\Delta(x) = w_0 \mathbb{1}\otimes\mathbb{1} \quad &+ \quad a_1 w_1 \sigma_1\otimes\sigma_1 + \frac{\alpha_1}{2} w_1 \sigma_1\otimes\sigma_2 + \frac{\gamma_1}{2} w_1 \sigma_1\otimes\sigma_3 \\
&+ \quad \frac{\alpha_1}{2} w_1 \sigma_2\otimes\sigma_1 + b_1 w_1 \sigma_2\otimes\sigma_2 + \frac{\beta_1}{2} w_1 \sigma_2\otimes\sigma_3 \\
&+ \quad \frac{\gamma_1}{2} w_1 \sigma_3\otimes\sigma_1 + \frac{\beta_1}{2} w_1 \sigma_3\otimes\sigma_2 + c_1 w_1 \sigma_3\otimes\sigma_3 \\
&+ \quad a_2 w_2 \sigma_1\otimes\sigma_1 + \frac{\alpha_2}{2} w_2 \sigma_1\otimes\sigma_2 + \frac{\gamma_2}{2} w_2 \sigma_1\otimes\sigma_3 \\
&+ \quad \frac{\alpha_2}{2} w_2 \sigma_2\otimes\sigma_1 + b_2 w_2 \sigma_2\otimes\sigma_2 + \frac{\beta_2}{2} w_2 \sigma_2\otimes\sigma_3 \\
&+ \quad \frac{\gamma_2}{2} w_2 \sigma_3\otimes\sigma_1 + \frac{\beta_2}{2} w_2 \sigma_3\otimes\sigma_2 + c_2 w_2 \sigma_3\otimes\sigma_3 \\
&+ \quad a_3 w_3 \sigma_1\otimes\sigma_1 + \frac{\alpha_3}{2} w_3 \sigma_1\otimes\sigma_2 + \frac{\gamma_3}{2} w_3 \sigma_1\otimes\sigma_3 \\
&+ \quad \frac{\alpha_3}{2} w_3 \sigma_2\otimes\sigma_1 + b_3 w_3 \sigma_2\otimes\sigma_2 + \frac{\beta_3}{2} w_3 \sigma_2\otimes\sigma_3 \\
&+ \quad \frac{\gamma_3}{2} w_3 \sigma_3\otimes\sigma_1 + \frac{\beta_3}{2} w_3 \sigma_3\otimes\sigma_2 + c_3 w_3 \sigma_3\otimes\sigma_3
\end{aligned}
$$

Calculating the last one, we obtain

$$\Delta(x) = \begin{pmatrix} w_0 + R & N - iP & N - iP & L - 2iM - O \\ N + iP & w_0 - R & L + O & -N + iP \\ N + iP & L + O & w_0 - R & -N + iP \\ L + 2iM - O & -N - iP & -N - iP & w_0 + R \end{pmatrix} \tag{15}$$

where

$$L = a_1 w_1 + a_2 w_2 + a_3 w_3, \quad M = \frac{\alpha_1}{2} w_1 + \frac{\alpha_2}{2} w_2 + \frac{\alpha_3}{2} w_3,$$
$$N = \frac{\gamma_1}{2} w_1 + \frac{\gamma_2}{2} w_2 + \frac{\gamma_3}{2} w_3, \quad O = b_1 w_1 + b_2 w_2 + b_3 w_3,$$
$$P = \frac{\beta_1}{2} w_1 + \frac{\beta_2}{2} w_2 + \frac{\beta_3}{2} w_3, \; R = c_1 w_1 + c_2 w_2 + c_3 w_3.$$

Theorem 2 *Let $\Delta : M_2(C) \to M_2(C) \otimes M_2(C)$ be a symmetric q-pure quasi q.q.o. with Haar state τ. Then Δ is not positive.*

Proof Let us prove from the contrary. Assume that Δ is positive. On the other hand, the q-purity of Δ implies that the conditions (i)–(iv) of Theorem 4.2 are satisfied.

Let us take $x = \mathbb{1} + \mathbf{a}\sigma$, then from (15) one gets

$$\Delta(x) = \begin{pmatrix} 1 + \langle \mathbf{c}, \mathbf{a} \rangle & -\frac{i}{2}\langle B, \mathbf{a} \rangle & -\frac{i}{2}\langle B, \mathbf{a} \rangle & 1 - \langle \mathbf{b}, \mathbf{a} \rangle \\ \frac{i}{2}\langle B, \mathbf{a} \rangle & 1 - \langle \mathbf{c}, \mathbf{a} \rangle & 1 + \langle \mathbf{b}, \mathbf{a} \rangle & \frac{i}{2}\langle B, \mathbf{a} \rangle \\ \frac{i}{2}\langle B, \mathbf{a} \rangle & 1 + \langle \mathbf{b}, \mathbf{a} \rangle & 1 - \langle \mathbf{c}, \mathbf{a} \rangle & \frac{i}{2}\langle B, \mathbf{a} \rangle \\ 1 - \langle \mathbf{b}, \mathbf{a} \rangle & -\frac{i}{2}\langle B, \mathbf{a} \rangle & -\frac{i}{2}\langle B, \mathbf{a} \rangle & 1 + \langle \mathbf{c}, \mathbf{a} \rangle \end{pmatrix}.$$

A simple algebra with (ii) of Theorem 4.1 shows that all eigenvalues of $\Delta(x)$ can be written as follows

$$\begin{aligned} \lambda_1 &= -2 + \frac{|\Gamma|^2}{2} + \frac{|A|^2}{2} \\ \lambda_2 &= 2 - \frac{|\Gamma|^2}{2} - \frac{|A|^2}{2} \\ \lambda_3 &= 2 + \frac{1}{2}\sqrt{\|A\|^4 - 2\|\Gamma\|^2\|A\|^2 + \|\Gamma\|^4 + \langle B, \mathbf{a} \rangle^2} \\ \lambda_4 &= 2 - \frac{1}{2}\sqrt{\|A\|^4 - 2\|\Gamma\|^2\|A\|^2 + \|\Gamma\|^4 + \langle B, \mathbf{a} \rangle^2}. \end{aligned}$$

Positivity of $\Delta(x)$ yields $\lambda_1 \geq 0, \lambda_2 \geq 0$, which means $\|A\|^2 + \|\Gamma\|^2 = 4$.

By considering elements $x = \mathbb{1} + \mathbf{b}\sigma$, $x = \mathbb{1} + \mathbf{c}\sigma$, respectively, and using the similar argument one finds

$$\|B\|^2 + \|A\|^2 = 4 \quad \|\Gamma\|^2 + \|B\|^2 = 4$$

Therefore, we conclude that

$$\|A\|^2 = 2, \; \|B\|^2 = 2, \; \|\Gamma\|^2 = 2.$$

Hence, again taking into account (ii) of Theorem 4.1 we find that

$$\langle \mathbf{a}, \mathbf{b} \rangle = 0, \ \langle \mathbf{a}, \mathbf{c} \rangle = 0, \ \langle \mathbf{b}, \mathbf{c} \rangle = 0.$$

This means that the vectors **a**, **b**, **c** are linearly independent. Therefore, one can write

$$A = \eta_1 \mathbf{a} + \mu_1 \mathbf{b} + \tau_1 \mathbf{c}$$
$$B = \eta_2 \mathbf{a} + \mu_2 \mathbf{b} + \tau_2 \mathbf{c}$$
$$\Gamma = \eta_3 \mathbf{a} + \mu_3 \mathbf{b} + \tau_3 \mathbf{c}$$

where $\eta_i^2 + \mu_i^2 + \tau_i^2 = 2, \ i = \overline{1,3}$.

From (iv) of Theorem 4.1 we find that

$$\eta_1 = 0, \ \eta_3 = 0, \ \mu_1 = 0, \ \mu_2 = 0, \ \tau_2 = 0, \ \tau_3 = 0.$$

This implies that

$$A = \tau_1 \mathbf{c}, \quad B = \eta_2 \mathbf{a}, \quad \Gamma = \mu_3 \mathbf{b}.$$

Hence, from (iii) of Theorem 4.1 we have

$$\langle A, B \rangle + \langle \mathbf{b}, \Gamma \rangle = 0 \ \Rightarrow \ \mu_3 = 0$$

which contradicts to $\mu_3 \neq 0$. This completes the proof.

By **C** we denote the unit circle in R^3, i.e.

$$\mathbf{C} = \left\{ \mathbf{p} = (p_1, p_2, 0) \in R^3 : p_1^2 + p_2^2 = 1 \right\}.$$

Using the same argument as in Theorem 4.1 one can prove the following

Theorem 3 *The operator V given by (8) maps* **C** *into itself if and only if the followings hold true*

(i) $\mathbf{a} = \mathbf{b}$, $A = 0$, $\|\mathbf{a}\| = 1$;
(ii) $\mathbf{a} = -\mathbf{b}$, $A = \pm 2(a_2, -a_1, 0)$, $\|\mathbf{a}\| = 1$, *where* $\mathbf{a} = (a_1, a_2, 0)$.

Theorem 4 *Let V given by (8) maps* **C** *into itself. Assume that (i) of Theorem 4.3 is satisfied. Then a symmetric quasi q.q.o.* $\Delta : M_2(C) \to M_2(C) \otimes M_2(C)$ *corresponding to V is not positive.*

5 Conclusion

In the present paper we studied quasi quantum quadratic operators (q.q.o) acting on the algebra of 2×2 matrices $M_2(C)$. We have introduced a weaker condition, called q-purity, than purity of the channel. To study q-pure channels, we have described all quasi q.q.o. acting on $M_2(C)$ with Haar state τ. Such a result allowed us to describe q-pure quadratic operators. Then we proved that q-pure symmetric quasi q.q.o. cannot be positive. We think that such a result will allow to check whether a given mapping from $M_2(C)$ to $M_2(C) \otimes M_2(C)$ is pure or not. On the other hand, our study is related to construction of pure quantum nonlinear channels.

Acknowledgement The authors acknowledge the MOHE grant FRGS11-022-0170 and the Junior Associate scheme of the Abdus Salam International Centre for Theoretical Physics, Trieste, Italy.

References

1. Bernstein, S.N.: Uchen. Zapiski NI Kaf. Ukr. Otd. Mat. **1**, 83–115 (1924)
2. Bratteli, O., Robertson, D.W.: Operator Algebras and Quantum Statistical Mechanics I. Springer, New York (1979)
3. Dohtani, A.: SIAM J. Appl. Math. **52**, 1707–1721 (1992)
4. Ganikhodzhaev, N.N.: Uzb. Matem. Zh. **3**, 8–20 (1997)
5. Ganikhodzhaev, N.N., Mukhamedov, F.M.: Izv. Math. **65**, 873–890 (2000)
6. Fisher, M.E., Goh, B.S.: J. Math. Biol. **4**, 265–274 (1977)
7. Franz, U., Skalski, A.: Colloq. Math. **113**, 13–23 (2008)
8. Hofbauer, J., Hutson, V., Jansen, W.: J. Math. Biol. **25**, 553–570 (1987)
9. Hofbauer, J., Sigmund, K.: Evolutionary Games and Population Dynamics. Cambridge University Press, Cambridge (1998)
10. Kesten, H.: Adv. in Appl. Probab. **2**, 1–82, 179–228 (1970)
11. Lyubich, Yu.I.: Mathematical structures in population genetics. Springer, Berlin (1992)
12. Majewski, W.A., Marciniak, M.: Rep. Math. Phys. **40**, 501–508 (1997)
13. Maksimov, V.M.: Theory Probab. Appl. **41**, 55–69 (1996)
14. Mukhamedov, F.M.: Method of Funct. Anal. and Topology **7**, 63–75 (2001)
15. Mukhamedov, F.M.: Izvestiya Math. **68**, 1009–1024 (2004)
16. Mukhamedov, F.: Proc. Inter. Symp. New Developments of Geometric Function Theory & its Appl pp. 425–430. In: Darus, M., Owa, S. (Eds) National Univ. Malaysia, Malaysia. (2008)
17. Mukhamedov, F., Abduganiev: A. Abst. Appl. Anal (Article ID 278606). **2013**, (2013)
18. Mukhamedov, F., Akin, H., Temir, S., Abduganiev, A.: Jour. Math. Anal. Appl. **376**, 641–655 (2011)
19. Nielsen, M.A., Chuang, I.L.: Quantum Computation and Quantum Information. Cambridge University Press, Cambridge (2000)
20. Plank, M., Losert, V.: J. Math. Phys. **36**, 3520–3543 (1995)
21. Ruskai, M.B., Szarek, S., Werner, E.: Lin. Alg. Appl. **347**, 159–187 (2002)
22. Soltan, P.M.: J. Noncommut. Geom. **4**, 1–28 (2010)
23. Udwadia, F.E., Raju, N.: Physica D. **111**, 16–26 (1998)
24. Ulam, S.M.: A collection of mathematical problems, Interscience. New York (1960)
25. Vallander, S.S.: Soviet Math. Dokl. **13**, 123–126 (1972)
26. Woronowicz, S.L.: Compact matrix pseudogroups. Comm. Math. Phys. **111**, 613–665 (1987)

On the Structure of the Essential Spectrum of Four-Particle Schrödinger Operators on a Lattice

Z. Muminov, F. Ismail and Z. Eshkuvatov

Abstract The four-particle discrete Schrödinger operator $H(K), K \in (-\pi, \pi]^3$ corresponding to the system of the four particles on the lattice $\mathbb{Z}^3$ with arbitrary "dispersion functions" not necessarily having compact support and interacting via short-range pair potentials, is described in the coordinate representation as bounded self-adjoint operator on the corresponding Hilbert space. We describe the location and structure of the essential spectrum of the four-particle discrete Schrödinger operator $H(K), K \in (-\pi, \pi]^3$ by means of the spectrum of the three-particle discrete Schrödinger operators and establish the resolvent equation.

1 Introduction

The essential spectrum of the many-particle Schrödinger operators in Euclidean space are sufficiently well studied by Hunziker, van Winter, Zhislin, and Simon. As is known, the theorem describing the essential spectrum for a system of many particles was named as the HWZ theorem which localizes the bottom of the essential spectrum.

Historically, the location of the essential spectrum of multiparticle Schrödinger operators called HVZ theorem, was provided by Hunziker [5], van Winter [10], and Zhislin [11]. Simon used the new approach (geometric deomposition) to prove the HVZ theorem (see e.g. [9]).

Alternative methods for the determination of the essential spectrum of generalized (continuous) Schrödinger operators involve C^* -algebra techniques used in Georgescu and Iftimovici [4]. The HVZ theorem for the Brown-Ravenhall operator with constant magnetic field and for a pseudo-relativistic operator were proven in Jakubassa-Amundsen [6], [7].

Z. Muminov (✉) · F. Ismail · Z. Eshkuvatov
Department of Mathematics, Faculty of Science,
University Putra Malaysia, Serdang, Malaysia
e-mail: zimuminov@gmail.com

F. Ismail
e-mail: fudziah@science.upm.edu.my

Z. Eshkuvatov
e-mail: ezaini@science.upm.edu.my

A. Kilicman et al. (eds.), *International Conference on Mathematical Sciences and Statistics 2013,* DOI 10.1007/978-981-4585-33-0_19,

In a lattice case, the location and structure of the essential spectrum of three-particle discrete Schrödinger operators with zero- or short range attractive pairs potentials have been investigated by S. N. Lakaev and his students Albeverio, Lakaev, Muminov [1], [2] and this porblem is considered for the a model "three-particle" discrete Schrödinger operators in Eshkabilov [3]. Four-particle discrete Schrödinger operators HVZ theorem was proven in Muminov [8] but in this case the structure of the essential spectrum have not been fully comprehended yet.

These results show that the HVZ thorem is proven for a variety of Hamiltonians.

2 Four-Particle Discrete Schrödinger Operator

Let $\mathbb{T}^3 = (\mathbb{R}/2\pi\mathbb{Z})^3 = (-\pi,\pi]^3$ be the three-dimensional torus (the first Brillouin zone, i.e., the dual group of $\mathbb{Z}^3$) equipped its Haar measure.

The free four-particle discrete Schrödinger operator $H_0(K)$ of four quantum particles on the three-dimensional lattice $\mathbb{Z}^3$ usually depends on the parameter $K \in \mathbb{T}^3$ and is associated with the following self-adjoint operator in the Hilbert space $\ell^2((\mathbb{Z}^3)^3)$ of ℓ^2-sequences $f(x), x \in (\mathbb{Z}^3)^3$:

$$H_0(K) = \hat{\Delta}_{x_1} + \hat{\Delta}_{x_2} + \hat{\Delta}_{x_3} + \hat{\Delta}_4(K),$$

with

$$\hat{\Delta}_{x_1} = \Delta_1 \otimes I_d \otimes I_d, \quad \hat{\Delta}_{x_2} = I_d \otimes \Delta_2 \otimes I_d, \quad \hat{\Delta}_{x_3} = I_d \otimes I_d \otimes \Delta_3,$$

where I_d is the identical operator on $\ell^2(\mathbb{Z}^3)$ and

$$\Delta_\alpha = \sum_{s\in\mathbb{Z}^3} \hat{\varepsilon}_\alpha(s) T(s), \quad \alpha = 1,2,3,$$

$$\hat{\Delta}_4(K) = \sum_{s\in\mathbb{Z}^3} \hat{\varepsilon}_4(s) e^{\mathrm{i}(s,K)} T(s) \otimes T(s) \otimes T(s).$$

Here, $T(y)$ is the shift operator in $\ell^2(\mathbb{Z}^3)$ by y, $y \in \mathbb{Z}^3$:

$$(T(y)f)(x) = f(x+y), \quad f \in \ell^2(\mathbb{Z}^3),$$

and the function $\hat{\varepsilon}_\alpha : \mathbb{Z}^3 \to \mathbb{C}$, called *dispersion function* of the particle α, $\alpha = \overline{1,4}$, is assumed to belong in $\ell^1(\mathbb{Z}^3)$ and also to satisfy the "self-adjointness" property

$$\hat{\varepsilon}_\alpha(s) = \overline{\hat{\varepsilon}_\alpha(-s)}, \quad s \in \mathbb{Z}^3.$$

In the physical literature, the symbol given by the Fourier series (without the factor $(2\pi)^{\frac{3}{2}}$)

$$\varepsilon_\alpha(p) = \sum_{s\in\mathbb{Z}^3} \hat{\varepsilon}_\alpha(s) e^{\mathrm{i}(s,p)}, \quad p \in \mathbb{T}^3,$$

being a real valued-function on $\mathbb{T}^3$, is called the *dispersion relations of normal modes* associated with the free particle α in question.

Remark 1 The free four-particle discrete Schrödinger operator $H_0(K)$ is a multi-dimensional Laurent–Toeplitz type operator defined by the function E_K : $(\mathbb{T}^3)^3 \to \mathbb{R}^1$:

$$E_K(k_1, k_2, k_3) = \sum_{\alpha=1}^{3} \varepsilon_\alpha(k_\alpha) + \varepsilon_4(K - k_1 - k_2 - k_3), \quad k_1, k_2, k_3 \in \mathbb{T}^3.$$

The total four-particle discrete Schrödinger operator $H(K)$ (in the coordinate representation) of the system of the four quantum particles, moving on three-dimensional lattice $\mathbb{Z}^3$, with the real-valued pair interactions V_{ij}, is a self-adjoint bounded operator in the Hilbert space $\ell^2((\mathbb{Z}^3)^3)$ of the form

$$H(K) = H_0(K) + V, \quad V = \sum_{1 \le i < j \le 3} V_{ij} + \sum_{1 \le i \le 3} V_{i4},$$

where V_{ij}, V_{j4}, $1 \le i < j \le 3$ are a multiplication operator in the form

$$(V_{ij} f)(x_1, x_2, x_3) = v_{ij}(x_i - x_j) f(x_1, x_2, x_3),\ 1 \le i < j \le 3,$$

$$(V_{i4} f)(x_1, x_2, x_3) = v_{i4}(x_i) f(x_1, x_2, x_3),$$

$$i = 1, 2, 3,\ f \in \ell^2((\mathbb{Z}^3)^3).$$

Hypothesis 1 *For the functions $v_{ij}(\,\cdot\,)$, it is assumed that* $\lim_{x\to\infty} v_{ij}(x) = 0$.

2.1 Cluster Decomposition

Definition 1 A partition D of the set $\{1, 2, 3, 4\}$ into nonintersecting subsets $C_1, C_2, \ldots, C_n$ is called a cluster decomposition.

Let $D : C_1 \cup C_2 \cup \cdots \cup C_n = \{1, 2, 3, 4\}$ be a cluster decomposition, and let the operator V_D be the sum of all potentials V_{ij} with i and j belonging to different clusters.

Definition 2 The operator

$$H_D(K) = H(K) + V_D, K \in \mathbb{T}^3,$$

acting in $\ell^2((\mathbb{Z}^3)^3)$ is called the channel operator of the corresponding cluster decomposition D.

Now we take clusters

$$D_\delta = \{\{\alpha,\beta,\gamma\},\{\delta\}\}, \quad D_{\alpha\beta} = \{\{\alpha,\beta\},\{\gamma,\delta\}\},$$

where

$$\{\alpha,\beta,\gamma,\delta\} = \{1,2,3,4\}.$$

Remark 2 There exist only four D_δ and three $D_{\alpha\beta}$ clusters.

Set

$$\Omega = \{12, 13, 14, 23, 24, 34\},$$

and let Ω_δ be the subset of Ω consisting of pairs corresponding to the cluster D_δ.

Let $|v_{\alpha\beta}|$ be the absolute value of the function $v_{\alpha\beta}$, $|v_{\alpha\beta}|^{\frac{1}{2}}$ be its square root, and set $v_{\alpha\beta}^{\frac{1}{2}} = |v_{\alpha\beta}|^{\frac{1}{2}}\operatorname{sgn} v_{\alpha\beta}$.

Let $V_{\alpha\beta}^{\frac{1}{2}}$ and $|V|_{\alpha\beta}^{\frac{1}{2}}$ (resp. $V_{\alpha 4}^{\frac{1}{2}}$ and $|V|_{\alpha 4}^{\frac{1}{2}}$) denote the multiplication operators corresponding to the functions $v_{\alpha\beta}^{\frac{1}{2}}(x_\alpha - x_\beta)$ and $|v_{\alpha\beta}|^{\frac{1}{2}}(x_\alpha - x_\beta)$ (resp. $v_{\alpha 4}^{\frac{1}{2}}(x_\alpha)$ and $|v_{\alpha 4}|^{\frac{1}{2}}(x_\alpha)$).

3 The HVJ Theorem and Faddeev Type Operator

Theorem 2 (The HVJ theorem) *For any $K \in \mathbb{T}^3$ we have*

$$\sigma_{ess}(H(K)) = \cup_{\alpha=1}^{4}\sigma(H_{D_\alpha}(K)) \bigcup \cup_{\alpha=1}^{3}\sigma(H_{D_{\alpha 4}}(K)).$$

Proof Set

$$\Sigma(K) = \cup_{\alpha=1}^{4}\sigma(H_{D_\alpha}(K)) \bigcup \cup_{\alpha=1}^{3}\sigma(H_{D_{\alpha 4}}(K)).$$

By the Weyl criterion, we can show that $\Sigma(K) \subset \sigma_{ess}(H(K))$.

We prove the converse inclusion $\sigma_{ess}(H(K)) \subset \Sigma(K)$ using the resolvent equation.

Let $R(K,z)$ and $R_0(K,z)$ be the resolvent of the operators $H(K)$ and $H_0(K)$.

Without lost of generality, we denote

$$H := H(K),\ H_0 := H_0(K),\ R(z) := R(K,z),\ R_0(z) := R_0(K,z).$$

The well known resolvent equation (see for instance [9]) has the form

$$R(z) = R_0(z) + R_0(z)\sum_{ij\in\Omega} V_{ij} R(z). \tag{1}$$

Multiplying (1) from the left by $|V|_{ij}^{\frac{1}{2}}$ and setting $\mathcal{X}_{ij}(z) \equiv |V|_{ij}^{\frac{1}{2}}R(z)$, we get the system of the equations

$$\mathcal{X}_{\alpha\beta}(z) = |V|_{\alpha\beta}^{\frac{1}{2}}R_0(z) + |V|_{\alpha\beta}^{\frac{1}{2}}R_0(z)\sum_{ij\in\Omega} V_{ij}^{\frac{1}{2}}\mathcal{X}_{ij}(z), \quad \alpha\beta \in \Omega.$$

Equivalently, we have the following system of three equations

$$\begin{aligned}(I - |V|_{\alpha\beta}^{\frac{1}{2}}R_0(z)V_{\alpha\beta}^{\frac{1}{2}})\mathcal{X}_{\alpha\beta}(z) &= |V|_{\alpha\beta}^{\frac{1}{2}}R_0(z) \\ &+ \sum_{ij\in\Omega\setminus\{\alpha\beta\}} |V|_{\alpha\beta}^{\frac{1}{2}}R_0(z)V_{ij}^{\frac{1}{2}}\mathcal{X}_{ij}(z), \qquad \alpha\beta \in \Omega.\end{aligned} \tag{2}$$

One can check that

$$W_{\alpha\beta}(z) = (I - |V|_{\alpha\beta}^{\frac{1}{2}}R_0(z)V_{\alpha\beta}^{\frac{1}{2}})^{-1}, \quad z \in \mathbb{C}\setminus\sigma(H_{D_{\alpha\beta}}(K))$$

exists and multiplying the equality (2) from the left by the operator $W_{\alpha\beta}(z)$ we get the following system of six equations

$$\mathcal{X}_{\alpha\beta} = W_{\alpha\beta}|V|_{\alpha\beta}^{\frac{1}{2}}R_0(z) + \sum_{\alpha'\beta'\in\Omega\setminus\{\alpha\beta\}} W_{\alpha\beta}|V|_{\alpha\beta}^{\frac{1}{2}}R_0(z)V_{\alpha'\beta'}^{\frac{1}{2}}\mathcal{X}_{\alpha'\beta'}, \quad \alpha\beta \in \Omega.$$

Set $\mathcal{H} = \oplus_{j=1}^{6}\ell^2((\mathbb{Z}^3)^3)$ and

$$T_{\alpha\beta,\alpha'\beta'}^{(\delta)}(z) = W_{\alpha\beta}|V|_{\alpha\beta}^{\frac{1}{2}}R_0(z)V_{\beta,\gamma,}^{\frac{1}{2}}, \quad \alpha\beta, \alpha'\beta' \in \Omega_\delta,\ \alpha\beta \neq \alpha'\beta'.$$

Lemma 1 *If for any $\alpha\beta, \alpha'\beta', ij, i'j' \in \Omega$ the equality $\{\alpha,\beta,\alpha',\beta',i,j,i',j'\} = \{1,2,3,4\}$ holds, then the operator $T_{\alpha\beta,\alpha'\beta'}^{(\delta)}(z)T_{ij,i'j'}^{(\delta)}(z)$ is compact in $\ell^2((\mathbb{Z}^3)^3)$.*

Let us define operators $\mathcal{T}^{(\delta)}(z)$ and $\mathcal{T}^{(0)}(z)$ acting $\mathcal{H}$ by the forms

$$\mathcal{T}^{(\delta)}(z) = (A_{\alpha\beta,\alpha'\beta'})_{\alpha\beta,\alpha'\beta'\in\Omega}$$

and

$$\mathcal{T}^{(0)}(z) = (B_{\alpha\beta,\alpha'\beta'})_{\alpha\beta,\alpha'\beta'\in\Omega},$$

where

$$A_{\alpha\beta,\alpha'\beta'} = \begin{cases} T_{\alpha\beta,\alpha'\beta'}^{\delta}(z), & \text{if} \quad \alpha\beta, \alpha'\beta' \in \Omega_\delta,\ \alpha\beta \neq \alpha'\beta' \\ 0, & \text{otherwise} \end{cases}$$

and

$$B_{\alpha\beta,\varsigma\tau} = \begin{cases} T_{\varsigma\tau,\alpha\beta}^{\delta}(z), & \text{if} \quad \alpha\beta, \varsigma\tau \in \Omega,\ \{\alpha,\beta,\varsigma,\tau\} = \{1,2,3,4\} \\ 0, & \text{otherwise.} \end{cases}$$

Lemma 2 *For any $z \in \mathbb{C} \setminus \sigma_{ess}(H(K))$ the inverse of operators $\mathbf{I} - \mathcal{T}^{(\delta)}(z)$, $\delta = \overline{0,4}$, exists and if*

$$(\mathbf{I} - \mathcal{T}^{(\delta)}(z))^{-1} = \mathbf{I} + \mathcal{D}^{(\delta)}(z)$$

then

$$\mathcal{D}^{(\delta)}(z) = (\mathbf{I} + \mathcal{D}^{(\delta)}(z))\mathcal{T}^{(\delta)}(z) \tag{3}$$

holds, and $\mathbf{I}$ is the identity operator in $\mathcal{H}$.

According to the definitions $\mathcal{T}^{(\delta)}(z)$, $\delta = \overline{0,4}$, we can write the system (3) of equations as

$$\mathbf{X}(z) = \mathbf{X}_0(z) + \sum_{\delta=0}^{4} \mathcal{T}^{(\delta)}(z)))\mathbf{X}(z),$$

where

$$\mathbf{X}(z) = (\mathcal{X}_{12}, \mathcal{X}_{13}, \mathcal{X}_{23}, \mathcal{X}_{14}, \mathcal{X}_{24}, \mathcal{X}_{34}),$$

$$\mathbf{X}_0(z) = (W_{12}|V|_{12}^{\frac{1}{2}} R_0(z), W_{13}|V|_{13}^{\frac{1}{2}} R_0(z), W_{23}|V|_{23}^{\frac{1}{2}} R_0(z),$$

$$W_{14}|V|_{14}^{\frac{1}{2}} R_0(z), W_{24}|V|_{24}^{\frac{1}{2}} R_0(z), W_{34}|V|_{34}^{\frac{1}{2}} R_0(z))$$

Applying this procedure four times, we have

$$\mathbf{X}(z) = (\mathbf{I} - \mathcal{T}^{(0)}(z))^{-1} \prod_{\delta=1}^{4} (\mathbf{I} + \mathcal{D}^{(4)}(z))\mathbf{X}_0(z)$$

$$+ (\mathbf{I} - \mathcal{T}^{(0)}(z))^{-1} \Bigg\langle \mathcal{D}^{(1)}(z) \Bigg(\mathcal{T}^{(0)}(z) + \mathcal{D}^{(2)}(z) \Bigg(\sum_{\delta=0}^{1} \mathcal{T}^{(\delta)}(z)$$

$$+\mathcal{D}^{(3)}(z) \Bigg(\sum_{\delta=0}^{2} \mathcal{T}^{(\delta)}(z) + \mathcal{D}^{(4)}(z) \Bigg(\sum_{\delta=0}^{3} \mathcal{T}^{(\delta)}(z) \Bigg)\Bigg)\Bigg)\Bigg)\Bigg\rangle \mathbf{X}(z),$$

that is

$$\mathbf{X}(z) = (\mathbf{I} - \mathcal{T}^{(0)}(z))^{-1} \prod_{\delta=0}^{4} (\mathbf{I} + \mathcal{D}^{(\delta)}(z))\mathbf{X}_0(z) + A(z)\mathbf{X}(z),$$

where $A(z)$ is the operator acting in $\mathcal{H}$ by the form

$$A(z) = (\mathbf{I} - \mathcal{T}^{(0)}(z))^{-1}$$

$$\times \Bigg\langle \mathcal{D}^{(1)}(z) \Bigg(\mathcal{T}^{(0)}(z) + \mathcal{D}^{(2)}(z) \Bigg(\sum_{\delta=0}^{1} \mathcal{T}^{(\delta)}(z)$$

$$+\mathcal{D}^{(3)}(z) \Bigg(\sum_{\delta=0}^{2} \mathcal{T}^{(\delta)}(z) + \mathcal{D}^{(4)}(z) \Bigg(\sum_{\delta=0}^{3} \mathcal{T}^{(\delta)}(z) \Bigg)\Bigg)\Bigg)\Bigg)\Bigg\rangle.$$

By the following assertion, we call $A(z)$ the Faddeev–Yakubovskiy type operator.

Lemma 3 *A number* $z \in \mathbb{C} \setminus \Sigma(K)$ *is an eigenvalue of the operator* $H(K)$ *if and only if the operator* $A(K; z)$ *has the eigenvalue 1 and their multiplicities coincide.*

Proof The proof of this lemma follows from the Lemma 1 and the equality 3.

From (1) we have the following representation for the resolvent

$$R(z) = R_0(z) + R_0(z) \sum_{\alpha\beta\in\Omega} V_{\alpha\beta}^{\frac{1}{2}} X_{\alpha\beta}(z). \tag{4}$$

The operator $A(z)$ is a compact operator-valued function on $\mathbb{C} \setminus \Sigma(K)$ and $\mathbf{I} - A(z)$ is invertible if z is real and either very negative or very positive. The analytic Fredholm theorem (see, e.g., Theorem $VI.14$ in [9]) implies that there is a discrete set $S \subset \mathbb{C} \setminus \Sigma(K)$ so that $(\mathbf{I} - A(z))^{-1}$ exists and is analytic in $\mathbb{C} \setminus (\Sigma(K) \cup S)$ and meramorphic in $\mathbb{C} \setminus \Sigma(K)$ with finite rank residues. Thus, the function $(\mathbf{I} - A(z))^{-1}(\mathbf{I} - \mathcal{T}^{(0)}(z))^{-1} \prod_{\delta=1}^{4} (\mathbf{I} + \mathcal{D}^{(4)}(z))\mathbf{X}_0(z) \equiv F(z)$ is analytic in $\mathbb{C} \setminus (\Sigma(K) \cup S)$ with finite rank residues at the points of S.

Let $z \notin S$, $\mathrm{Im} z \neq 0$, then by (1), (4) we have $F(z) = X(z)$. In particular,

$$R(z)(H(K) - zI) = (R_0(z) + R_0(z) \sum_{\alpha\beta\in\Omega} V_{\alpha\beta}^{\frac{1}{2}} X_{\alpha\beta}(z))(H(K) - zI) = I.$$

By analytic continuation, this holds for any $z \notin \Sigma(K) \cup S$. Thus, for any such z, the operator $H(K) - zI$ has a bounded inverse. Therefore, $\sigma(H(K)) \setminus \Sigma(K)$ consists of isolated points and only the frontier points of $\Sigma(K)$ are possible their limit points. Finally, since $R(z)$ has finite rank residues at $z \in S$, we conclude that $\sigma(H(K)) \setminus \Sigma(K)$ belongs to the discrete spectrum $\sigma_d(H(K))$ of $H(K)$, which completes the proof of Theorem 1.

Remark 3 We recall that for the four-particle continuous Schödinger operators, the essential spectrum coincides with the semi-axis $[\Sigma, \infty)$.

Acknowledgement This work was partially supported by University Putra Malaysia under Research University Grant Scheme (RUGS), and the project code is 9190883. The first named author gratefully acknowledges the hospitality of the Abdus Salam International Center for Theoretical Physics (ICTP) during his stay in the winter of 2011, and he thanks S. N. Lakaev for useful discussions. This work was also partially supported by the Fundamental Science Foundation of Uzbekistan.

References

1. Albeverio, S., Lakaev, S.N., Muminov Z.I.: Schrödinger operators on lattices. The Efimov effect and discrete spectrum asymptotics. Ann. Henri Poincaré **5**, 1–30 (2004)
2. Albeverio, S., Lakaev, S.N., Muminov Z.I.: On the structure of the essential spectrum for the three-particle Schrödinger operators on lattices. Math. Nachr. **280**(7), 1–18 (2007)
3. Eshkabilov, Yu.Kh: The efimov effect for a model "three-particle" discrete Schrödinger operator. Theo. Math. Phys. **164**(1), 896–904 (2010)
4. Georgescu, V., Iftimovici, A.: Crossed products of C^*-algebras and spectral analysis of quantum Hamiltonians. Commun. Math. Phys. **228**, 519–560 (2002)

5. Hunziker, W.: On the spectra of Schrödinger multiparticle Hamiltonians. Helv. Phys. Acta. **39**, 451–462 (1966)
6. Jakubassa-Amundsen, D.H.: Localization of the essential spectrum for relativistic N-electron ions and atoms. Doc. Math. **10**, 417–445 (2005)
7. Jakubassa-Amundsen, D.H.: The HVZ theorem for a pseudo-relativistic operatorar. Ann. Henri Poincaré. **8**, 337–360 (2007)
8. Muminov, M.I.: A Hunziker-van Winter-Zhislin theorem for a four-particle lattice schrödinger operator. Theor. Math. Phys. **148**(3), 1236–1250 (2006)
9. Reed, M., Simon, B.: Analysis of Operators, Vol. IV of Methods of Modern Mathematical Physics. Academic Press, New York (1978)
10. van Winter, Clasine: Theory of finite systems of particles I. The Green function. Mat. Fys. Dan. Vid. Selsk. **2**(8), 1–60 (1964)
11. Zhislin, G.M.: A study of the spectrum of the Schrödinger operator for a system of several particles. Trudy Moskov. Mat. Obsc. **9**, 81–120 (1960)

Symmetrizers for Runge–Kutta Methods

N. Razali and R. P. K. Chan

Abstract L-stable symmetrizers for symmetric Runge–Kutta methods are constructed to preserve the asymptotic error expansion in even powers of the stepsize and to provide necessary damping for the numerical solution of stiff initial value ordinary differential equations. The process is called symmetrization and has the effect of dampening down undesirable oscillations that may arise when the problem is stiff. The symmetry property can be exploited by Richardson extrapolation in increasing the order by two at a time. The effect of the damping is to stabilize the order behaviour of the extrapolation. In this paper, we discuss two types of symmetrizers; the well-known one-step smoothing formula of Gragg, and the new two-step smoothing formula. We construct these symmetrizers for the implicit midpoint and trapezoidal rules and investigate the active mode of application with extrapolation. Finally, we present numerical results in a constant stepsize setting that show two-step symmetrizers having certain advantages over one-step symmetrizers for stiff linear and nonlinear problems.

1 Introduction

The use of Richardson extrapolation [11] in accelerating the convergence of a sequence has been studied by many researchers. This technique has been successfully applied in numerical quadrature [12] and the numerical solution of ordinary differential equations. It is more efficient if the method has an asymptotic error expansion [13] in even powers of the stepsize, since the order can be increased by two at a time. Gragg [8] first proved that the explicit midpoint rule has this property. He also introduced the concept of smoothing to suppress the effects of the parasitic oscillatory component in the numerical solution and showed the power of smoothing applied

N. Razali (✉)
Fundamental Engineering Unit, Faculty of Engineering and Built Environment,
Universiti Kebangsaan Malaysia, Bangi Selangor Darul Ehsan, Selangor, Malaysia
e-mail: nraz865@aucklanduni.ac.nz

R. P. K. Chan
Department of Mathematics, Faculty of Science, University of Auckland,
Auckland, New Zealand
e-mail: chan@math.auckland.ac.nz

A. Kilicman et al. (eds.), *International Conference on Mathematical Sciences and Statistics 2013,* DOI 10.1007/978-981-4585-33-0_20,

with extrapolation. The GBS extrapolation algorithm [2] was successfully developed for the numerical solution of nonstiff problems. Dahlquist and Lindberg [7], Lindberg [9], Bader and Deuflhard [1] then extended the application of extrapolation to the solution of stiff problems.

Later, Chan [3], [4] generalized smoothing for arbitrary symmetric Runge–Kutta methods. The process is called *symmetrization* and the method used for this purpose is called a symmetrizer. It is constructed so as to preserve the symmetry property and to provide the necessary damping for stiff problems. In 2011, Chan and Gorgey [6] reported on a study that compared the active and passive modes of application of the one-step symmetrization. They analyzed the efficiency of implementing symmetrization with and without extrapolation on higher order methods, providing evidence that one-step symmetrizers can restore the classical order of particular members of the Gauss and Lobatto IIIA family of methods [5], [6].

In Sect. 2, we construct two-step symmetrizers for the implicit midpoint rule and the implicit trapezoidal rule, and in Sect. 3 we discuss the implementation of active symmetrization with active extrapolation. We present numerical results for stiff and nonstiff cases in Sect. 4 and in Sect. 5 we present our conclusion.

2 Symmetry and Symmetrization

An s-stage Runge–Kutta method $\mathcal{R} = (A, b, c)$ is defined by

$$Y_i = y_n + h \sum_{j=1}^{s} a_{ij} f(x_n + c_j h, Y_j), \quad i = 1, \dots, s, \tag{1}$$

$$y_{n+1} = y_n + h \sum_{i=1}^{s} b_i f(x_n + c_i h, Y_i), \tag{2}$$

where A is the $s \times s$ Runge–Kutta matrix, b and c are the $s \times 1$ vector of weights and abscissas respectively. The method $\mathcal{R}$ is symmetric if $\mathcal{R}^* = \mathcal{R}$, where $\mathcal{R}^*$ denotes the adjoint of $\mathcal{R}$. This leads to the algebraic characterization of symmetry given by

$$PA + AP = eb^T, \quad Pb = b, \quad Pc = e - c.$$

Here e is the $s \times 1$ vector of units, and P is the $s \times s$ permutation matrix that reverses the order of the stages with the (i, j)th element given by the Kronecker $\delta_{i,s+1-j}$. These conditions assume that $b^T e = 1$ and $Ae = c$ hold. A symmetric method has an h^2-asymptotic error expansion [12].

The one-step symmetrized value of the IMR and the ITR satisfying the damping and order-one conditions is given by

$$\widetilde{y}_n = \frac{1}{4}\left(y_{n-1} + 2y_n + y_{n+1}\right).$$

This is the well-known smoothing formula [8].

2.1 Two-Step Symmetrization

Let $\mathcal{R}_n = \frac{1}{n}\mathcal{R}^n$ denote the composition of n-steps of a symmetric Runge–Kutta method with stepsize $h = H/n$. For two-step symmetrization, the last two symmetric steps are replaced by $\widetilde{\mathcal{R}}$. If $\widetilde{\mathcal{R}}$ satisfies $\widetilde{\mathcal{R}} \circ \widetilde{\mathcal{R}}^* = \mathcal{R}^4$, then the composite method $\widehat{\mathcal{R}}_n = \frac{1}{n}(\mathcal{R}^{n-2} \circ \widetilde{\mathcal{R}})$ satisfies $\widehat{\mathcal{R}}_n = \widehat{\mathcal{R}}_{-n}$ for all n, that is, the h^2-symmetry is preserved. This symmetrizer is a composition of four symmetric steps but with different weights generated by

$$\widetilde{\mathcal{R}} = (\widetilde{A}, \widetilde{b}, \widetilde{c}) \equiv \left(\begin{bmatrix} A & 0 & 0 & 0 \\ eb^T & A & 0 & 0 \\ eb^T & eb^T & A & 0 \\ eb^T & eb^T & eb^T & A \end{bmatrix}, \begin{bmatrix} b - Pv \\ b - Pu \\ u \\ v \end{bmatrix}, \begin{bmatrix} c \\ e + c \\ 2e + c \\ 3e + c \end{bmatrix} \right)$$

A two-step symmetrizer carries twice the number of parameters (u and v) compared to a one-step symmetrizer (u) [5]. This gives flexibility especially for higher order methods in satisfying the order and other desirable conditions. From (2), we define the symmetrized step by

$$\begin{aligned} \widetilde{Y}^{[n]} &= \widetilde{e} y_{n-2} + h \widetilde{A} \widetilde{F}\left(x_{n-2} + \widetilde{c} h, \widetilde{Y}^{[n]}\right), \\ \widetilde{y}_n &= \widetilde{b}^T \widetilde{A}^{-1} \widetilde{Y}^{[n]}. \end{aligned} \tag{3}$$

The stability function $\widetilde{R}(z) = 1 + z\widetilde{b}^T(\widetilde{I} - z\widetilde{A})^{-1}\widetilde{e}$ reduces to

$$\begin{aligned} \widetilde{R}(z) &= R(z)^2 \left(1 + 2z^2 u^T (I - z^2 A^2)^{-1} c\right) \\ &\quad + R(z)\left(R(z)^2 z v^T (I - zA)^{-1} e - z v^T (I + zA)^{-1} e\right). \end{aligned} \tag{4}$$

If A is nonsingular then since $R(\infty) = \pm 1$, we have

$$\widetilde{R}(\infty) = 1 - 2u^T A^{-1} e - R(\infty) 2 v^T A^{-1} e,$$

and so $\widetilde{R}(\infty) = 0$ gives the damping condition

$$u^T A^{-1} e + R(\infty) v^T A^{-1} e = \frac{1}{2}. \tag{5}$$

In the case that A is singular, we restrict our consideration to the family of Lobatto IIIA methods. The special structure of the matrix A with the first row of zeros allows us to partition the vectors and matrices as follows:

$$A = \begin{bmatrix} 0 & 0 \\ a & \overline{A} \end{bmatrix}, \quad b = \begin{bmatrix} b_1 \\ \overline{b} \end{bmatrix}, \quad c = \begin{bmatrix} 0 \\ \overline{c} \end{bmatrix}, \quad I = \begin{bmatrix} 1 & 0 \\ 0 & \overline{I} \end{bmatrix},$$

$$e = \begin{bmatrix} 1 \\ \overline{e} \end{bmatrix}, \quad u = \begin{bmatrix} u_1 \\ \overline{u} \end{bmatrix}, \quad v = \begin{bmatrix} v_1 \\ \overline{v} \end{bmatrix}.$$

Thus the stability function then gives, as $z \to \infty$,

$$\widetilde{R}(\infty) = 1 - 2\overline{u}^T\overline{A}^{-2}\overline{c} + (-1)^s 2\overline{v}^T\overline{A}^{-2}\overline{c} - 2\overline{b}^T\overline{A}^{-3}\overline{c}(v_1 - \overline{v}^T\overline{A}^{-1}a) + O(1/z) \tag{6}$$

and so $\widetilde{R}(\infty) = 0$ gives the damping condition

$$\overline{u}^T\overline{A}^{-2}\overline{c} - (-1)^s\overline{v}^T\overline{A}^{-2}\overline{c} + \overline{b}^T\overline{A}^{-3}\overline{c}(v_1 - \overline{v}^T\overline{A}^{-1}a) = \frac{1}{2}. \tag{7}$$

In the case of the IMR, the parameter u and v are chosen to satisfy (5) with $R(\infty) = -1$ and the order-two condition, namely,

$$u^Tc + v^T(e + c) = 0. \tag{8}$$

This yields $u = \frac{3}{16}$ and $v = -\frac{1}{16}$ and thus (3) becomes

$$\widetilde{y}_n = \frac{1}{16}(-y_{n-2} + 4y_{n-1} + 10y_n + 4y_{n+1} - y_{n+2}). \tag{9}$$

As for the ITR, since the matrix A is singular, the damping condition (7) together with order-two condition (8) yields $u = \frac{1}{16}$ and $v = -\frac{1}{32}$ and the formula for the update turns out to be also given by (9). The symmetrized update at step n in this case involves two steps beyond the nth step.

3 Implementation

The implementation of the symmetrizers on linear and nonlinear problems have been discussed in [10]. In this section, we will discuss the implementation of symmetrization together with extrapolation. There are a few possible combinations of active/passive symmetrization with active/passive extrapolation. In this paper we will only focus on active symmetrization with active extrapolation.

3.1 Active Symmetrization

In the active mode, each time a symmetrized value is computed, it is then used to propagate the numerical solution. Symmetrization can be performed at every step, every two steps or three steps etc. This implementation is reported in [6], but for two-step symmetrization the symmetrized update involves two steps beyond the n-th step.

Table 1 Notation for numerical experiments.

MX/TX	IMR/ITR applied with active extrapolation
1AMX/1ATX	one-step active symmetrization with active extrapolation
2AMX/2ATX	two-step active symmetrization with active extrapolation

3.2 Active Symmetrization with Active Extrapolation

The diagram below shows active extrapolation at x_2 with active symmetrization. The horizontal lines represent symmetrized solutions while the vertical lines denote applying active extrapolation using two symmetrized values with stepsize h and $h/2$. The symmetrized extrapolated value is used to propagate the solution.

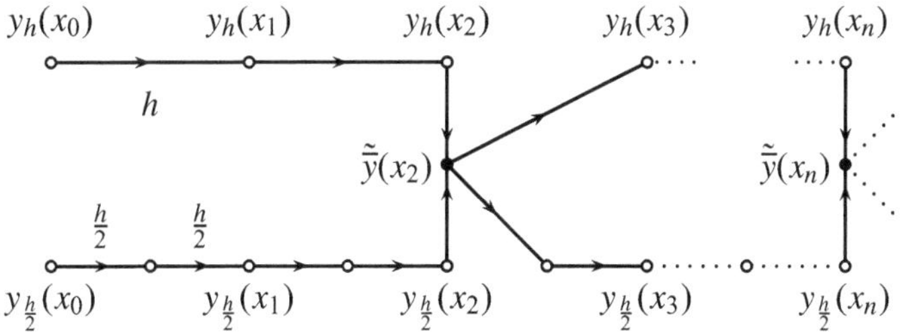

4 Results and Discussion

In this section we discuss the accuracy and efficiency of the IMR and the ITR with active symmetrization applied with active extrapolation on linear and nonlinear problems. For the Prothero–Robinson and the Kaps problems, the experiments were done on nonstiff and stiff cases with $\lambda = -1$ and $\lambda = -10^6$ respectively. Table 1 shows the abbreviation used to explain the results.

Problem 1 Prothero-Robinson

$$y' = \lambda(y - g(x)) + g'(x), \quad y(x) = g(x)$$

integrated to $X = 5$ with stepsize $h = 0.1$.

Problem 1 is a linear problem with $g(x) = e^{-x}$ and the initial value $y(0) = 1$. In terms of accuracy of the IMR, the base method is more accurate compared to the other methods for the nonstiff case but in the stiff case, 2AMX is more accurate than the other methods (See Fig. 1). For the ITR, the base method is more accurate compared to other methods in the nonstiff case but in the stiff case, 2ATX is more accurate (See Fig. 2).

In terms of the efficiency, for the IMR, the base method is most efficient in the nonstiff case while in the stiff case 2AMX is most efficient as seen in Fig. 3. For the ITR, we observed the same behaviour as seen in Fig. 4.

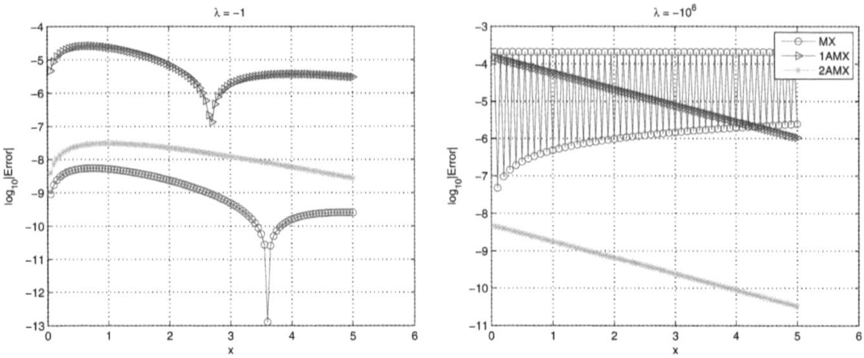

Fig. 1 PR problem for $g(x) = e^{-x}$: Accuracy diagram of IMR with active symmetrizations and extrapolation

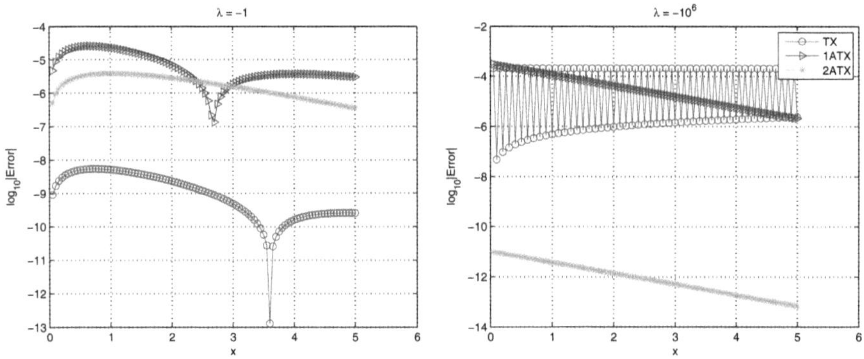

Fig. 2 PR problem for $g(x) = e^{-x}$: Accuracy diagram of ITR with active symmetrizations and extrapolation

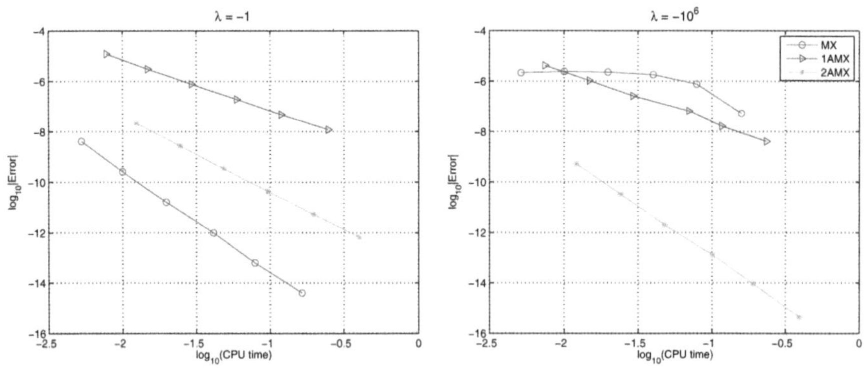

Fig. 3 PR problem for $g(x) = e^{-x}$: Efficiency diagram of IMR with active symmetrizations and extrapolation

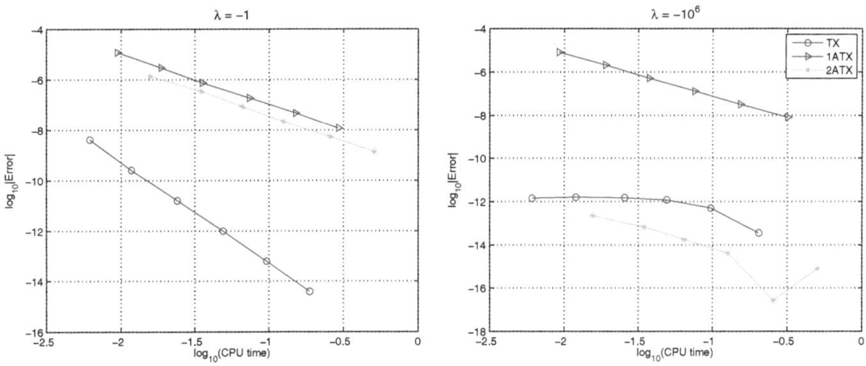

Fig. 4 PR problem for $g(x) = e^{-x}$: Efficiency diagram of ITR with active symmetrizations and extrapolation

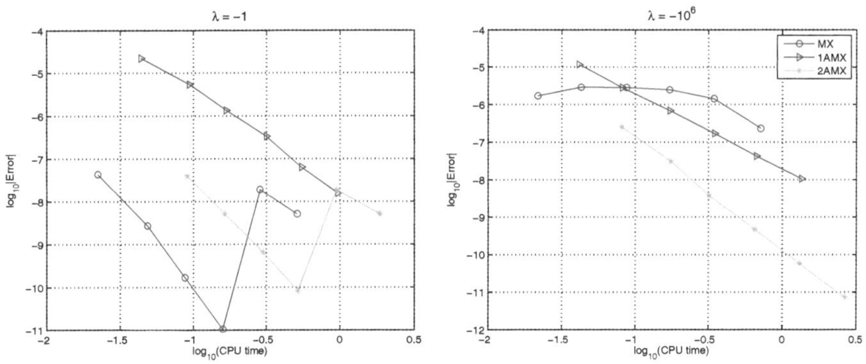

Fig. 5 Kaps problem: Efficiency diagram of IMR with active symmetrizations and extrapolation

Problem 2 Kaps

$$y_1' = (\lambda - 2)y_1 - \lambda y_2^2, \quad y_1(0) = 1,$$
$$y_2' = y_1 - y_2(1 + y_2), \quad y_2(0) = 1,$$

integrated to $X = 3$ with stepsize $h = 0.1$.

Problem 2 is a nonlinear problem of two dimensions. For the IMR, the base method is more efficient than the other methods in the nonstiff case while in the stiff case 2AMX is more efficient. These can be seen in Fig. 5. For the ITR, the base method is more efficient in both nonstiff and stiff cases as seen in Fig. 6.

Table 2, summarizes the order behaviour of all symmetrizers in the nonstiff and stiff cases while Table 3 summarizes the order behaviour of active symmetrization applied with extrapolation. The obvious observation is that the two-step symmetrizers attained order-two behaviour in the active symmetrization compared to the order-one behaviour of one-step symmetrizers. Thus, this explains the accuracy that two-step symmetrizers achieved for the few problems shown in Figs. 1 and 2.

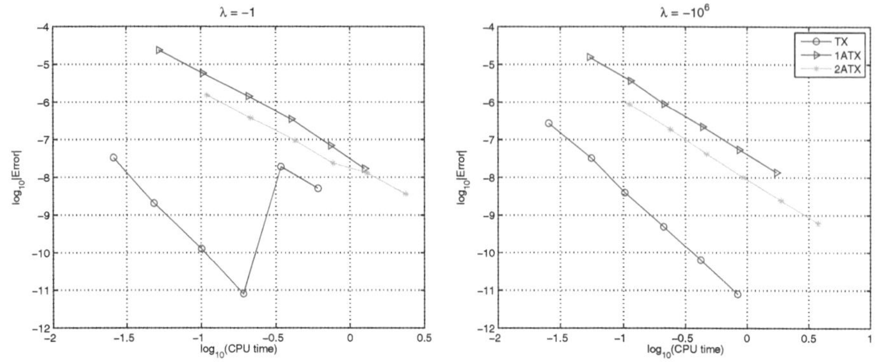

Fig. 6 Kaps problem: Efficiency diagram of ITR with symmetrizations and active extrapolation

Table 2 Order behaviour of symmetrizers in active mode for nonstiff/stiff problems

Linear scalar problem	Active	Nonlinear Kaps problem	Active
One-step (IMR)	1/2	One-step (IMR)	1/1
Two-step (IMR)	2/2	Two-step (IMR)	2/2
One-step (ITR)	1/2	One-step (ITR)	1/1
Two-step (ITR)	2/4	Two-step (ITR)	2/2

Table 3 Table of order behaviour of active symmetrization with extrapolation for nonstiff/stiff problem

Linear scalar problem	Active
One-Step (IMR)	2/2
Two-Step (IMR)	3/4
One-Step (ITR)	2/4
Two-Step (ITR)	3/6

5 Conclusion

By using active symmetrization together with active extrapolation, the two-step symmetrizer for the IMR and the ITR can show order behaviour higher than the one-step symmetrizer as seen in Table 2. We thus conclude that two-step symmetrization is significantly more accurate than one-step symmetrization in both the nonstiff and stiff cases. Even though the two-step symmetrizer requires twice as much computational cost per step our results show that in all the cases tested it is more efficient than the one-step symmetrizer. We note, however, that it might not be as efficient as the base method in some cases (see, for example, the Kaps problem). With active symmetrization, the two-step symmetrizer for ITR shows super-convergent order-4 behaviour and when applied with active extrapolation the method shows order-6 behaviour. The promising behaviour of two-step symmetrizers for low order symmetric methods provides motivation for extending the study of two-step symmetrizers to symmetric methods of higher order. A variable stepsize, variable order extrapolation code based on the IMR/ITR with one and two step symmetrization is under development.

Acknowledgements This paper is part of a PhD research study entitled 'Two-Step Symmetrization of Runge–Kutta methods' from the Department of Mathematics, Faculty of Science, The University of Auckland with support funding of the Press Account and a PhD scholarship courtesy by the Ministry of Higher Education Malaysia and UKM, Malaysia.

References

1. Bader, G., Deuflhard, P.: A semi-implicit mid-point rule for stiff systems of ordinary differential equations. Numer. Math. **41**, 373–398 (1983)
2. Bulirsch, R., Stoer, J.: Numerical treatment of ordinary differential equations by extrapolation methods. Numer. Math. **8**, 1–13 (1966)
3. Chan, R.P.K.: Extrapolation of Runge–Kutta Methods for Stiff Initial Value Problems. PhD Thesis, University of Auckland (1989)
4. Chan, R.P.K.: Generalized symmetric Runge–Kutta methods. Computing. **50**, 31–49 (1993)
5. Chan, R.P.K., Gorgey, A.: Order-4 symmetrized Runge–Kutta methods for stiff problems. J. Qual. Meas. Anal. **7**(1), 53–66 (2011)
6. Chan, R.P.K., Gorgey, A.: Active and passive symmetrization of Runge–Kutta Gauss methods. Appl. Numer. Math. **67**, 64–77 (2013). doi:10.1016/j.apnum.2011.06.013
7. Dahlquist, G., Lindberg, B.: On some implicit one-step methods for stiff differential equations, Report TRITA-NA-7302 (1973)
8. Gragg, W.B.: On extrapolation algorithms for ordinary initial value problems. SIAM J. Num. Anal. **2**, 384–403 (1965)
9. Lindberg, B.: On smoothing and extrapolation for the trapezoidal rule. BIT Numer. Math. **11**, 29–52 (1971)
10. Razali, N., Chan, R.P.K.: On two-step symmetrization of Runge–Kutta methods for stiff problems. Proceedings of International Malaysia Research Conference in Auckland, iMaRCA 2012, 151–173 (2013a)
11. Richardson, L.F., Gaunt, J.A.: The deferred approach to the limit. Philos. Trans. Roy. Soc. London Ser. A. **226**, 299–361 (1927)
12. Romberg, W.: Vereinfachte numerische integration, Norske Vid. Selsk. Forhandlinger (Trondheim). **28**, 30–36 (1955)
13. Stetter, H.J.: Asymptotic expansions for the error of discretization algorithms for non-linear functional equations. Numer. Math. **7**, 18–31 (1965)

A Method of Estimating the p-adic Sizes of Common Zeros of Partial Derivative Polynomials Associated with a Complete Cubic Form

S. S. Aminudin, S. H. Sapar and K. A. Mohd Atan

Abstract Let $x = (x_1, x_2, \ldots, x_n)$ be a vector in the space Q^n with Q field of rational numbers and q be a positive integer, f a polynomial in x with coefficient in Q. The exponential sum associated with f is defined as $S(f; q) = \Sigma_{xmodq} e^{((2if(x))/q)}$, where the sum is taken over a complete set of residues modulo q. The value of $S(f; q)$ depends on the estimate of cardinality $|V|$, the number of elements contained in the set $V = \{xmodq | f_x \equiv 0 mod q\}$ where f_x is the partial derivative of f with respect to x. To determine the cardinality of V, the p-adic sizes of common zeros of the partial derivative polynomials need to be obtained. In this paper, we estimate the p-adic sizes of common zeros of partial derivative polynomials of $f(x, y)$ in $Q_p[x, y]$ with a complete cubic form by using Newton polyhedron technique. The polynomial is of the form $f(x, y) = ax^3 + bx^2y + cxy^2 + dy^3 + \frac{3}{2}ax^2 + bxy + \frac{1}{2}cy^2 + sx + ty + k$.

1 Introduction

In our discussion, we use notations the completion of algebraic closure of Q_p the field of rational p-adic numbers (Ω_p) and the p-adic size of x which means the highest power of p dividing x ($ord_p x$). It follows that for rational number x and y, $ord_p x = \infty$ if and only if $ord_p x = 0$; $ord_p(xy) = ord_p x + ord_p y$ and $ord_p(x + y) \geq \min\{ord_p x, ord_p y\}$, with equality if $ord_p x \neq ord_p y$.

Loxton and Vaughan [5] are the researchers who investigate the exponential sums $S(f; q) = \Sigma_{xmodq} e^{((2if(x))/q)}$ where f is a nonlinear polynomial in Z[x]. They showed that the number of common zeros of the partial derivative polynomials of f with respect to x modulo q gives the estimation of S(f; q).

S. S. Aminudin (✉) · K. A. M. Atan
Institute for Mathematical Research, Universiti Putra Malaysia,
Serdang 43400, Malaysia
e-mail: syaheera.siti@yahoo.com

S. H. Sapar
Department of Mathematics, Faculty of Science, Universiti Putra Malaysia,
Serdang 43400, Malaysia
e-mail: sitihas@putra.upm.edu.my

K. A. M. Atan
e-mail: kamel@inform.upm.edu.my

A. Kilicman et al. (eds.), *International Conference on Mathematical Sciences and Statistics 2013,* DOI 10.1007/978-981-4585-33-0_21,

Loxton and Smith [4] found that the p-adic sizes of common zeros to partial derivative polynomials associated with f in the neighborhood of points in the product space Ω_p^n, $n > 0$ can estimate the cardinality of V. Their result is the estimation of $ord_p(x - \xi_i)$ that will lead to a derivation of estimate of $N(f, p^\alpha)$.

The estimations for lower degree two-variable polynomials by using Newton polyhedron technique are found by many researchers such as Mohd. Atan [6] who defines the p-adic sizes of common zeros to two polynomials in $Q_p[x, y]$, Chan [1] who estimates the cardinality $N(f, p^\alpha)$ of the set of solutions to congruence equations modulo a prime power and also Heng [2] who found a better estimate for the exponential sums under various conditions on the coefficients of $f(x, y)$. However, results for the polynomials of higher degrees are less complete.

Our approach entails the work developed by Mohd. Atan [7] who presented the p-adic Newton polyhedral method of finding the p-adic order of polynomials in $\Omega_p[x, y]$ which is an analogue of Newton polygon defined by Koblitz [3]. Sapar and Mohd. Atan [8] improved the result.

Sapar and Mohd. Atan [9] estimate a method of the p-adic sizes of common zeros of partial derivative polynomials associated with a quintic form for prime $p > 5$. Recently, Yap et al. [10] showed that the p-adic sizes of common zeros of partial derivative polynomials associated with a cubic form can be found explicitly on the overlapping segment of the indicator diagrams associated with the polynomials by using Newton polyhedron technique.

Our work involves application of the Newton polyhedron technique at the point of intersection in the combination of Indicator diagrams to determine explicitly the p-adic sizes of the component (ξ, η) a common root of partial derivative polynomials of $f(x, y)$ in $Q_p[x, y]$ of complete cubic form.

2 p-adic Orders of Zeros of a Polynomial

Sapar and Mohd. Atan [8] proved that for every point of intersection of the indicator diagrams, there exist common zeros of both polynomials in $Q_p[x, y]$ whose p-adic orders correspond to point (μ_1, μ_2) as mentioned in Theorem 1 below:

Theorem 1 *Let p be a prime. Suppose f and g are polynomials in $Q_p[x, y]$. Let (μ_1, μ_2) be a point of intersection of the indicator diagrams associated with f and g at the vertices or simple points of intersections. Then, there are ξ and η in Ω_p^2 satisfying $f(\xi, \eta) = g(\xi, \eta) = 0$ and $ord_p\xi = \mu_1$, $ord_p\eta = \mu_2$.*

Our investigation concentrates on the p-adic sizes of common zeros of partial derivative associated with a polynomial $f(x, y) = ax^3 + bx^2y + cxy^2 + dy^3 + \frac{3}{2}ax^2 + bxy + \frac{1}{2}cy^2 + sx + ty + k$. First, we have the following assertion:

Lemma 1 *Let $p > 3$ be a prime, a, b, c and d in Q_p and λ_1, λ_2 zeros of $k(\lambda) = (3bd - c^2)\lambda^2 + (9ad - bc)\lambda + (3ac - b^2)$. Let $\alpha_1 = \frac{b+\lambda_1 c}{3a+\lambda_1 b}$ and $\alpha_2 = \frac{b+\lambda_2 c}{3a+\lambda_2 b}$.*

Therefore,

$$(\alpha_1 - \alpha_2) = \frac{-\sqrt{\Delta}}{(3ac - b^2)},$$

where $\Delta = (9ad - bc)^2 - 4(3bd - c^2)(3ac - b^2)$.

Proof Let $\lambda_i = \dfrac{-(9ad - bc) \pm \sqrt{(9ad - bc)^2 - 4(3bd - c^2)(3ac - b^2)}}{2(3bd - c^2)}$.

Given that $\alpha_i = \frac{b+\lambda_i c}{3a+\lambda_i b}$, i = 1, 2. It can be shown that

$$\alpha_1 - \alpha_2 = \frac{(\sqrt{(9ad - bc)^2 - 4(3bd - c^2)(3ac - b^2)})(3ac - b^2)}{(3bd - c^2)(3a + \lambda_1 b)(3a + \lambda_2 b)} \tag{1}$$

By simplifying Eq. (1), we obtain

$$\alpha_1 - \alpha_2 = \frac{-\sqrt{\Delta}}{(3ac - b^2)},$$

where $\Delta = (9ad - bc)^2 - 4(3bd - c^2)(3ac - b^2)$ as asserted. □

Throughout the following discussion,

$$\alpha_1 = \frac{b + \lambda_1 c}{3a + \lambda_1 b}, \quad \alpha_2 = \frac{b + \lambda_2 c}{3a + \lambda_2 b} \tag{2}$$

where λ_1, λ_2 are the zeros of $k(\lambda) = (3bd - c^2)\lambda^2 + (9ad - bc)\lambda + (3ac - b^2)$.$\alpha_1 \neq \alpha_2$ since $\lambda_1 \neq \lambda_2$.

Lemma 2 *Suppose* (U, V) *in* Ω_p^2. *Let* $p > 3$ *be a prime, a, b, c and d coefficients of* α_1 *and* α_2 *as in (2) in* Q_p. *Therefore*

$$\alpha_1 V - \alpha_2 U = \frac{[(9ad - bc)(V - U) - \sqrt{\Delta}(V + U)]}{2(3ac - b^2)}$$

where $\Delta = (9ad - bc)^2 - 4(3bd - c^2)(3ac - b^2)$.

Proof We have

$$\alpha_1 V - \alpha_2 U = \left(\frac{b + \lambda_1 c}{3a + \lambda_1 b}\right) V - \left(\frac{b + \lambda_2 c}{3a + \lambda_2 b}\right) U \tag{3}$$

Suppose

$$\lambda_1 = \frac{-(9ad - bc) + \sqrt{\Delta}}{2(3bd - c^2)} \text{ and } \lambda_2 = \frac{-(9ad - bc) - \sqrt{\Delta}}{2(3bd - c^2)}.$$

where $\Delta = (9ad - bc)^2 - 4(3bd - c^2)(3ac - b^2)$.

By substituting λ_1, λ_2 in Eq. (3), it can be shown that

$$\alpha_1 V - \alpha_2 U = \frac{[(9ad - bc)(V - U) - \sqrt{\Delta}(V + U)]}{2(3ac - b^2)}$$

where $\Delta = (9ad - bc)^2 - 4(3bd - c^2)(3ac - b^2)$ as asserted. □

Lemma 3 *Suppose* (U, V) *in* Ω_p^2 *with* $U = x + \alpha_1 y$ *and* $V = x + \alpha_2 y$ *where* α_1 *and* α_2 *as in (2). Let* $p > 3$ *be a prime, a, b, c and d coefficients of* α_1 *and* α_2*, in* Q_p *and* $ord_p bc > ord_p ad$*. Then,* $ord_p x \geq W$ *or* $ord_p x \geq W - \frac{1}{2}\varepsilon$ *and* $ord_p y \geq W + ord_p \frac{a}{b}$ *or* $ord_p y \geq W + ord_p \frac{a}{b} - \frac{1}{2}\varepsilon$ *or* $ord_p y \geq W + ord_p \frac{b}{c}$ *or* $ord_p y \geq W + ord_p \frac{b}{c} - \frac{1}{2}\varepsilon$ *in an exceptional case with* $W = \min\{ord_p V, ord_p U\}$ *and some* $\varepsilon \geq 0$ *which can be specified explicitly.*

Proof From $U = x + \alpha_1 y$ and $V = x + \alpha_2 y$, we have

$$x = \frac{\alpha_1 V - \alpha_2 U}{(\alpha_1 - \alpha_2)}, \quad y = \frac{V - U}{(\alpha_1 - \alpha_2)} \tag{4}$$

Thus, by substituting Lemma 1 and Lemma 2 into (4), we have

$$ord_p x = ord_p[(9ad - bc)(V - U) - \sqrt{\Delta}(V + U)] - ord_p \sqrt{\Delta} \tag{5}$$

$$ord_p y = ord_p(V - U) + ord_p(3ac - b^2) - ord_p \sqrt{\Delta} \tag{6}$$

where $\Delta = (9ad - bc)^2 - 4(3bd - c^2)(3ac - b^2)$.

From (5), we have to consider two cases.

Case I: $\min\{ord_p(9ad - bc)(V - U), ord_p \sqrt{\Delta}(V + U)\} = ord_p \sqrt{\Delta}$
Case II: $\min\{ord_p(9ad - bc)(V - U), ord_p \sqrt{\Delta}(V + U)\} = ord_p(9ad - bc)(V - U)$

For Case I, it is easy to prove that $ord_p x \geq W$, where $W = \min\{ord_p V, ord_p U\}$. For Case II, by substituting Δ, we have

$$\begin{aligned} ord_p x = {} & ord_p(V - U) + ord_p(9ad - bc) \\ & - ord_p \sqrt{(9ad - bc)^2 - 4(3bd - c^2)(3ac - b^2)} \end{aligned}$$

$$\begin{aligned} ord_p y = {} & ord_p(V - U) + ord_p(3ac - b^2) \\ & - ord_p \sqrt{(9ad - bc)^2 - 4(3bd - c^2)(3ac - b^2)}. \end{aligned}$$

For the above two equations, we will consider another two cases:

(i) $ord_p(9ad - bc)^2 \neq ord_p 4(3bd - c^2)(3ac - b^2)$
(ii) $ord_p(9ad - bc)^2 = ord_p 4(3bd - c^2)(3ac - b^2)$.

Since $ord_p bc > ord_p ad$, $p > 3$ and by considering all cases, it can be proved that

$$ord_p x \geq W \text{ or } ord_p x \geq W - \frac{1}{2}\varepsilon \text{ and } ord_p y \geq W + ord_p \frac{a}{b} \text{ or}$$

$$ord_p y \geq W + ord_p \frac{a}{b} - \frac{1}{2}\varepsilon \text{ or } ord_p y \geq W + ord_p \frac{b}{c} \text{ or } ord_p y \geq W + ord_p \frac{b}{c} - \frac{1}{2}\varepsilon$$

where $W = \min\{ord_p \mathrm{V}, ord_p \mathrm{U}\}$ and $\varepsilon = ord_p(m - n)$ as asserted. □

Lemma 4 *Suppose (U, V) in Ω_p^2 and $U = x + \alpha_1 y$, $V = x + \alpha_2 y$ where α_1 and α_2 as in (2). Let $p > 3$ be a prime, a, b, c, d, s and t in Q_p, $ord_p bc > ord_p ad$, $\delta = \max\{ord_p a, ord_p b, ord_p c, ord_p d\}$ and $ord_p s, ord_p t \geq \alpha > 2\delta$. If $ord_p U = ord_p \frac{s+\lambda_1 t}{3a+\lambda_1 b}$ and $ord_p V = ord_p \frac{s+\lambda_2 t}{3a+\lambda_2 b}$ then $ord_p x \geq \alpha - \delta$ or $ord_p x \geq \alpha - \delta - \frac{1}{2}\varepsilon$ and $ord_p y \geq \alpha - \delta$ or $ord_p y \geq \alpha - \delta - \frac{1}{2}\varepsilon$ or $ord_p y \geq \alpha - 2\delta$ or $ord_p y \geq \alpha - 2\delta - \frac{1}{2}\varepsilon$ for some $\varepsilon \geq 0$.*

Proof Since $U = x + \alpha_1 y$, $V = x + \alpha_2 y$, $ord_p bc > ord_p ad$ and from Lemma 3, we have

$$ord_p x \geq W \text{ or } ord_p x \geq W - \frac{1}{2}\varepsilon \text{ and } ord_p y \geq W + ord_p \frac{a}{b} \text{ or } ord_p y \geq W + ord_p \frac{a}{b} - \frac{1}{2}\varepsilon \text{ or } ord_p y \geq W + ord_p \frac{b}{c} \text{ or } ord_p y \geq W + ord_p \frac{b}{c} - \frac{1}{2}\varepsilon$$

where $W = \min\{ord_p V, ord_p U\}$ and $\varepsilon = ord_p(m-n)$. Now,

$$ord_p U = ord_p \frac{s+\lambda_1 t}{3a+\lambda_1 b}, ord_p V = ord_p \frac{s+\lambda_2 t}{3a+\lambda_2 b}. \tag{7}$$

Since $ord_p bc > ord_p ad$, $p > 3$, $\delta = \max\{ord_p a, ord_p b, ord_p c, ord_p d\}$ and $ord_p s$, $ord_p t \geq \alpha > 2\delta$, it can be proved that

$$ord_p x \geq \alpha - \delta \text{ or } ord_p x \geq \alpha - \delta - \frac{1}{2}\varepsilon \text{ and } ord_p y \geq \alpha - \delta \text{ or }$$
$$ord_p y \geq \alpha - \delta - \frac{1}{2}\varepsilon \text{ or } ord_p y \geq \alpha - 2\delta \text{ or } ord_p y \geq \alpha - 2\delta - \frac{1}{2}\varepsilon$$

for some $\varepsilon \geq 0$ as asserted. □

The next theorem gives the p-adic sizes of common zeros of partial derivative polynomials associated with a polynomial $f(x, y)$ in $Q_p[x, y]$, in terms of the coefficients of its dominant terms.

Theorem 2 *Let $f(x, y) = ax^3+bx^2y+cxy^2+dy^3+\frac{3}{2}ax^2+bxy+\frac{1}{2}cy^2+sx+ty+k$ be a polynomial in $Q_p[x, y]$ with $p > 3$. Let $\alpha > 0$, $\delta = \max\{ord_p a, ord_p b, ord_p c, ord_p d\}$ and $ord_p bc > ord_p ad$. If $ord_p f_x(0,0), ord_p f_y(0,0) \geq \alpha > 2\delta$ then there exists (ξ, η) in Ω_p^2 such that $f_x(\xi, \eta) = 0$, $f_y(\xi, \eta) = 0$ and $ord_p \xi \geq \alpha - \delta$ or $ord_p \xi \geq \alpha - \delta - \frac{1}{2}\varepsilon$ and $ord_p \eta \geq \alpha - \delta$ or $ord_p \eta \geq \alpha - \delta - \frac{1}{2}\varepsilon$ or $ord_p \eta \geq \alpha - 2\delta$ or $ord_p \eta \geq \alpha - 2\delta - \frac{1}{2}\varepsilon$ for some $\varepsilon \geq 0$.*

Proof Let $g = f_x$ and $h = f_y$ and λ be a constant where,

$(g + \lambda h)(x, y) = (3a + \lambda b)x^2 + (2b + 2\lambda c)xy + (c + 3\lambda d)y^2 + (3a + \lambda b)x + (b + \lambda c)y + (s + \lambda t)$.

That is,

$$\frac{(g + \lambda h)(x, y)}{3a + \lambda b} = x^2 + \left(\frac{2b + 2\lambda c}{3a + \lambda b}\right)xy + \left(\frac{c + 3\lambda d}{3a + \lambda b}\right)y^2 + x + \left(\frac{b + \lambda c}{3a + \lambda b}\right)y + \frac{s + \lambda t}{3a + \lambda b}. \tag{8}$$

By factoring Eq. (8), we have

$$\frac{(g+\lambda h)(x,y)}{3a+\lambda b}=\left(x+\frac{b+\lambda c}{3a+\lambda b}y\right)^2+\left(x+\frac{b+\lambda c}{3a+\lambda b}y\right)+\left(\frac{s+\lambda t}{3a+\lambda b}\right). \quad (9)$$

if $\frac{c+3\lambda d}{3a+\lambda b}-\left(\frac{b+\lambda c}{3a+\lambda b}\right)^2=0$. Thus, we have

$$(3bd-c^2)\lambda^2+(9ad-bc)\lambda+(3ac-b^2)=0. \quad (10)$$

Let λ_1, λ_2 be the zeros of the Eq. (10) whose expressions are given in Lemma 1. $\lambda_1 \neq \lambda_2$ since $ord_p bc > ord_p ad$ implies $bc \neq ad$. Now, let

$$U = x+\frac{b+\lambda_1 c}{3a+\lambda_1 b}y \quad (11)$$

$$V = x+\frac{b+\lambda_2 c}{3a+\lambda_2 b}y \quad (12)$$

$$F(U,V)=(g+\lambda_1 h)(x,y) \quad (13)$$

$$G(U,V)=(g+\lambda_2 h)(x,y) \quad (14)$$

Substituting U and V in (9) for i = 1, 2 gives the following polynomials in (U,V):

$$F(U,V)=(3a+\lambda_1 b)U^2+(3a+\lambda_1 b)U+(s+\lambda_1 t) \quad (15)$$

$$G(U,V)=(3a+\lambda_2 b)V^2+(3a+\lambda_2 b)V+(s+\lambda_2 t). \quad (16)$$

The combination of the indicator diagrams associated with the Newton polyhedron of (15) and (16) is as follows:

From Fig. 1 and by Theorem 1, there exists $(\hat{U}\hat{V})$ in Ω_p^2 such that $F(\hat{U}\hat{V})=0, G(\hat{U}\hat{V})=0$ and $ord_p\hat{U}=\mu_1$, $ord_p\hat{V}=\mu_2$ with $\mu_1 = ord_p\left(\frac{s+\lambda_1 t}{3a+\lambda_1 b}\right)$ and $\mu_2 = ord_p\left(\frac{s+\lambda_2 t}{3a+\lambda_2 b}\right)$.

Let $U=\hat{U}$ and $V=\hat{V}$ in (11) and (12). Thus, there exists (x_0, y_0) in Ω_p^2 such that

$$\hat{U}=x_0+\alpha_1 y_0, \quad \hat{V}=x_0+\alpha_2 y_0.$$

More specifically, x_0, y_0 are given by

$$x_0=\frac{\alpha_1 V-\alpha_2 U}{(\alpha_1-\alpha_2)}, \quad y_0=\frac{V-U}{(\alpha_1-\alpha_2)}$$

with $\alpha_1=\frac{b+\lambda_1 c}{3a+\lambda_1 b}, \alpha_2=\frac{b+\lambda_2 c}{3a+\lambda_2 b}$ and λ_1, λ_2 the zeros of $k(\lambda)=(3bd-c^2)\lambda^2+(9ad-bc)\lambda+(3ac-b^2)$, and $\alpha_1 \neq \alpha_2$ since $\lambda_1 \neq \lambda_2$. Since $ord_p\hat{U}=ord_p\frac{s+\lambda_1 t}{3a+\lambda_1 b}$ and $ord_p\hat{V}=ord_p\frac{s+\lambda_2 t}{3a+\lambda_2 b}$, $ord_p x_0 \geq \alpha-\delta$ or $ord_p x_0 \geq \alpha-\delta-\frac{1}{2}\varepsilon$ and $ord_p y_0 \geq \alpha-\delta$

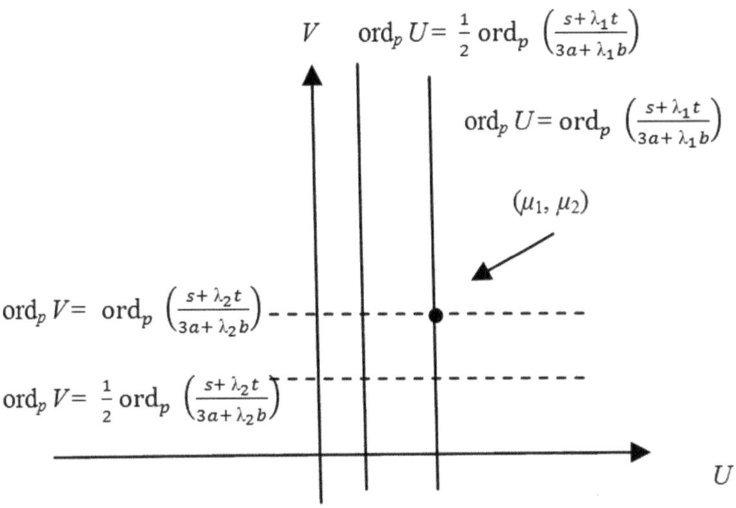

Fig. 1 The Indicator diagrams associated with F(U, V) (bold line) and G(U, V) (broken line).

or $ord_p y_0 \geq \alpha - \delta - \frac{1}{2}\varepsilon$ or $ord_p y_0 \geq \alpha - 2\delta$ or $ord_p y_0 \geq \alpha - 2\delta - \frac{1}{2}\varepsilon$. Let $x_0 = \xi$ and $y_0 = \eta$. Since $F(\hat{U}\hat{V}) = 0$ and $G(\hat{U}\hat{V}) = 0$, by back substitution in (15) and (16) we would have $g(\xi, \eta) = f_x(\xi, \eta) = 0$ and $h(\xi, \eta) = f_y(\xi, \eta) = 0$.

Thus, $ord_p\xi \geq \alpha - \delta$ or $ord_p\xi \geq \alpha - \delta - \frac{1}{2}\varepsilon$ and $ord_p\eta \geq \alpha - \delta$ or $ord_p\eta \geq \alpha - \delta - \frac{1}{2}\varepsilon$ or $ord_p\eta \geq \alpha - 2\delta$ or $ord_p\eta \geq \alpha - 2\delta - \frac{1}{2}\varepsilon$, where (ξ, η) is a common zero of f_x and f_y, $\delta = \max\{ord_p a, ord_p b, ord_p c, ord_p d\}$ for some $\varepsilon \geq 0$. □

3 Conclusion

If $p > 3$ with coefficients in Q_p such that $\alpha > 0$, $\delta = \max\{ord_p a, ord_p b, ord_p c, ord_p d\}$ and $ord_p bc > ord_p ad$. If $ord_p f_x(0,0), ord_p f_y(0,0) \geq \alpha > 2\delta$ then there exists (ξ, η) in Ω_p^2 such that $f_x(\xi, \eta) = 0$, $f_y(\xi, \eta) = 0$ and $ord_p\xi \geq \alpha - \delta$ or $ord_p\xi \geq \alpha - \delta - \frac{1}{2}\varepsilon$ and $ord_p\eta \geq \alpha - \delta$ or $ord_p\eta \geq \alpha - \delta - \frac{1}{2}\varepsilon$ or $ord_p\eta \geq \alpha - 2\delta$ or $ord_p\eta \geq \alpha - 2\delta - \frac{1}{2}\varepsilon$ for some $\varepsilon \geq 0$. The p-adic sizes of common zeros that we obtained in this project can be used to find the cardinality V. Through cardinality, exponential sums can also be defined.

Acknowledgement We would like to acknowledge with thanks the financial support from Graduate Research Fellowship Grant of Universiti Putra Malaysia that has enabled us to carry out this research.

References

1. Chan, K.L., Mohd Atan, K.A.: On the estimate to solutions of congruence equations associated with a quartic form. J. Phys. Sci. **8**, 21–34 (1997)
2. Heng, S.H., Mohd Atan, K.A.: An estimation of exponential sums associated with a cubic form. J. Phys. Sci. **10**, 1–21 (1999)
3. Koblitz, N.: p-adic Numbers, p-adic Analysis, and Zeta-Functions. Springer-Verlag, New York (1977)
4. Loxton, J.H., Smith, R.A.: Estimate for multiple exponential sums. J. Aust. Math. Soc. **33**, 125–134 (1982)
5. Loxton, J.H., Vaughan, R.C.: The estimate of complete exponential sums. Canad. Math. Bull. **28**(4), 440–454 (1985)
6. Mohd Atan, K.A.: Newton polyhedral method of determining p-adic orders of zeros common to two polynomials in Qp [x, y]. Pertanika. **9**(3), 375–380 (1986)
7. Mohd Atan, K.A., Loxton, J.H.: Newton polyhedra and solutions of congruences. In: Loxton, J.H., Poorten, A. Van der. (eds.) In Proceeding of Diophantine Analysis, pp. 67–82. Cambridge University Press (1986)
8. Sapar, S.H., Mohd Atan, K.A.: Estimate for the cardinality of the set of solution to congruence equations. J. Technol. **36**(C), 13–40 (2002)
9. Sapar, S.H., Mohd Atan, K.A.: A method of estimating the p-adic sizes of common zeros of partial derivative polynomials associated with a quintic form. World Sci. **5**, 541–554 (2009)
10. Yap, H.K., Sapar, S.H., Mohd Atan, K.A.: Estimation of p-adic sizes of common zeros of partial derivative associated with a cubic form. Sains Malaysiana. **40**(8), 921–926 (2011)

Feynman Graph Representation to Stochastic Differential Equations Driven by Lévy Noise

Boubaker Smii

Abstract Stochastic differential equations driven by Lévy noise are intensively studied. But so far there seems to be no recipe to find out what kind of noise it is, given the general structure of the equation. This can be obtained by recalling a graphical representation of the solution of the stochastic differential equations driven by Lévy noise. The graphs introduced are called generalized Feynman graphs and a numerical value will be assigned to each graph. Our graphs' formalism can be applied to different kinds of stochastic differential equations. As an example, a graphical representation of the generalized Ornstein–Uhlenbeck process will be given in this work.

1 Introduction and Motivation

We consider a stochastic differential equation (SDE):

$$\begin{cases} \frac{\partial}{\partial t} X_t(x) = -m X_t(x) + \eta(t), \ m > 0 \\ X_0(x) = X_0, \quad (t, X_0) \in [0, \infty] \times \mathbf{R}^d \end{cases} \tag{1}$$

where $\eta = \eta(t)$ is a Lévy noise.

Equation (1) is called generalized Ornstein–Uhlenbeck (OU) equation, which is studied in [4] and others, but so far only few results, e.g., [5], find out the type of the noise which gives the general structure of Eq. (1).

The construction of Lévy noise η followed in this work is based on Bochner–Minlos theorem, See, e.g., [5]. However, by replacing the continuous time t by a discrete one, we will introduce the generalized random process, and therefore, we will give our approach for the construction of the Lévy noise.

A graphical calculus will be applied to the generalized OU-process in this paper. We will introduce a special types of graphs called rooted trees with two types of leaves, e.g. [5], [8], [9]. An analytic value will be assigned to each rooted trees, hence a graphical representation of the OU process will be given by a divergent series, but it can be given the meaning of asymptotic expansions, see,. e.g, [2], [3].

B. Smii (✉)
Department of Mathematics and Statistics,
King Fahd University of Petroleum and Minerals, Dhahran, 31261 Saudi Arabia
e-mail: boubaker@kfupm.edu.sa

A. Kilicman et al. (eds.), *International Conference on Mathematical Sciences and Statistics 2013*, DOI 10.1007/978-981-4585-33-0_22,

Before we go over to describe the contents of the present paper, let us mention that our study of SDEs of type (1) can be extended to other classes of nonlinear SDEs, e.g., for force F of gradient type, $F = \nabla X$, we obtain a nonlinear SDE. One can ask the same questions as before, and again, expansion into graphs is possible. Generalizing further, we pass from SDEs driven by Lévy noise to stochastic partial differential equations.

Graphical representations of the OU process seems to be of great importance. They can be used for modeling biological processes such as neuronal response, See, e.g., [1], [2], mathematical finance, See, e.g., [4]. Relations to certain problems in the study of statistics of processes described by Lévy noises should also be mentioned, see, e.g., [5], [6].

Let us now describe the single sections of the article:

The next section will be reserved to an approach of construction of Lévy noise.

In section 3, we recall a theorem which determines the solution of Eq. (1). The distribution of the generalized OU process will be given and the particular case when the noise η is of Gaussian type will be deduced.

Section 4 will be devoted to graphical representations. We will introduce rooted trees with two types of leaves, and then develop an algorithm which gives a numerical value to each rooted tree. The graphical representation of the generalized OU process will be then given.

2 Construction of the Lévy Noise

This section is devoted to the construction of the Lévy noise η. We will introduce the Lévy characteristic ψ. By taking the time to be discrete, the Fourier transform of the joint distribution of η will converge to a function that depends on ψ. The generalized random process will be given as well. Hence, by the use of the well-known Bochner–Minlos theorem, see, e.g., [5], we recall a theorem which gives the Lévy noise.

It is well known that an infinite divisible probability distribution P is a probability distribution that satisfies the existence of a probability distribution P_n such that $P = P_n^{(n)} = P_n * \cdots * P_n$ (n times).

Let P be an infinitely divisible probability distribution. By Lévy–Khinchine theorem, e.g., [7], the Fourier transform or characteristic function of P satisfies: $C_P(\xi) = \int_{\mathbf{R}^d} e^{i\langle\zeta,\xi\rangle}\, dP(\zeta) = e^{\psi(\xi)}$, $\xi \in \mathbf{R}^d$, where ψ is the Lévy characteristic of P, see, e.g., [7], which is uniquely represented by:

$$\psi(t) = i\langle \mathrm{a}, \xi\rangle - \langle \xi, \mathrm{D}\xi\rangle + z\int_{\mathbf{R}^d\setminus\{0\}} \left(e^{i\langle\zeta,\xi\rangle} - 1\right) dM(\zeta), \quad \forall\, \xi \in \mathbf{R}^d, \tag{2}$$

with M satisfying $\int_{\backslash\mathbf{R}^d\setminus\{0\}} \min(1, \mathrm{s}^2)\mathrm{dM}(\zeta) < \infty$. Here $\mathrm{a} \in \backslash\mathrm{R}^d, \mathrm{z} >= 0$ and D a real and positive semi-definite $\mathrm{d} \times \mathrm{d}$ matrix.

We can explain the meaning of Eq. (2) as follows:
If $D = M = 0$, then $\psi(t) = i\langle a, \xi\rangle$, and the process $\{X(t),\ t \geq 0\}$ is simply deterministic motion. Here a is called the drift, and its physical interpretation is the velocity of this motion.

Now if only $M = 0$ so that $\psi(\xi) = i\langle a, \xi\rangle - \langle \xi, D\xi\rangle$, and therefore, $C_P(\xi) = e^{[i\langle a,\xi\rangle - 1\langle \xi, D\xi\rangle]}$, this is a characteristic function of a Gaussian random variable $X(t)$ having mean $\langle a, \xi\rangle$ and covariance $D\xi$. The process $\{X(t),\ t \geq 0\}$ in this case is known as Brownian motion, e.g., [7].

The last case is when $a \neq 0$, $D \neq 0$, and $M \neq 0$. In this case the process $\{X(t),\ t \geq 0\}$ can be represented by $X(t) = \langle b, \xi\rangle + \sqrt{c}\, B(t) + N(t)$, where b is the drift, $c > 0$, $B(t)$ the Brownian motion, and $N(t)$ will be recognized as a Poisson process with intensity λ taking values in $\{nh;\ h > 0,\ n \in \mathbf{N}\}$.

In the following, we denote by F the probability distribution on $\mathbf{R}_+$ of a given stochastic process.

Definition 1 A generalized stochastic process $\eta(t)$ is a stochastic process such that the following proprieties are satisfied:

- $F(\eta(t)) = F(\eta(s)),\ \forall t, s \geq 0$
- $\eta(t)$ is independent of $\eta(s)$, i.e., $F(\eta(t)\eta(s)) = F(\eta(t)) \otimes F(\eta(s))$. Here $\otimes$ is the convolution product.

Now the stochastic process η will be taken to be of general type, i.e, a generalized stochastic process.

In order to construct a generalized Lévy noise, let us start with a model where the continuous time $\{t \in \mathbf{R}_+\}$ is replaced by discrete time $\{t \in \frac{1}{n}\mathbf{N}\}$. In this case, one can model the noise $\{\eta^{(n)}(t)\}_{t \in \frac{1}{n}\mathbf{N}}$ by a collection of independently and i.i.d. random variables such that the "global random fluctuations" that the noise gives should not depend on the lattice scale $\frac{1}{n}$ fixed in our model, i.e., to model the noise independently of the lattice scale, we need that:

$$\begin{aligned} F(\eta^{(1)}(1)) &= F\left(\eta^{(n)}\left(\frac{1}{n}\right) + \eta^{(n)}\left(\frac{2}{n}\right) + \cdots + \eta^{(n)}\left(\frac{n}{n}\right)\right) \\ &= \left(F(\eta^{(n)}(1))\right)^{*n}, \end{aligned} \tag{3}$$

where $F(\eta^{(1)}(1))$ is the random fluctuation of the unit interval.

Lemma 1 *The Fourier transform of $\eta^{(n)}(t)$ is given by*

$$\mathcal{F}(\eta^{(n)})(f)(t) = e^{\frac{\psi(f(t))}{n}}, \quad f(t) \in \mathbf{R}^d \tag{4}$$

Proof for the proof we refer to Sato, see [7] □

Proposition 1 *The Fourier transform of the joint distribution of $\eta^{(n)}\left(\frac{1}{n}\right), \eta^{(n)}\left(\frac{2}{n}\right)$, ... converges to $e^{\int_{\mathbf{R}_+} \psi(f(t))\,dt}$, as $n \longrightarrow \infty$.*

Proof We have:

$$\mathcal{F}\left(\eta^{(n)}\left(\frac{1}{n}\right).\,\eta^{(n)}\left(\frac{2}{n}\right)\ldots.\ldots\right)\left(f\left(\frac{1}{n}\right).\,f\left(\frac{2}{n}\right)\ldots.\ldots\right)=\prod_{t\in\frac{1}{n}\mathbf{N}} e^{\frac{\psi(f(t))}{n}}=\exp\left(\sum_{t\in\frac{1}{n}\mathbf{N}}\frac{\psi(f(t))}{n}\right). \tag{5}$$

It is now easy to see that the last expression converges to $e^{\int_{\mathbf{R}_+}\psi(f(t))\,dt}$, as $n \longrightarrow \infty$.
□

Now it turns out that, we can not define clearly $\eta(t)$ as a random function of t, since the limit:

$$F\left((\eta^{(n)}(1))\right)=\delta_0 \tag{6}$$

shows that the fluctuations of the random variable $\eta^{(n)}(t)$ vanish for $n \longrightarrow \infty$.

The solution of this problem lies in the fact that $\eta(t)$ is not a stochastic process in the classical sense, but rather a distribution function in t. In the following we will present our approach for the construction for the Lévy noise η.

Let $\mathcal{S}(\mathbf{R}^d)$ be the Schwartz space of all rapidly decreasing functions on $\mathbf{R}^d$ endowed with the Schwartz topology. Its topological dual is the space of tempered distribution noted by $\mathcal{S}'(\mathbf{R}^d)$. We denote by $\langle .,.\rangle$ the dual pairing between $\mathcal{S}(\mathbf{R}^d)$ and $\mathcal{S}'(\mathbf{R}^d)$.

Definition 3 Let $(\Omega, \mathcal{B}, P)$ be a probability space. A generalized random process is a mapping:

$$\phi : (\Omega, \mathcal{B}, P) \longrightarrow \mathcal{S}'(\mathbf{R}, \mathbf{R}^d)$$

such that $\omega \longrightarrow \langle \phi(w), f\rangle$ is measurable and for $f_n \longrightarrow f$ in $\mathcal{S}(\mathbf{R}, \mathbf{R}^d)$, $F(\phi(f_n)) \longrightarrow F(\phi(f))$ in low, i.e.,

$$\begin{aligned}\lim_{n\longrightarrow\infty}\int_\Omega F(\phi(f_n))\,dP &= \lim_{n\longrightarrow\infty}\int_{\mathbf{R}^d} F(x)\,F(\phi(f_n))(x)\,dx\\ &= \int_{\mathbf{R}^d} F(x)\,F(\phi(f))(x)\,dx\\ &= \int_\Omega F(\phi(f))\,dP.\end{aligned}$$

Theorem 1 *(P. Lévy) Let X and X_n, $n \in N$ be stochastic processes. Then: $X_n \longrightarrow X$ in low if and only if $\mathcal{F}(X_n)(k) \longrightarrow \mathcal{F}(X)(k)$, $\forall\, k \in \mathbf{R}^d$.*

We note by $\mathcal{B}$ the σ-algebra generated by the cylinder sets of $\mathcal{S}'(\mathbf{R}^d)$. Then $(\mathcal{S}'(\mathbf{R}^d), \mathcal{B})$ is a measurable space.

We define a characteristic functional on $\mathcal{S}(\mathbf{R}^d)$, as a functional $\mathcal{C} : \mathcal{S}(\mathbf{R}^d) \longrightarrow \mathbf{C}$ such that:

1. $\mathcal{C}(0) = 1$.
2. $\mathcal{C}$ is continuous on $\mathcal{S}(\mathbf{R}^d)$.
3. $\mathcal{C}$ is positive-definite, i.e., $\forall z_1, ..., z_n \in \mathbf{C}, \forall n \in \mathbf{N}, f_1, ..., f_n \in \mathcal{S}(\mathbf{R}^d)$

$$\sum_{l,j=0}^{n} z_l \bar{z}_j \mathcal{C}(f_l - f_j) \geq 0 . \tag{7}$$

In the following, we choose $(\Omega, \mathcal{B}) = (\mathcal{S}', \mathcal{B}(\mathcal{S}'))$.

By the well-known Bochner–Minlos theorem, we know that for a given a characteristic functional $\mathcal{C}$, there exists a unique (up to equivalence in low) random process ϕ such that:

$$\mathcal{C}(f) = \mathbf{E}[e^{i\,\phi(f)}] = \int_{\mathcal{S}'} e^{i\,\phi(f)(\omega)}\, dP(\omega), \quad \forall\, f \in \mathcal{S}(\mathbf{R}, \mathbf{R}^d). \tag{8}$$

Theorem 2 *Let ψ be a Lévy characteristic given by the representation (2). Then there exists a unique measure η on $\mathcal{S}'$ such that:*

$$\mathcal{C}(f) = \int_{\mathcal{S}'} e^{i\,\langle \omega, f\rangle}\, d\eta(\omega) = \exp\left(\int_{\mathbf{R}^d} \psi(f(x))\, dx\right), \quad x \in \mathbf{R}^d. \tag{9}$$

Proof It is easy to see that the right hand side of Eq. (9) is a characteristic equation. To conclude it suffices to apply Bochner–Minlos theorem. □

Definition 5 The measure η given by Theorem 2 is called white noise measure with Lévy characteristic ψ, and $(\mathcal{S}'(\mathbf{R}^d), \mathcal{B}, \eta)$ will be the generalized white noise space associated with ψ. The associated coordinate process:

$$\eta : S(\mathbf{R}^d) \times (S'(\mathbf{R}^d), \mathcal{B}, \eta) \longrightarrow \mathbf{C}, \;\; \eta(f, \xi) = \langle f, \xi\rangle \;\forall f \in \mathcal{S},\; \xi \in \mathcal{S}', \tag{10}$$

is called a Lévy noise.

3 Generalized Ornstein–Uhlenbeck Processes

In this section, the generalized stochastic process η will be taken as a Lévy noise in the sense of Definition 5.

When there is no confusion, we denote for simplicity $X(t)$ by X_t.

Definition 6 The stochastic process $(X_t)_{t\geq 0}$ verifying Eq. (1), where η is a Lévy noise is called the generalized Ornstein–Uhlenbek (OU) process.

Theorem 3 *The solution of the generalized OU Eq. (1) is given by:*

$$X_t(x) = X_0\, e^{-m\,t} + \int_0^t e^{-m\,(t-s)}\, \eta(s)\, ds \tag{11}$$

Proof The solution of the homogeneous equation associated with Eq. (1) is given by $X_{t,C} = C\, e^{-mt}$. Now by the method of variation of parameters, we get:

$$X_t(x) = \int_0^t e^{-m\,(t-s)}\, \eta(s)\, ds + C\, e^{-mt} \tag{12}$$

The result now follows by taking $X_0(x) = C = X_0$. □

Definition 7 Let $(\Omega, \mathcal{B}, P)$ be a probability space, and $\mathbf{F}$ be the distribution of a Lévy OU process. The Fourier transform (or characteristic function) of the Lévy OU process X_t is given by:

$$\begin{aligned} \mathcal{F}(X_t)(x) = \langle\, e^{i\,<X_t,x>} \rangle_P \quad &= \int_\Omega e^{i\,<X_t(w),x>}\, dP(w) \\ &= \int_{\mathbf{R}^d} e^{i<y,x>}\, d\mathbf{F}(y) \end{aligned} \tag{13}$$

Lemma 1 *The Fourier transform of the Lévy OU process X_t is represented as*

$$\mathbf{E}(e^{i\langle\, X_t, \xi\rangle}) = e^{t\,\psi(\xi)}, \quad \forall t > 0, \quad \xi \in \mathbf{R}^d, \tag{14}$$

where ψ is the characteristic function given by Eq. (2).

Proof For the proof we refer to Sato, see, [7]. □

Lemma 2 *Let $\{\eta_t,\ t \geq 0\}$ be a Lévy process on $\boldsymbol{R}^d$. Then there exists a temporally homogeneous transition function $P_t(x, .)$ on $\boldsymbol{R}^d$ such that:*

$$\int_{\mathbf{R}^d} e^{i\langle z,y\rangle}\, P_t(x,y)dy = \exp\left[i\, e^{-mt}\langle x, a\rangle + \int_0^t \psi\left(e^{-ms}\, a\right) ds \right], \quad a \in \mathbf{R}^d \tag{15}$$

Proof For the proof we refer to Sato, see, [[7], p. 105]. □

Let X_t^j, j standing for jumps, be the Lévy OU process associated to jumps only, i.e., X_t^j satisfy the stochastic differential equation:

$$\begin{cases} \frac{\partial}{\partial t} X_t^j(x) = \eta_t^j(x),\ t > 0 \\ X_0^j(x) = 0 \end{cases} \tag{16}$$

where η_t^j is the Lévy noise associated to the jumps and corresponding to the third term on the r.h.s of Eq. (2).

Let P_t^j be the transition probabilities of the X_t^j, then P_t^j satisfy:

$$\int_{\mathbf{R}^d} e^{i\langle z,y\rangle}\, P_t^j(x,y)dy = \exp\left[\int_0^t \tilde{\psi}(e^{-ms}\, x)ds \right], \quad x \in \mathbf{R}^d, \tag{17}$$

where

$$\tilde{\psi}(x) = z \int_{\mathbf{R}^d\setminus\{0\}} \left(e^{i\langle s,x\rangle} - 1\right) dM(s),\ z \geq 0. \tag{18}$$

Theorem 4 *The distribution* $\mathbf{F}$ *of the generalized OU process verify:*

$$\mathbf{F}(X_t) = \mathbf{F}\left(e^{-mt}X_0\right) * \mathbf{F}\left(\eta_{-t}\right) \tag{19}$$

Proof The Fourier transform of X_t is given by:

$$\begin{aligned}\mathcal{F}(X_t)(x) &= \exp\left(i\,\langle X_0,\, x\rangle e^{-mt} + \int_{\backslash \mathrm{R}^{\mathrm{d}}} \psi(\mathbf{1}_{s\leq t}e^{-m(t-s)}x)ds\right)\\ &= \exp\left(i\,\langle X_0,\, x\rangle e^{-mt} + \int_0^t \psi(e^{-m(t-s)}x)ds\right)\end{aligned}$$

□

Corollary 1

1. *Let* η *be a Gaussian noise, then the distribution of the OU process, known as "Mehler's formula," is given by:*

$$\mathbf{F}(X_t)(dx) = K_t\, \exp\left\{-\frac{1}{2}\frac{m}{1-e^{-2mt}}\langle\left(x-e^{-mt}X_0\right),\, D^{-1}\left(x-e^{-mt}X_0\right)\rangle\right\}dx \tag{20}$$

where $K_t = (det A)^{-\frac{1}{2}}\left(\frac{2\pi}{m}\left(1-e^{-2mt}\right)\right)^{-\frac{d}{2}}$ and $A = \left(\frac{D}{m}\left(1-e^{-2mt}\right)\right)^{-1}$.

2. *The distribution of Brownian motion is given by:*

$$\mathbf{F}(B_t)(dx) = N\,\exp\left\{-\frac{1}{4t}\langle(x-X_0),\, D^{-1}(x-X_0)\rangle\right\}dx \tag{21}$$

where $N = (det A)^{-\frac{1}{2}}(4\pi)^{-\frac{d}{2}}$.

Proof

1. If the noise η is taken to be Gaussian noise, then $\psi(\xi) = -\langle\xi,\, D\xi\rangle$. Thus:

$$\begin{aligned}\int_0^t \psi\left(e^{-m(t-s)}\xi\right)ds &= -\langle\xi,\,\mathrm{D}\xi\rangle\int_0^t e^{-2m(t-s)}ds\\ &= -\langle\xi,\, D\xi\rangle\frac{1}{2m}\left(1-e^{-2mt}\right)\end{aligned}$$

Hence by applying the Fourier integral, one obtains:

$$\begin{aligned}\mathbf{F}(X_t)(dx) &= \frac{\sqrt{det A}}{(2\pi)^{\frac{d}{2}}}\exp\left\{-\frac{1}{2}\langle(x-e^{-mt}X_0),\, A(x-e^{-mt}X_0)\rangle\right\}dx\\ &= K_t\,\exp\left\{-\frac{1}{2}\frac{m}{1-e^{-2mt}}\langle(x-e^{-mt}X_0),\, D^{-1}(x-e^{-mt}X_0)\rangle\right\}dx.\end{aligned}$$

2. If $m \longrightarrow 0$, the r.h.s. of Eq. (20) converges to

$$(det A)^{-\frac{1}{2}}(4\pi)^{-\frac{d}{2}}\,\exp\left\{-\frac{1}{4t}\langle(x-X_0),\, D^{-1}(x-X_0)\rangle\right\}dx, \tag{22}$$

which is the distribution of the Brownian motion. □

4 Graphical Representation of the Generalized OU Process

This section will be reserved to the graphical applications to the generalized OU process. To do this, we will introduce a special type of graph called rooted trees with two types of leaves. We start with some useful notations and standard definitions.

Let $G(t, x)$ be the Green function which satisfies :

$$\begin{cases} \frac{\partial G(t,x)}{\partial t} = -mG(t, x) + \delta(x), \\ G(t, \mathbf{x}) = 0, \quad t < 0 \end{cases} \tag{23}$$

Here $\delta(x)$ is the Dirac distribution.

Proposition 2 *Let $X_0(x) = X_0 = f(x)$, the analytic solution of the generalized OU Eq. (1) is given by:*

$$X_t = G * \eta(x) + G \star f(x) \tag{24}$$

Here $*$ and $\star$ are two convolution products acting on (t, x) and x respectively.

Proof From the generalized OU equation we obtain

$$\left(\frac{\partial}{\partial t} + m\right) X_t(x) = \eta(x) \tag{25}$$

Now the result holds by applying the Green function given by Eq. (23). □

Definition 8 A tree is a graph in which any two vertices are connected by exactly one simple path. In other words, any connected graph without cycles is a tree.[1] A tree is called a rooted tree if one vertex has been designated the root, in which case the edges have a natural orientation, towards or away from the root. Each vertex of a tree T that is connected with only one edge is called a leaf.

In our model, we consider a rooted tree T with two types of leaves. The set of all trees with two types of leaves and i inner vertices will be denoted by $\mathcal{T}(i)$.

The following algorithm named as "Feynman rule" gives a numerical value to each rooted tree with two types of leaves.

Definition 9 Let $\mathcal{R}(T, \eta)$ be the analytic value of a given rooted tree $T \in \mathcal{T}(i)$. Then $\mathcal{R}$ is obtained as follows:

1. Assign $r = (t, x)$ to the root of the tree T.
2. For each edge of the tree T multiply by G, where G is the Green function given by equation (23).
3. For each leaf of type one multiply by the noise η.
4. For each leaf of type two multiply by the initial condition f.
5. Integrate with respect to the Lebesgue measure dx.

For the convenience of the reader, definitions (8) and (9) are explained in Table 1 and Fig. 1.

[1] The various kinds of trees used as data structures in computer science are not really trees in this sense, but rather, types of ordered directed trees.

Table 1 Different types of vertices

Root	Inner vertices	Leaf 1	Leaf 2	Edge
$\times$	$\bullet$	$\diamond$	$\otimes$	$\longrightarrow$

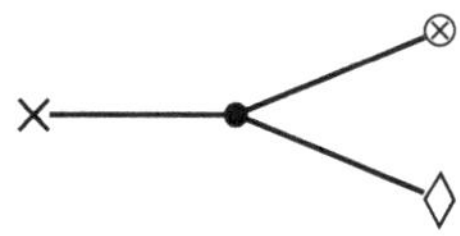

Fig. 1 Construction of a rooted tree with two types of leaves and one inner vertex

Theorem 5 *Let $T \in \mathcal{T}(i)$, $i \in N$. The solution of the generalized OU equation (1) is given by a sum over all rooted trees that are evaluated according to the rule given in definition (9), i.e.,*

$$X_t^i = \sum_{T \in \mathcal{T}(i)} \mathcal{R}(T, \eta) \tag{26}$$

Proof The proof is done by induction on i. For $i = 0$ we have:

$$\begin{aligned} X_t^i &= G * \eta(x) + G \star f(x) \\ &= \times\!-\otimes + \times - \diamond \\ &= \mathcal{R}(T_1, \eta) + \mathcal{R}(T_2, \eta) \end{aligned} \tag{27}$$

Suppose equation (26) is true for all $m < i$. Then:

$$\begin{aligned} X_t^i &= G * \sum_{i \geq 0} X_t^m(x, \eta) \\ &= G * \sum_{m \geq 0} \sum_{T \in \mathcal{T}(m)} \mathcal{R}(T, \eta) \\ &= \sum_{T \in \mathcal{T}(i)} \mathcal{R}(T, \eta) \end{aligned} \tag{28}$$

□

Remark The graphical representations introduced previously can be applied for more complicated OU equations, e.g., by replacing the r.h.s. of equation (1) by $-m\,X^p + \eta$, p $\in$ \N. In this case, our graph formalism still valid, and the rooted trees will contain more inner vertices. Hence, in the algorithm given by definition 9, one should integrate over all vertices.

Acknowledgements This work was supported by King Fahd University of Petroleum and Minerals under the project $\sharp\, IN101025$. The author gratefully acknowledges this support.

References

1. Albeverio, S., Dipersio, L., Mastrogiacomo, E., Smii, B.: Invariant measures for stochastic differential equations driven by Lévy noise. Preprint (2013)
2. Albeverio, S., Mastrogiacomo, E., Smii, B.: Small noise asymptotic expansions for stochastic PDE's driven by dissipative nonlinearity and Lévy noise. Stoch. Proc. Appl. **123**, 2084–2109 (2013)
3. Albeverio, S., Smii, B.: Asymptotic expansions for SDE's with small multiplicative noise. Preprint (2013)
4. Brandorff-Nielsen, O.E., Shephard, N.: Non-Gaussian Ornstein–Uhlenbeck-based models and some of their uses in financial economics. J. R. Stat. Soc. Ser. B Stat. Methodol. **63**(2), 167–241 (2001)
5. Gottschalk, H., Smii, B.: How to determine the law of the solution to a SPDE driven by a Lévy space-time noise. J. Math. Phys. **43**, 1–22 (2007)
6. Gottschalk, H., Smii, B., Thaler, H.: The Feynman graph representation of general convolution semigroups and its applications to Lévy statistics. Bernoulli **14**(2), 322–351 (2008)
7. Sato, K.-I.: Lévy Processes and Infinite Divisibility. Cambridge University Press, Cambridge (1999)
8. Smii, B.: A linked cluster theorem of the solution of the generalized Burger equation. Appl. Math. Sci. (Ruse) **6**(1), 21–38 (2012)
9. Smii, B.: A graphical representation of the truncated moments of a mixed noise. Preprint (2013)

Some Properties of the Concurrent Grammars

Gairatzhan Mavlankulov, Mohamed Othman, Mohd Hasan Selamat and Sherzod Turaev

Abstract Petri nets are becoming one of the most important mathematical tools in Computer Science. In this paper, we study some mathematical properties of concurrent grammars which are controlled by Petri nets under parallel firing strategies, where transitions of Petri nets fire simultaneously in different modes. We propose a notion of concurrent context-free grammar which is a similar case of the context-free Petri nets under parallel firing strategy, where parallel firing modes of context-free Petri nets were converted to rule applications in context-free grammars and we investigate their properties.

1 Introduction

The rapid evolution of computing, communication, network, and sensor technologies has brought about the proliferation of computer-integrated systems, mostly technological and often highly complex. These systems, indispensable for our modern life, include air traffic control systems; automated manufacturing systems; computer and communication networks; embedded and networked systems; and software systems. The challenge for the researchers and engineers is the planning, modeling, analysis, verification, control, scheduling, and control implementation. Petri nets are increasingly becoming one of the most important mathematical tools to handle the above problems. One of the fundamental approaches in this area is to consider Petri nets as language generators. If the transitions in a Petri net are labeled with a set of symbols, a sequence of transition firing generates a string of symbols. The set of strings generated by all possible firing sequences defines a language called a Petri net language. With different kinds of labeling functions and different kinds of final marking sets, various classes of Petri net languages were introduced and investigated by Hack [8] and Peterson [10].

G. Mavlankulov (✉) · M. Othman · M. H. Selamat
Faculty of Computer Science and Information Technology, Universiti Putra Malaysia, Seri Kembagan, Malaysia

S. Turaev
Kulliyyah of Information and Communication Technology, International Islamic University Malaysia, Gombak, Malaysia

A. Kilicman et al. (eds.), *International Conference on Mathematical Sciences and Statistics 2013*, DOI 10.1007/978-981-4585-33-0_23,

Recently, in [2], [3], [4], [5], [11] different variants of a Petri net controlled grammar were introduced, one of which is a context-free grammar equipped with a Petri net, whose transitions are labeled with rules of the grammar or the empty string, and the associated language consists of all terminal strings which can be derived in the grammar. The sequence of rules in every terminal derivation corresponds to some occurrence sequence of transitions of the Petri net which is enabled at the initial marking and finished at a final marking of the net. It can be considered as a mathematical model for the study of concurrent systems appearing in systems biology and automated manufacturing systems. The distinguished feature of all of these variants is that the transitions of a Petri net fire sequentially. The concept of maximal parallelity in Petri nets was studied by Burkhard in [1]. Another different viewpoint on the parallel firing of transitions in Petri nets was taken by Farwer, Kudlek and Rolke in [6], [7], where Turing Machines (called Concurrent Turing Machines) with Petri nets as a finite control were introduced. The variant of the Concurrent Finite Automation (CFA) has been defined and studied in [7]. In [9], some modes of firing transitions in Petri nets were compared and classes of languages specified by them were investigated. In this paper, we introduce a new variant of theoretical models for parallel computation using Petri nets under parallel firing strategies, called grammars controlled by Petri nets under parallel firing strategies (concurrent grammars), i.e., the transitions of a Petri net fire simultaneously in different modes.We investigate some properties of the concurrent context-free languages and compare them with other known grammars. For instance,we show some examples of concurrent context-free grammars which can generate non-context free languages. As noted, these languages cannot be generated by Petri Net controlled grammars in the sequential case.

2 Preliminaries

Let $\mathbb{N}$ be the set of all nonnegative integers and $\mathbb{N}^k$ be the set of all vectors of k nonnegative integers. The cardinality of a set X is denoted by $|X|$.

2.1 Grammars and Languages

Let Σ be an *alphabet* which is a finite nonempty set of symbols. A *string* over the alphabet Σ is a finite sequence of symbols from Σ. The *empty* string is denoted by λ. The set of all strings over the alphabet Σ is denoted by Σ^*. A subset of Σ^* is called a *language*. The *length* of a string w, denoted by $|w|$, is the number of occurrences of symbols in w. The number of occurrences of a symbol a in a string w is denoted by $|w|_a$.

A *context-free grammar* is a quadruple $G = (V, \Sigma, S, R)$ where V and Σ are the disjoint finite sets of *nonterminal* and *terminal* symbols, respectively, $S \in V$ is the *start* symbol, and $R \subseteq V \times (V \cup \Sigma)^*$ is a finite set of *(production) rules*.

Usually, a rule (A, x) is written as $A \to x$. A rule of the form $A \to \lambda$ is called an *erasing rule*. $x \in (V \cup \Sigma)^+$ *directly derives* $y \in (V \cup \Sigma)^*$, written as $x \Rightarrow y$, iff there is a rule $r = A \to \alpha \in R$ such that $x = x_1Ax_2$ and $y = x_1\alpha x_2$. The rule $r : A \to \alpha \in R$ is said to be *applicable* in sentential form x, if $x = x_1Ax_2$, where $x_1, x_2 \in (V \cup \Sigma)^*$. The reflexive and transitive closure of $\Rightarrow$ is denoted by $\Rightarrow^*$. A derivation using the sequence of rules $\pi = r_1r_2\cdots r_n$ is denoted by $\stackrel{\pi}{\Rightarrow}$ or $\stackrel{r_1r_2\cdots r_n}{\Longrightarrow}$. The *language* generated by G is defined by $L(G) = \{w \in \Sigma^* \mid S \Rightarrow^* w\}$. The family of context-free languages is denoted by $\mathcal{CF}$. Context-free grammar is called *linear* if all production rules have a form $R \subseteq V \times (\Sigma^*V\Sigma^* \cup \Sigma^*)$. The family of linear languages is denoted by $\mathcal{LIN}$.

2.2 Multisets

A *multiset* over an alphabet Σ is a mapping $\mu : \Sigma \to \mathbb{N}$. The set Σ is called the *basic set* of a multiset ν and the elements of Σ are called the *basic elements* of a multiset μ. A multiset μ over an alphabet $\Sigma = \{a_1, a_2, \ldots a_n\}$ can be denoted by

$$\mu = (\mu(a_1)a_1, \mu(a_2)a_2, \ldots, \mu(a_n)a_n)$$

where $\mu(a_i)$, $1 \leq i \leq n$, is the multiplicity of a_i, or as a vector

$$\mu = (\mu(a_1), \mu(a_2), \ldots, \mu(a_n)),$$

or as the set in which each basic element $a \in \Sigma$ occurs $\mu(a)$ times

$$\mu = \{\underbrace{a_1, \ldots, a_1}_{\mu(a_1)}, \underbrace{a_2, \ldots, a_2}_{\mu(a_2)}, \ldots, \underbrace{a_n, \ldots, a_n}_{\mu(a_n)}\}.$$

The empty multiset is denoted by ϵ, that is $\epsilon(a) = 0$ for all $a \in \Sigma$. The set of all multisets over Σ is denoted by $\Sigma^\oplus$. Since Σ is finite, $\Sigma^\oplus = \mathbb{N}^{|\Sigma|}$. The power (or cardinality) of a multiset $\mu = (\mu(a_1), \mu(a_2), \ldots, \mu(a_n))$ denoted by $|\mu|$, is $\sum_{i=1}^{n} \mu_i$. A multiset μ is a *set* if and only if $\mu(a) \leq 1$ for all $a \in \Sigma$.

For two multisets μ and ν over the same alphabet Σ, we define

- The *inclusion* $\mu \sqsubseteq \nu$ by

$$\mu \sqsubseteq \nu \text{ if and only if } \mu(a) \leq \nu(a) \text{ for all } a \in \Sigma;$$

- The *sum* $\mu \oplus \nu$ by

$$(\mu \oplus \nu)(a) = \mu(a) + \nu(a) \text{ for each } a \in \Sigma,$$

and we denote the sum of multisets $\mu_1, \mu_2, \ldots, \mu_k$ by $\bigoplus\sum_{i=1}^{k} \mu_i$, i.e.,

$$\bigoplus\sum_{i=1}^{k} \mu_i = \mu_1 \oplus \mu_2 \oplus \cdots \oplus \mu_k;$$

- The *difference* $\mu \ominus \nu$ by

$$(\mu \ominus \nu)(a) = \max\{0, \mu(a) - \nu(a)\} \text{ for each } a \in \Sigma.$$

2.3 Petri Nets

A Petri net is a triple (P, T, δ) where P and T are finite disjoint sets of *places* and *transitions*, respectively, a mapping $\delta : T \to P^{\oplus} \times P^{\oplus}$ is a mapping which assigns to each transition $t \in T$ a pair $\delta(t) = (\alpha, \beta)$. Graphically, a Petri net is represented by a bipartite directed graph with the node set $P \cup T$ where places are drawn as *circles* and transitions as *boxes*. For each transition $t \in T$ with $\delta = (\alpha, \beta)$, the multiplicities $\alpha(p)$, $\beta(p)$ of a place $p \in P$ give the number of arcs from p to t and from t to p, respectively. A multiset $\mu \in P^{\oplus}$ is called a *marking*. For each $p \in P$, $\mu(p)$ gives the number of *tokens* in p. Graphically, tokens are drawn as small solid *dots* inside circles.

A *place/transition net* (*p/t net* for short) is a quadruple $N = (P, T, \delta, \mu_0)$ where (P, T, δ) is a Petri net and $\iota \in P^{\oplus}$ is the *initial marking*.

A transition $t \in T$ with $\delta(t) = (\alpha, \beta)$ is *enabled* at a marking $\mu \in P^{\oplus}$ if and only if $\alpha \sqsubseteq \mu$. In this case, t can *occur* (*fire*). Its occurrence transforms the marking μ into the marking $\mu' \in P^{\oplus}$ defined by $\mu' = \mu \ominus \alpha \oplus \beta$. We write $\mu \xrightarrow{t}$ to denote that t may fire in μ, and $\mu \xrightarrow{t} \mu'$ to indicate that the firing of t in μ leads to μ'. A finite sequence $t_1 t_2 \cdots t_k$, $t_i \in T, 1 \leq i \leq k$, is called *an occurrence sequence* enabled at a marking μ and finished at a marking μ_k if there are markings $\mu_1, \mu_2, \ldots, \mu_{k-1}$ such that

$$\mu \xrightarrow{t_1} \mu_1 \xrightarrow{t_2} \ldots \xrightarrow{t_{k-1}} \mu_{k-1} \xrightarrow{t_k} \mu_k.$$

In short, this sequence can be written as $\mu \xrightarrow{t_1 t_2 \cdots t_k} \mu_k$ or $\mu \xrightarrow{\nu} \mu_k$ where $\nu = t_1 t_2 \cdots t_k$. For each $1 \leq i \leq k$, marking μ_i is called *reachable* from marking μ. $\mathcal{R}(N, \mu) \subseteq P^{\oplus}$ denotes the set of all reachable markings from a marking μ.

Let $N = (P, T, \delta, \iota)$ be a p/t net and $F \subseteq \mathcal{R}(N, \iota)$ be a set of markings which are called *final markings*. An occurrence sequence ν of transitions is called *successful* for F if it is enabled at the initial marking ι and finished at a final marking τ of F. If F is understood from the context, we say that ν is a *successful occurrence sequence*.

A *labeled Petri net* is a tuple $K = (\Delta, N, \gamma, F)$ where Δ is an alphabet, $N = (P, T, \delta, \iota)$ is a p/t net, $\gamma : T \to \Delta \cup \{\lambda\}$ is a transition labeling function, and $F \subseteq \mathcal{R}(N, \iota)$.

The labeling function γ is extended to occurrence sequences in a natural way, i.e., if $\nu t \in T^*$ is an occurrence sequence, then $\gamma(\nu t) = \gamma(\nu)\gamma(t)$ and $\gamma(\lambda) = \lambda$. For an occurrence sequence $\nu \in T^*$, $\gamma(\nu)$ is called a *label sequence*.

A *Petri net language* of K with respect to a transition labeling function γ and a final marking set F is defined by

$$L(K) = \{\gamma(\nu) \in \Delta^* \mid \iota \xrightarrow{\nu} \mu \text{ where } \nu \in T^* \text{ and } \mu \in F\}.$$

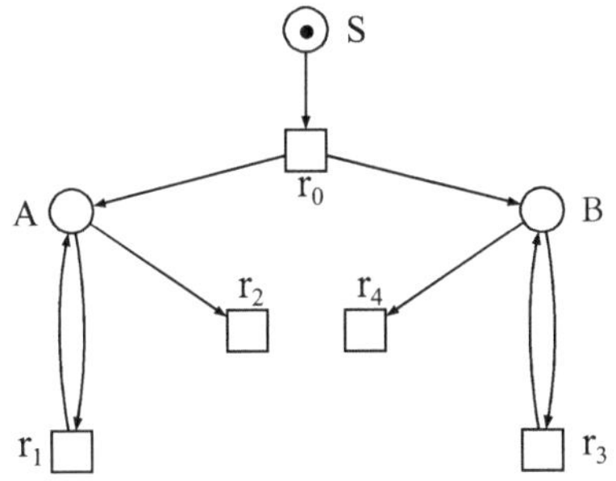

Fig. 1 A context-free Petri net

2.4 *Context-Free Petri Nets*

A context-free Petri net is a Petri net $N = (P, T, F, \phi, \beta, \gamma, \iota)$ where
• Labeling function $\beta : P \to V$ and $\gamma : T \to R$ are bijections;
• There is an arc from place p to transition t if and only if $\gamma(t) = A \to \alpha$ and $\beta(p) = A$. The weight of the arc (p, t) is 1;
• There is an arc from transition t to place p if and only if $\gamma(t) = A \to \alpha$ and $\beta(p) = \chi$ where $|\alpha|_\chi > 0$. The weight of the arc (t, p) is $|\alpha|_\chi$;
• The initial marking ι is defined by $\iota(b^{-1}(S)) = 1$ and $\iota(p) = 0$ for all $p \in P - \{\beta^{-1}(S)\}$.

Example 1 Let G_1 be a context-free grammar with the rules:

$$r_0 : S \to AB,\ r_1 : A \to aAb,\ r_2 : A \to ab,\ r_3 : B \to cB,\ r_4 : B \to c$$

(the other components of the grammar can be seen from these rules). Figure 1 illustrates a cf Petri net N_1 with respect to the grammar G_1. Obviously, $L(G_1) = \{a^n b^n c^m \mid n, m \geq 1\}$.

3 Definitions, Examples, Results

3.1 *Multisteps*

Definition 1 Let $G = (V, \Sigma, S, R)$ be context-free grammar. $K = (\Delta, N, \gamma, F)$, $N = (P, T, \delta, \iota)$ be a labeled Petri net such that $\Delta = R$. Let $A = \{t_1, t_2, \dots, t_k\} \subseteq T$ with $\delta(t_i) = (\alpha_i, \beta_i)$ for $1 \leq i \leq k$. The transitions of a multiset $\nu \in A^\oplus$ are *simultaneously/parallelly enabled/firable* at a marking $\mu \in \mathcal{R}(N, \iota)$ if and only if

$$\sum_{i=1}^{k} \nu(t_i)\alpha_i \sqsubseteq \mu.$$

Then, the transitions of ν *parallelly fire* resulting in the new marking μ' defined by

$$\mu' = \mu \ominus \sum_{i=1}^{k} \nu(t_i)\alpha_i \oplus \sum_{i=1}^{k} \nu(t_i)\beta_i.$$

A multiset ν whose transitions fire parallelly is called a *multistep*. We write $\mu \xrightarrow[m]{\nu} \mu'$ to denote that the a multistep ν at μ leads to μ'.

Let $X = \{t_1, t_2, \ldots, t_k\} \subseteq T$ with $t_i = (\alpha_i, \beta_i)$, $1 \leq i \leq k$, and let a multistep $\nu \in X^{\oplus}$ be enabled at a marking $\mu \in P^{\oplus}$. We will define some special types(modes) of multisteps with respect to the basic sets and multisets.

1. The multistep ν is said to be in k *mode* if $|\nu| = k$. Similarly, ν is called in $\leq k$ mode ($\geq k$ mode) if $|\nu| \leq k(|\nu| \geq k)$.
2. Let $A \in V$.The multistep ν is said to be in *A-nonterminal labeled mode* if $T_A = \{t \in T : \gamma(t) = A \to \alpha$ for some $A \to \alpha \in$R$\}$.
3. Let $r \in R$.The multistep ν is said to be in *r-rule labeled mode* if $T_r = \{t \in T : \gamma(t) = r\}$
4. The multistep ν is said to be in *wide* mode if $\nu(t) > 0$ for all $t \in X$ and
   - $X = T$ or
   - For all $\nu^{'} \in Y^{\oplus}$, where $X \subset Y \subseteq T$,

$$\bigoplus_{t \in Y} \nu^{'}(t)\alpha \not\sqsubseteq \mu.,$$

5. The multistep ν is said to be in *global* mode if and only if for all $\eta \in X^{\oplus}$,

$$\bigoplus_{i=1}^{k} \eta(t_i)\alpha_i \sqsubseteq \mu \text{ imply } \eta = \nu.$$

6. The multistep ν is said to be a in *step* mode if ν is a set, i.e., $\nu \subseteq X$.

3.2 Concurrent Context-Free Grammar and Language

In this paper, we will consider context-free Petri nets under parallel strategy. First, we convert parallel firing modes in Petri net to rule applications in context-free grammars.

Definition 2 Let $R' = \{r_1, r_2, \cdots r_n\} \subseteq R$, where $r_i = A_i \to \alpha_i (1 \leq i \leq n)$ are applicable rules in the sentential form x. Multiset $\Re_{app(x)} \subseteq R^{'\oplus} = \{\rho(r_1)r_1, \rho(r_2)r_2 \cdots \rho(r_t)r_t\} (t \leq n)$ is said to be *parallelly applicable* in the sentential form x if x can be represented as $x = x_1 A_{i_1} x_2 A_{i_2} \cdots x_k A_{i_k} x_{k+1}$ where $\{A_{i_j}, 1 \leq j \leq k\} = \{\rho(r_1)A_1, \rho(r_2)A_2, \cdots \rho(r_t)A_t, t \leq k\}$. A set of all multisets of parallelly applicable rules in the sentential form x is denoted by $\Re^{'}_{app(x)}$.

Definition 3 Let $x = x_1A_1x_2A_2 \cdots x_mA_mx_{m+1}$ and $y = x_1u_1x_2u_2 \cdots x_mu_mx_{m+1}$, where $x_i \in (V \cup \Sigma)^*(1 \le i \le m+1), A_j \in V^*, u_j \in (V \cup \Sigma)^*(1 \le j \le m)$, and $\{r_i : r_i = A_i \to u_i, 1 \le i \le m\} \subseteq R$. We say that x *directly derives* y

(i) In *a multistep mode*, denoted by m, if a multiset $m \in \mathfrak{R}'_{app(x)}$

(ii) In *a step mode*, denoted by s, if $s \in \mathfrak{R}'_{app(x)}$ and $s \subseteq R'$.

(iii) In k *mode*, denoted by mode k, if $k \in \mathfrak{R}'_{app(x)}$ and $|k| \le k$.

(iv) In *a nonterminal labeled mode*, denoted by n, if $n \in \mathfrak{R}'_{app(x)}$ and $n = \{r : r = A_i \to u_i\}$, where $A_j = A_i$ for any $1 \le j \le m$;

(v) In *a rule labeled mode*, denoted by r, if $r \in \mathfrak{R}'_{app(x)}$ and $r = \{r : r = A_i \to u_i\}$, where $A_j = A_i$ and $u_j = u_i$ for any $1 \le j \le m$

(vi) In *a global mode*, denoted by g, if $g \in \mathfrak{R}'_{app(x)}$ and $g \cup r \notin \mathfrak{R}'_{app(x)}$ for any $r \in R'$

(vii) In *a wide mode*, denoted by w, if $w \in \mathfrak{R}'_{app(x)}$ and
- The multiset w consists all rules $r_i \in R'$
  or
- The multiset $(\rho \cup r_i) \notin \mathfrak{R}'_{app(x)}$ for any $r_i \in R'(\notin w)$ and $\rho = \{\rho_1(r_1), \rho_2(r_2), \cdots \rho_t(r_t)\} \sqsubseteq w$, where $\rho_i(r_i) \ge 1$ for all $1 \le i \le t$.

It is also of interest to consider some combined cases of these modes. $ws, wg, wk, wn, ng, nk, rg, rk, kg$ denote *wide step, wide global, wide k, wide nonternimal labeled, nonterminal labeled global, nonternimal labeled k, rule labeled global, rule labeled* k and k *global* modes respectively.

Let $F = \{m, s, k, n, r, g, w, ws, wg, wk, wn, ng, nk, rg, rk, kg\}$. We use a general notion $x \underset{f}{\Rightarrow}$ if x *directly derives* y in f mode, where $f \in F$. The reflexive and transitive closure of $\underset{f}{\Rightarrow}$ is denoted by $\overset{*}{\underset{f}{\Rightarrow}}$.

Definition 4 A *concurrent context-free grammar in* f *mode* is a tuple $\mathcal{G} = (V, \Sigma, S, R, f)$ where $G = (V, \Sigma, S, R)$ is a context-free grammar and $f \in F$.

Definition 5 The language $L(\mathcal{G})$ generated by *concurrent context-free grammar in f mode* is defined by $L(\mathcal{G}) = \{w \in \Sigma^* | \, S \overset{*}{\underset{f}{\Rightarrow}} w\}$.

The family of languages generated by concurrent context-free grammars in f mode is denoted by $f\mathcal{CF}$, where $f \in F$.

3.3 Examples

Example 2 Let $\mathcal{G}_1 = (V, \Sigma, S, R, rg)$ be concurrent context-free grammar, where $R = \{S \to AA, A \to aA, A \to a\}$, for all $a \in \Sigma$.

It can be easily seen that the grammar generates the language $L(\mathcal{G}_1) = \{ww : w \in \Sigma^*\}$ which is not context-free language. For example, if $\Sigma = \{a, b\}$, the set of labeled rules will be as

r_1: $S \to AA$
r_2: $A \to aA$
r_3: $A \to bA$
r_4: $A \to a$
r_5: $A \to b$.
The derivation steps for generating the word **aaabaaab** would be $S \overset{r_1}{\Rightarrow} AA \overset{r_2}{\Rightarrow} aAaA \overset{r_2}{\Rightarrow} aaAaaA \overset{r_2}{\Rightarrow} aaaAaaaA \overset{r_5}{\Rightarrow} aaabaaab$

Example 3 Let $\mathcal{G}_2 = (V, \Sigma, S, R, rg)$ be context-free grammar, where R=$\{r_1 : S \to SS, r_2 : S \to a\}$ and $\Sigma = \{a\}$.

We get a doubled number of **S** s in each derivation step by using the rule r_1. $S \underset{rg}{\overset{r_1}{\Rightarrow}} S^2 \underset{rg}{\overset{r_1}{\Rightarrow}} S^4 \underset{rg}{\overset{r_1}{\Rightarrow}} S^8 \underset{rg}{\overset{r_1}{\Rightarrow}} \ldots . \underset{rg}{\overset{r_1}{\Rightarrow}} S^{2^k}$.

Application of the rule r_2 in any step replaces all **S**s with a s; consequently, $S \underset{*}{\overset{rg}{\Rightarrow}} a^{2^k}$.

Therefore, $L(\mathcal{G}_2) = \{a^{2^n} : n \geq 0\}$ which is not context-free.

3.4 Results

In this section, we investigate some properties of the concurrent context-free grammars. Based on the previous definitions and examples, the instance results are as follows.
Let $F = \{m, s, k, n, r, g, w, ws, wg, wk, wn, ng, nk, rg, rk, kg\}$ be the set of modes.

Theorem 1 *$\mathcal{LIN} = f\mathcal{LIN}$, where $f \in F$.*

Proof By definition, every rule of the linear grammar has a form $R \subseteq V \times (\Sigma^* V \Sigma^* \cup \Sigma^*)$, therefore in every derivation step a sentential form of the grammar has at most one nonterminal. Hence, all modes of the concurrent linear grammar has the same derivation step with linear grammar.

We cite the following theorems without proofs.

Theorem 2 *$\mathcal{CF} = x\mathcal{CF}$, where $x \in \{s, m, k, n, r\}$.*

Theorem 3 *$rg\mathcal{CF} - \mathcal{CF} = \emptyset$.*

4 Conclusion

We have introduced a new variant of theoretical models for parallel computation using Petri nets under parallel firing strategies called grammars controlled by Petri nets under parallel firing strategies (i.e., concurrent grammars), which are natural formal models of concurrent, asynchronous, distributed, parallel, nondeterministic, and stochastic systems. Various concurrent grammars were defined with respect to classes of Petri nets, firing modes, labeling strategies, and final marking sets. We consider a context-free Petri net under parallel strategy and define parallel firing modes. Moreover, we convert these firing modes to the rule application in context-free grammar, and hence define concurrent context-free grammars. Some properties of the concurrent context-free grammars are also investigated. This kind of approach can be used as theoretical models for parallel computing devices.

Acknowledgments This work has been supported by Ministry of Higher Education of Malaysia via Fundamental Research Grant Scheme FRGS/1/11/SG/UPM/01/1.

References

1. Hans-Dieter, B. Ordered firing in Petri nets. Inf. Process. Cybern. **17**(2/3) 71–86 (1989)
2. Dassow, J., Turaev, S.: k-Petri net controlled grammars. In: Hutchison, D., et al. (eds.) Language and Automata Theory and Applications. Lecture Notes in Computer Science, vol. 5196, pp. 209–220. Springer, Berlin (2008) (LATA 2008. Revised Papers)
3. Dassow, J., Turaev, S.: Grammars controlled by special Petri nets. In: Hutchison, D., et al. (eds.) Language and Automata Theory and Applications. Lecture Notes in Computer Science, vol. 5457, pp. 326–337. Springer, Berlin (2009) (Third International Conference, LATA 2009)
4. Dassow, J., Turaev, S.: Petri net controlled grammars: the power of labeling and final markings. Rom. J. Inf. Sci. Tech. 12(2), 191–207 (2009)
5. Dassow, J., Mavlankulov, G., Othman, M., Turaev, S., Selamat, M.H., Stiebe, R.: Grammars controlled by Petri nets. In: Pawlewski, P. (ed.) Petri Nets. Intech, Rijeka (2012) (ISBN 978-953-51-0700-2.)
6. Farwer, B., Jantzen, M., Kudlek, M., Rolke, H., Zetzsche, G.: Petri net controlled finite automata. Fund. Inform. 85(1–4), 111–121 (2008)
7. Farwer, B., Kudlek, M., Rolke., H.: Concurrent Turing machines. Fund. Inform. 79(3–4), 303–317 (2007)
8. Hack, M.: Petri net languages. MIT, Cambridge (1975) (Computation Structures Group Memo, Project MAC 124)
9. Jantzen, M., Zetzsche, G.: Labeled step sequences in Petri nets. In: van Hee Kees, M., Rüdiger, V. (eds.) Applications and Theory of Petri Nets. Lecture Notes in Computer Science, vol. 5062, pp. 270–283. Springer, Berlin (2008)
10. Peterson, J.L.: Petri net theory and modeling of systems. Prentice-Hall, Englewood Cliffs (1981)
11. Turaev, S.: Petri net controlled grammars. In: Matyska, L., et al. (eds.) Third Doctoral Workshop on Mathematical and Engineering Methods in Computer Science, MEMICS 2007, pp. 233–240, Prestige Hotel, Znojmo, Czechia, 26–28 Oct 2007 (ISBN 978-80-7355-077-6.)

Statistical Analysis on LBlock Block Cipher

Nik Azura Nik Abdullah, Kamaruzzaman Seman and Norita Md Norwawi

Abstract In this research paper, we present the statistical analysis on full round lightweight block cipher; LBlock. This block cipher has a fixed block size of 64-bit, utilizes an 80-bit key and executes in 32 rounds. To determine the randomness of ciphertext produced by this algorithm, NIST Statistical Test Suite is used. Inputs (plaintext and key) for this block cipher algorithm are generated using nine different types of data. From the analysis done, it is concluded that the LBlock block cipher is not random based on 1 % significance level only.

1 Introduction

LBlock block cipher is a most recent block cipher used in lightweight cryptography that was proposed by Wenling Wu and Lei Zhang in 2011 and was published at ACNS 2011. This algorithm can be implemented efficiently not only in hardware environment but also in software platforms and can achieve competitive performances compared with other known ultra-lightweight block ciphers. Wu and Zhang claimed that their security analysis showed that the full-round LBlock encryption can provide enough security margins against known cryptanalytic techniques, such as differential cryptanalysis, linear cryptanalysis, impossible differential cryptanalysis, related-key attack, etc. [7].

The Advanced Encryption Standard (AES) has been the most widely-accepted block cipher encryption algorithm when protecting personal data, network traffic, corporate IT infrastructures, etc. The National Security Agency (NSA) of the US Government declared that the design and strength of AES algorithm are sufficient to

N. A. N. Abdullah (✉)
Cyber Technology Research Department, CyberSecurity Malaysia,
Kuala Lumpur, Malaysia
e-mail: azura@cybersecurity.my

K. Seman · N. M. Norwawi
Faculty of Science and Technology, University Science Islam Malaysia (USIM),
Negeri Sembilan, Malaysia
e-mail: drkzaman@usim.edu.my

N. M. Norwawi
e-mail: norita@usim.edu.my

A. Kilicman et al. (eds.), *International Conference on Mathematical Sciences and Statistics 2013,* DOI 10.1007/978-981-4585-33-0_24,

protect classified information up to top secret level [8]. When the National Institute of Standards and Technology (NIST) first announced the request for candidate algorithm nominations for AES, it was mentioned that one of the characteristics required is that the algorithm can be implemented as a random number generator [4]. Therefore, the suitability of the algorithm to act as a random number generator is considered to be an important requirement when designing a block cipher algorithm. This is to ensure that the ciphertext produced from the algorithm will be uniformly distributed should all of the plaintext blocks used during encryption are different. To determine whether the tested block cipher algorithm fulfills this requirement, a randomness test using statistical analysis can be applied. It is important to note here that passing a fixed set of statistical analysis is necessary, but it is not sufficient enough for a block cipher algorithm to be secure.

Over the past couple of years, some work has been conducted in order to evaluate the security of LBlock block cipher against various cryptanalysis attacks but statistical analysis has never been performed before.

This chapter presents the statistical analysis results for full round LBlock block cipher using NIST Statistical Test Suite, and is organized as follows:

Section 2 gives a short description of LBlock block cipher. Section 3 explains in detail the nine different categories of data used to generate inputs (plaintext and key) for LBlock block cipher. Section 4 explains the statistical analysis used to determine the randomness of ciphertext produced from LBlock block cipher, which is the NIST Statistical Test Suite. Section 5 presents the experimental setup in conducting experiment towards full round LBlock block cipher. Section 6 presents the empirical results and analysis. Section 7 demonstrates the summary and conclusion of this research.

2 Description of LBlock Block Cipher

LBlock block cipher is based on a modified Fiestel structure with a block size of 64-bit. This algorithm accepts an 80-bit key and executes in 32 rounds. It also employs S-box layers in both encryption algorithm and key scheduling; confusion function acts as a nonlinear layer, and diffusion function acts as a linear layer.

2.1 Encryption Algorithm

In the encryption algorithm of LBlock block cipher, the 64-bit plaintext, M is processed by first dividing it into two separate sequences: X_1 is the left 32-bit half and X_0 is the right 32-bit half. This is denoted as:

$$M\ (64\text{-bit}) = X_1\ \ (32\text{-bit})\ ||X_0\ (32\text{-bit}) \tag{1}$$

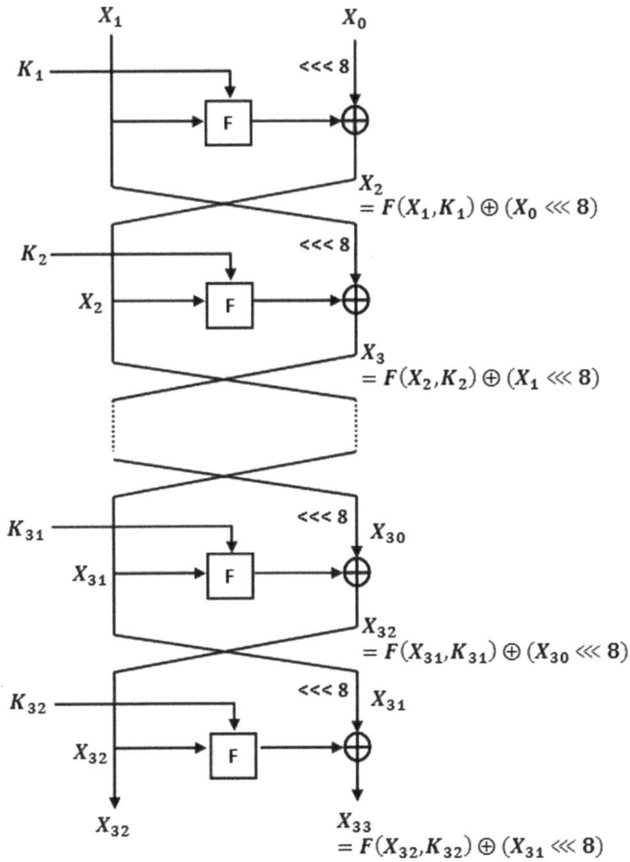

Fig. 1 Encryption of LBlock block cipher

The encryption process continues in 32 rounds for $2 \leq i \leq 33$ as follows:

$$X_i = F\,(X_{i-1}, K_{i-1}) \,\oplus (X_{i-2} <<< 8) \tag{2}$$

where F and K_i are denoted as the round function and the 32-bit round subkey respectively, meanwhile ||, $\oplus$ and $<<<$ are denoted as concatenation of two binary strings, exclusive-OR operation and a left cyclic shift operation respectively.

The produced ciphertext, C, from this encryption process is a concatenation of X_{32} and X_{33}, and is presented as:

$$C = X_{32}||X_{33} \tag{3}$$

Figure 1 shows the encryption process of LBlock block cipher in 32 rounds. Figure 2 shows the round function that is used in each round.

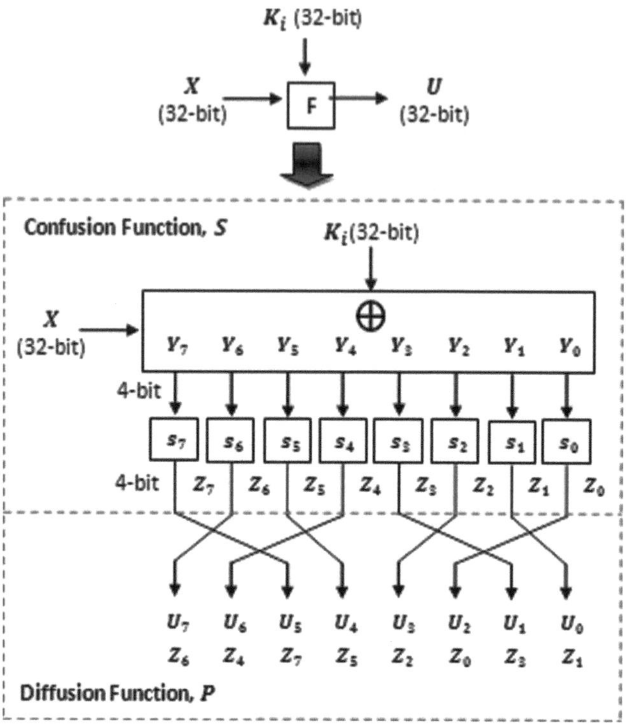

Fig. 2 Round function, F which includes confusion function S and diffusion function P

Table 1 Content of S-boxes s_0, s_1, s_2, s_3, s_4, s_5, s_6, and s_7

S-boxes	Content
s_0	14, 9, 15, 0, 13, 4, 10, 11, 1, 2, 8, 3, 7, 6, 12, 5
s_1	4, 11, 14, 9, 15, 13, 0, 10, 7, 12, 5, 6, 2, 8, 1, 3
s_2	1, 14, 7, 12, 15, 13, 0, 6, 11, 5, 9, 3, 2, 4, 8, 10
s_3	7, 6, 8, 11, 0, 15, 3, 14, 9, 10, 12, 13, 5, 2, 4, 1
s_4	14, 5, 15, 0, 7, 2, 12, 13, 1, 8, 4, 9, 11, 10, 6, 3
s_5	2, 13, 11, 12, 15, 14, 0, 9, 7, 10, 6, 3, 1, 8, 4, 5
s_6	11, 9, 4, 14, 0, 15, 10, 13, 6, 12, 5, 7, 3, 8, 1, 2
s_7	13, 10, 15, 0, 14, 4, 9, 11, 2, 1, 8, 3, 7, 5, 12, 6

Round Function, F Constructed from two other functions; confusion function, S, and diffusion function, P. This is defined as follows:

$$U = F\,(X, K_i) = P(S(X\ \oplus K_i)) \tag{4}$$

Confusion Function, S The nonlinear layer of the round function F that consists of eight 4×4 S-boxes, namely, s_0, s_1, s_2, s_3, s_4, s_5, s_6, and s_7 (shown in Table 1). This function is defined as follows:

$$Y = Y_7\,||Y_6||\,Y_5\,||Y_4||\,Y_3\,||Y_2||\,Y_1||Y_0 \quad \rightarrow \quad Z = Z_7\,||Z_6||\,Z_5\,||Z_4||\,Z_3\,||Z_2||\,Z_1||Z_0 \tag{5}$$

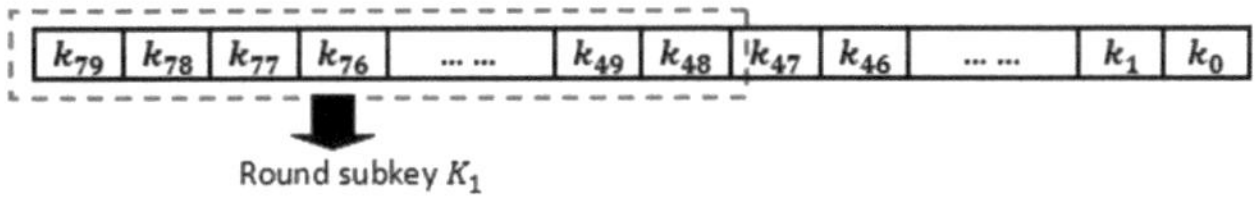

Fig. 3 Step 1 of key scheduling

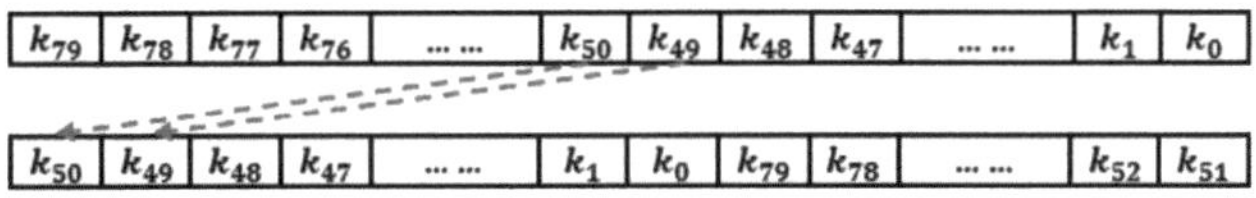

Fig. 4 Step 2(a) of key scheduling

Table 2 Content of S-boxes s_9 and s_8

S-boxes	Content
s_9	8, 7, 14, 5, 15, 13, 0, 6, 11, 12, 9, 10, 2, 4, 1, 3
s_8	11, 5, 15, 0 7, 2, 9, 13, 4, 8, 1, 12, 14, 10, 3, 6

where $Z_7 = s_7\,(Y_7),\quad Z_6 = s_6\,(Y_6),\quad Z_5 = s_5\,(Y_5),\quad Z_4 = s_4\,(Y_4),$

$$Z_3 = s_3(Y_3),\quad Z_2 = s_2(Y_2),\quad Z_1 = s_1(Y_1),\quad Z_0 = s_0(Y_0).$$

Diffusion Function, P The linear layer of the round function F that permutes inputs Z to produce outputs U. This function is defined as follows:

$$Z = Z_7\,|Z_6|\,|Z_5|\,|Z_4|\,|Z_3|\,|Z_2|\,|Z_1|\,|Z_0 \rightarrow U = U_7\,||U_6||\,U_5\,||U_4||\,U_3\,||U_2||\,U_1||U_0 \tag{6}$$

where $U_7 = Z_6,\quad U_6 = Z_4,\quad U_5 = Z_7,\quad U_4 = Z_5,$

$$U_3 = Z_2,\quad U_2 = Z_0,\quad U_1 = Z_3,\quad U_0 = Z_1.$$

2.2 *Key Scheduling*

The LBlock block cipher accepts an 80-bit master key, $K = k_{79}, k_{78}, k_{77}, \ldots, k_1, k_0$, which is stored in a key register. After 32 rounds of execution, this algorithm will produce 32 round subkeys denoted as K_i for $1 \le i \le 31$.

Step 1: Round subkey, K_1, is the leftmost 32 bits of the master key, K (Fig. 3).

Step 2: For $1 \le i \le 31$, the key register is updated as follows:

a. Contents in key register are shifted 29 bits to the left with a cyclic shift operation (Fig. 4).
b. Contents in key register at positions 79, 78, 77, 76 and 75, 74, 73, 72 are replaced with contents from two 4 × 4 S-boxes s_9 and s_8 (shown in Table 2), respectively.

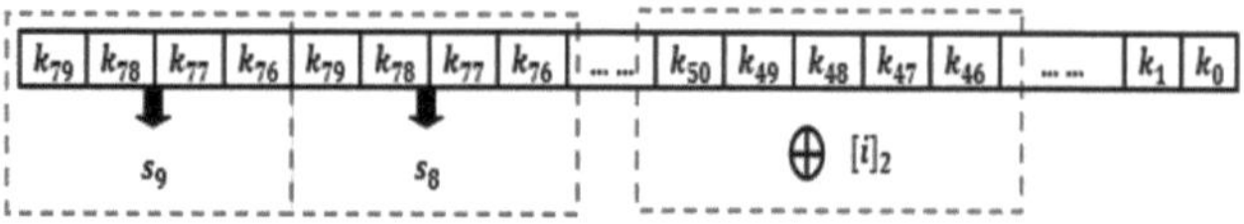

Fig. 5 Steps 2(b) and 2(c) of key scheduling

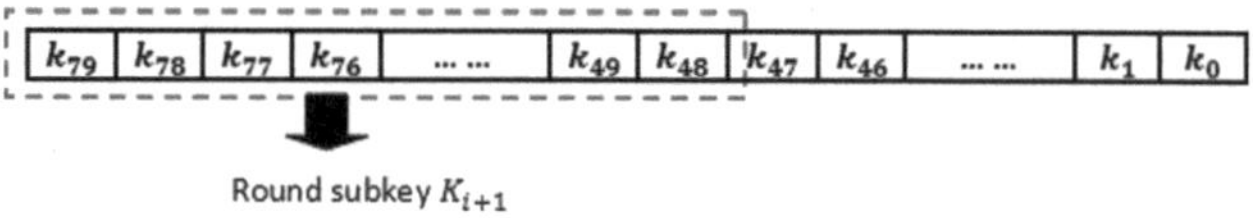

Fig. 6 Step 2(d) of key scheduling

c. Contents in key register at positions 50, 49, 48, 47, and 46 are XORed with binary form of current round (Fig. 5).

$$[k_{50}, k_{49}, k_{48}, k_{47} \text{ and } k_{46}] \oplus [i]_2 \tag{7}$$

d. Round subkey, K_{i+1}, is the leftmost 32 bits of the current key register (Fig. 6).

3 LBlock Block Cipher Data Categories

During the AES Round 2 analysis period, NIST conducted randomness testing to determine the suitability of the five finalist candidates to act as a random number generator. Juan Soto and Lawrence Bassham have specified nine different categories of data required to conduct this test [2], [3], [5], [6]. These data categories are: strict key avalanche (SKA), strict plaintext avalanche (SVA), plaintext/ciphertext correlation (PCC), cipher block chaining mode (CBCM), random plaintext/random key (RPRK), low density key (LDK), high density key (HDK), low density plaintext (LDP), and high density plaintext (HDP).

NIST Statistical Test Suite requires a large sequence of bits. The nine data categories are used to generate inputs (plaintext and key) for LBlock block cipher to produce ciphertexts that will be concatenated in a certain way. Figure 7 illustrates this process. In this experiment, 100 samples are generated using each data category. The number of blocks produced for every sample is depending on the block size or key size of the LBlock block cipher.

Strict Key Avalanche (SKA)^{A} and Strict Plaintext Avalanche (SPA)^{B} The strict key {Plaintext} avalanche data category is used to examine the sensitivity of LBlock block cipher to changes in the 80-bit key {plaintext}. Each sample for this data category utilizes plaintext {key} that is set to all zero, and 196 blocks of random 80-bit base keys {245 blocks of random 64-bit base plaintext}. Each base key {base

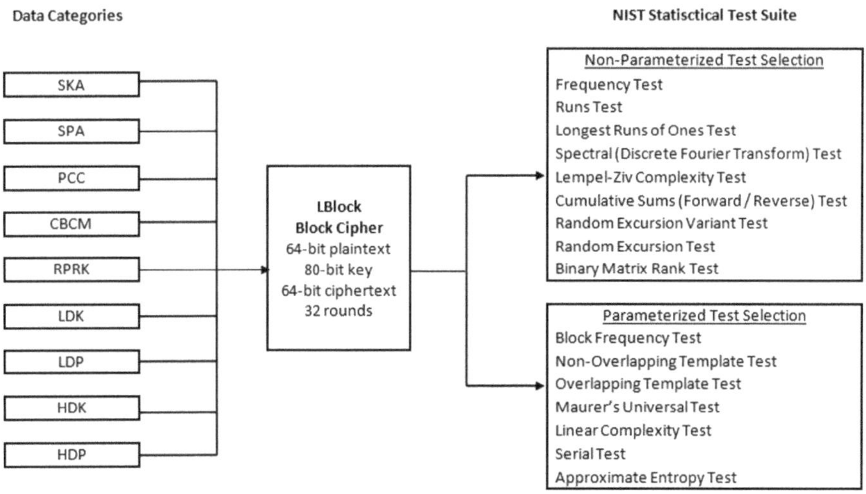

Fig. 7 The process for conducting statistical analysis towards LBlock block cipher

plaintext} is flipped at the ith bit, for $1 \leq i \leq 80$ $\{1 \leq i \leq 64\}$ giving a total of 15,680 perturbed keys {perturbed plaintexts}. The all-zero plaintext {key} is then encrypted using each perturbed key {perturbed plaintexts}. All resultant ciphertexts are XORed with the ciphertext resulting from the encryption using its corresponding base key {base plaintext}. Output product of the XOR operation is called a derived block.

Plaintext/Ciphertext Correlation (PCC)[C] The plaintext/ciphertext correlation data category is used to examine the correlation between plaintext/ciphertext pairs using electronic codebook (ECB) mode of operation. Each sample for this data category utilizes 15,625 blocks of random 64-bit plaintext and one random 80-bit key. Each plaintext block is encrypted using the random 80-bit key. The resultant ciphertext is XORed with its corresponding plaintext. Output product of the XOR operation is called a derived block.

Ciphertext Block Chaining Mode (CBCM)[D] The ciphertext block chaining mode data category is computed using CBC mode of operation. Each sample for this data category utilizes plaintext that is set to all zero (PT), a random 80-bit key (k), and an all-zeroes initialization vector (IV). Derived blocks for this data category are ciphertext blocks computed in CBC mode of operation. The first ciphertext block (CT_1) is defined as $CT_1 = E_k\ (IV \oplus PT_0)$, whereas subsequent ciphertext blocks are defined as $CT_{i+1} = E_k\ (CT_i \oplus PT_i)$ for $1 \leq i \leq 15{,}625$. To construct a large sequence of bits to be inserted into the NIST Statistical Test Suite, derived blocks are concatenated.

Random Plaintext/Random Key (RPRK)[E] The random plaintext/random key data category is used to examine the randomness of ciphertext based on random plaintext and random key. Each sample for this data category utilizes 15,625 blocks of

Table 3 Number of derived block and derived bits per sample for all nine data categories

Data categories	Derived blocks	Derived bits
SKA, SPA	15,680	1,003,520
PCC, CBCM, RPRK	15,625	1,000,000
LDK, HDK, LDP, HDP	3,241	207,424

random 64-bit plaintext and one random 80-bit key. Each plaintext block is encrypted using the random 80-bit key. Derived blocks for this data category are ciphertext blocks computed in ECB mode of operation.

Low Density Keys (LDK)^{F} and High Density Keys (HDK)^{G} The low {high} density keys data category is formed based on low-density {high-density} 80-bit keys. Each sample for this data category utilizes 3,241 blocks of random 64-bit plaintext. The first plaintext block is encrypted using an all-zeroes {all-ones} 80-bit key. Plaintext blocks 2–81 are encrypted using an 80-bit key with a single '1' {'0'} in each of the 80-bit positions of the key, and all other key bits are set to '0' {'1'}. Plaintext blocks 82–3,241 are encrypted using an 80-bit key with two '1's {'0's} in each combination of two bit positions of the key, and all other key bits are set to '0' {'1'}. Derived blocks for this data category are ciphertext blocks computed in ECB mode of operation.

Low Density Plaintext (LDP)^{H} and High Density Plaintext (HDP)^{I} The low {high} density plaintext data category is formed based on low-density {high-density} 64-bit plaintext blocks. Each sample for this data category utilizes 2,081 blocks of random 80-bit keys. The first plaintext block is encrypted using an all zeroes {all-ones} 64-bit plaintext. Plaintext blocks 2–65 consist of one bit of '1' {'0'}, and all other bits are '0' {'1'}. Plaintext blocks 66–2,081 consist of two bits of '1' {'0'}, and all other bits are '0' {'1'}. The '1's {'0's} appear in each combination of two-bit positions of the plaintext block. Derived blocks for this data category are ciphertext blocks computed in ECB mode of operation.

To construct a large sequence of bits to be inserted into the NIST Statistical Test Suite, derived blocks produced from each data categories are concatenated. The number of derived blocks and derived bits per sample produced from each data categories are listed in Table 3.

4 NIST Statistical Test Suite

To test the randomness of binary sequences produced by either hardware- or software-based cryptographic random or pseudorandom number generators, National Institute of Standards and Technology (NIST) developed a statistical package known as NIST Statistical Test Suite. This statistical package focuses on a variety of different types of nonrandomness that could exist in a sequence [1], [5], [6]. It consists of 16 tests separated into two categories. Nine tests categorized as nonparameterized test selection do not require users to enter any parameter in obtaining the p values for each test,

Table 4 Minimum number of bits recommended by NIST for all 16 tests

	Statistical test	Derived blocks
Nonparameterized test selection	Frequency	$n \geq 100$
	Runs	$n \geq 100$
	Longest runs of ones	$n \geq 750{,}000$
	Discrete Fourier transform (DFT)	$n \geq 1{,}000$
	Lempel–Ziv complexity	$n \geq 10^6$
	Cumulative sums	$n \geq 100$
	Random excursion variant	$n \geq 10^6$
	Random excursion	$n \geq 10^6$
	Binary matrix rank	$n \geq 38{,}912$
Parameterized Test Selection	Block frequency	$n \geq 100$
	Nonoverlapping templates	Not specified
	Overlapping template	$n \geq 10^6$
	Maurer's universal	$n \geq 387{,}480$
	Linear complexity	Not specified
	Serial	Not specified
	Approximate entropy	Not specified

whereas the other seven tests categorized as the parameterized test selection require users to define one or two parameter value(s). All 16 tests are listed according to their category in Table 4. NIST has recommended a minimum number of bits required for each test. This is also presented in Table 4.

5 Experimental Setup

The randomness testing activity was conducted on a full round of LBlock block cipher using NIST Statistical Test Suite by analyzing ciphertext produced. Inputs (plaintext and key) for this algorithm were generated using nine different categories of data. Out of 16 tests in the NIST Statistical Test Suite, 11 provided only 1 p value, whereas 2 tests provided 2 p values and the other 3 tests provided 8, 18, and 148 p values [2]. Due to the limitation of minimum number of bits required, only 5 categories of data were able to undergo all 16 tests, whereas the other 4 categories were able to undergo only 11 tests. This can be seen by comparing Tables 3 and 4. Table 5 lists the number of p values produced by all tests according to each data category.

The parameterized test selection requires users to define one or two parameter value(s). To explain the parameters that need to be entered, the following abbreviation is used: block length (M or L), length of bit string (n), nonoverlapping blocks ($N = n/M$), template length (m), number of degrees of freedom (K), theoretical probabilities (π_i), and number of block in the initialization sequence (Q).

The requirements to be considered in choosing the parameter(s) are as per listed below [1]:

Table 5 Breakdown of the 1,585 p values obtained per sample during experimentation

Non-parameterized Test Selection	**Data categories**								
	A	**B**	**C**	**D**	**E**	**F**	**G**	**H**	**I**
**Frequency	1	1	1	1	1	1	1	1	1
**Runs	1	1	1	1	1	1	1	1	1
Longest runs of ones	1	1	1	1	1	1	1	1	1
Spectral DFT	1	1	1	1	1	1	1	1	1
*Lempel–Ziv complexity	1	1	1	1	1	–	–	–	–
**Cumulative sums	2	2	2	2	2	2	2	2	2
*Random excursion variant	18	18	18	18	18	–	–	–	–
*Random excursion	8	8	8	8	8	–	–	–	–
Binary matrix rank	1	1	1	1	1	1	1	1	1
Parameterized Test Selection	**Data categories**								
	A	**B**	**C**	**D**	**E**	**F**	**G**	**H**	**I**
**Block frequency	1	1	1	1	1	1	1	1	1
Non-overlapping	148	148	148	148	148	148	148	148	148
*Overlapping	1	1	1	1	1	–	–	–	–
Maurer's universal	1	1	1	1	1	–	–	–	–
Linear complexity	1	1	1	1	1	1	1	1	1
Serial	2	2	2	2	2	2	2	2	2
Approximate entropy	1	1	1	1	1	1	1	1	1

* tests which require minimum bit length of 10^6 bits.
** tests which require only 100 bits.

- Block frequency test: M is selected such that $n \geq MN$, $M \geq 20$, $M > 0.01n$ and $N < 100$. Therefore, M chosen is 20,000 for all categories of data.
- Nonoverlapping template test: $N = 8$ has been specified, it is recommended that $m = 9$ or $m = 10$, $N \leq 100$, and $M > 0.01n$ where $N = [n/M]$. Therefore, m chosen is 10 for all categories of data.
- Overlapping template test: K, M, and N have been fixed in the test code such that $n \geq 10^6$, it is recommended that $m = 9$ or $m = 10$, $n \geq MN$, N (min π_i), $\lambda = (M - m + 1)/2^m \approx 2$, $m \approx \log_2 M$, and $K \approx 2\lambda$. Therefore, m chosen is 10. Note that this test is only suitable for categories of data which produce sequences larger than 10^6 bits.
- Maurer's universal test: the value of L must be in the range of $6 \leq L \leq 16$, $Q = 10.^{2L}$ and $n \geq (Q + K)L$ where $K = [n/L] - Q \approx 1000.^{2L}$. Therefore, L and Q chosen are 7 and 1,280, respectively. Note that this test is only suitable for categories of data which produce sequences larger than 904,960 bits.
- Linear complexity test: the value of M must be in the range $500 \leq M \leq 5000$ and $N \geq 200$. Therefore, M chosen is 2,000 for categories of data with $n \geq 10^6$ and 600 for categories of data with $n < 10^6$.
- Serial and approximate entropy tests: m and n are selected such that $m < [\log_2 n] - 2$. Therefore, m chosen is 2.

Table 6 Number of maximum rejection for random excursion variant test with 18 p values per sample and random excursion test with 8 p values per sample

		SKA	SPA	PCC	CBCM	RPRK
Number of samples evaluated		72	60	62	58	61
Random Excursion Variant Test	Number of p values	1,296	1,080	1,116	1,044	1,098
	Maximum rejection	23	20	21	20	20
Random Excursion Test	Number of p values	576	480	496	464	488
	Maximum rejection	12	11	11	11	11

6 Empirical Results and Analysis

This research paper considers full round LBlock block cipher algorithm to be tested under 9 categories of data with each having 100 samples. During experimentation, a total of 900 binary sequences (i.e., 100 samples $\times$ 9 data categories) were evaluated. Only 5 categories of data were evaluated using all 16 NIST tests, which constructed 8,000 test values (i.e., 100 samples $\times$ 5 data categories $\times$ 16 tests), whereas the other 4 categories of data were evaluated using 11 tests, which constructed 4,400 test values (i.e., 100 samples $\times$ 4 data categories $\times$ 11 tests). Therefore, there were 12,400 test values constructed altogether. As explained in Table 5, one sample produces 1,585 p values, thus the whole experiment examined 158,500 p values in total.

The maximum number of binary sequences that were expected to be rejected at the chosen significance level was computed using formula (8) [3]:

$$s\left(\alpha + 3\sqrt{\frac{\alpha(1-\alpha)}{s}}\right) \tag{8}$$

where α is the significance level that was set at 0.01, and s is the number of samples used in the experiment. As the experiment consists of 100 samples, the rejection rate should not exceed 3.9850 (~ 3) sequences. The nonoverlapping template test produces 14,800 p values (148 p values/sample). Therefore the rejection rate for this test should not exceed 184.3136 (~ 184) sequences.

Only samples with the number of cycles exceeding 500 were evaluated for random excursion variant and random excursion tests. Tests were not applicable for samples with insufficient number of cycles. Therefore, these 2 tests did not use all 100 samples. Table 6 presents the number of p values and number of maximum rejections, which corresponds to the number of samples acceptable for these 2 tests.

The actual test results from NIST Statistical Test Suite based on 1 % significance level are presented in Table 7. Tests which exceeded the maximum number of rejection rate are highlighted in . indicates that test is not applicable for that particular data category.

There are some cases in which the Lempel–Ziv complexity test produces p value equal to 1. This occurs for SKA and SVA data categories.

Table 7 Number of samples rejected in full round testing of LBlock Block Cipher

Non-parameterized Test Selection		Data Categories								
		A	B	C	D	E	F	G	H	I
Frequency		2	0	0	3	3	0	3	2	0
Runs		0	3	2	0	0	4	2	4	0
Longest Runs of Ones		0	1	1	0	0	5	2	0	2
Spectral DFT		0	1	1	0	0	0	0	0	0
Lempel-Ziv Complexity		0	0	0	0	2				
Cumulative Sums	Fwd	2	0	2	0	2	0	1	2	0
	Rvs	2	0	1	2	2	1	2	2	0
Random Excursion Variant		10	30	9	37	11				
Random Excursion		8	2	7	14	5				
Binary Matrix Rank		2	0	1	1	0	1	1	2	0
Parameterized Test Selection		**Data Categories**								
		A	**B**	**C**	**D**	**E**	**F**	**G**	**H**	**I**
Block Frequency		0	2	4	0	1	2	1	0	0
Non-Overlapping		126	143	143	139	150	180	171	172	180
Overlapping		0	0	0	2	2				
Maurer's Universal		2	0	0	0	1				
Linear Complexity		0	1	1	2	1	2	1	0	0
Serial	p-value1	2	3	2	2	2	3	1	2	0
	p-value2	0	3	2	0	0	4	2	4	0
Approximate Entropy		2	1	0	3	2	2	0	2	2

7 Conclusion

We have presented the statistical analysis on full round LBlock block cipher using NIST Statistical Test Suite. To determine whether the algorithm tested is random or non-random, the significance level was set at 1 %.

Overall results show that SKA, CBCM, LDK and HDP fail one test; RPRK and LDP fail two tests. Therefore, this algorithm is nonrandom based on the 1 % significance level.

Acknowledgement A special thanks goes to Cyber Technology Research department of CyberSecurity Malaysia for the efforts they have put in conducting the statistical test, and Faculty of Science and Technology, University Sains Islam Malaysia for all the support they have given in completing this study.

References

1. Rukhin, A., Soto, J., Nechvatal, J., Smid, M., Barker, E., Leigh, S., Levenson, M., Vangel, M., Banks, D., Heckert, A., Dray, J., Vo, S.: A statistical test suite for random and pseudorandom number generators for cryptographic applications in NIST Special Publication 800-2 (2001)
2. Soto, J.: Randomness testing of the AES candidate algorithms in NIST Interagency Reports, NISTIR 6390 (1999)
3. Soto, J., Bassham, L.: Randomness testing of the advanced encryption standard finalist candidates in NIST Interagency Reports, NISTIR 6483 (1999)
4. Hathaway, L.: National policy on the use of the advanced encryption standard (AES) to protect national security systems and national security information in The Committee on National Security Systems (CNSS), June 2003
5. Abdullah, N.A.N., Lot, N.H., Rani, H.A.: Analysis on lightweight block cipher, KTANTAN. In: Proceedings of Information Assurance and Security (IAS), 2011 7th International Conference, pp. 46–51, Malacca, 5–8 Dec 2011
6. Lot, N.H., Abdullah, N.A.N., Rani, H.A.: Statistical analysis on KATAN block cipher. In: Proceedings of Research and Innovation in Information Systems (ICRIIS), Seri Pacific Hotel, Kuala Lumpur, 23–24 Nov 2011
7. Wu, W., Zhan, L.: LBlock: A lightweight block cipher. In: Hutchison, D., et al. (eds.) Applied Cryptography and Network Security. Lecture Notes in Computer Science, vol. 6715, pp. 327–344. Springer, Berlin (2011)
8. National Institute of Standards and Technology (NIST). Announcing request for candidate algorithm nominations for the advanced encryption standard. In: Federal Register, The Daily Journal of the United States Government, 12 September 1997, vol. 62. https://www.federalregister.gov/articles/1997/09/12/97-24214/announcing-request-for-candidate-algorithm-nominations-for-the-advanced-encryption-standard (1997) 12 February 2014

A New Threshold-Authenticated Encryption Scheme

Mohd Saiful Adli Mohamad, Eddie Shahril Ismail and Haslinda Ibrahim

Abstract An authenticated encryption scheme is a message recovery scheme that provides the authenticity property. In an authenticated encryption scheme, the recipients not only verify the message authentication, but they also could recover the message. In this paper, we propose a new authenticated encryption scheme based on two hard number theoretical problems: factoring and discrete logarithm. In our new scheme, t out of n signers/senders are required to sign and at the same time encrypt a message, while k out of l recipients cooperate to verify and recover the original message. We also show that our scheme is secure against some cryptographic attacks and requires reasonable number of operations in both signature/encryption and verification/decryption phases.

1 Introduction

Authenticated encryption is a cryptographic scheme that provides the properties of digital signature and cryptosystem in a single scheme. In an authenticated encryption scheme, the recipients could not only recover the message, but also verify the message signature. By combining these two in a single scheme, the cost of communication and operation can be reduced, as compared with performing both schemes separately [1].

The first authenticated encryption scheme was proposed by Nyberg and Rueppel [2]. In their paper, they presented a modified version of the Digital Signature Algorithm to obtain message recovery. However, the scheme proposed by Nyberg and Rueppel only allows single signer and single recipient in both the signature/encryption and verification/decryption phases. Later on, Hsu and Wu [3] proposed an authenticated encryption scheme with (t, n) shared verification. It applied the idea of threshold cryptography [4], [5], which is based on the concept of Shamir's secret sharing [6], for the verification phase.

M. S. A. Mohamad (✉) · H. Ibrahim
School of Quantitative Sciences, UUM College of Arts and Sciences,
Universiti Utara Malaysia, 06010 Sintok, Kedah, Malaysia
e-mail: msadli@uum.edu.my

E. S. Ismail
School of Mathematical Sciences, Faculty of Sciences and Technology,
Universiti Kebangsaan Malaysia, 43600 Bangi, Selangor, Malaysia

A. Kilicman et al. (eds.), *International Conference on Mathematical Sciences and Statistics 2013,* DOI 10.1007/978-981-4585-33-0_25,

Recently, researches in this area have attracted more attention. Wang et al. [7] presented an authenticated encryption scheme for group communications based on the discrete logarithm problem and Hsu et al. [8] proposed an improvement of Wang's scheme. Then, Chen et al. [1] introduced such a scheme based on elliptic curve cryptosystem. In the schemes in [1], [7], [8], (t, n) signers participate in the signature and encryption phases, while (k, l) recipients are required to verify and recover the original message.

All the schemes reviewed here were developed based on a single-number theoretical problem. However, if the problem becomes solvable someday, then the single-problem schemes will no longer be secure. In this paper, an authenticated encryption scheme based on factoring [9] and discrete logarithm problems [10] is proposed. The security of this scheme is based on the difficulty to solve both problems simultaneously.

2 The Proposed Authenticated Encryption Scheme

In this paper, an authenticated encryption scheme based on two hard number theoretical problems, namely factoring and discrete logarithm, is proposed. This scheme consists of the following phases:

1. Generating parameters and keys
2. Signing/encrypting message
3. Verifying/decrypting message

2.1 Phase 1: Generating Parameters and Keys

Before it generates the secret and public keys, the system authority (SA) sets the following parameters:

1. p is a 1024-bit prime number.
2. $N = ab$ is a factor of $p - 1$, where a and b are two safe primes.
3. $\bar{N} = \bar{a}\bar{b}$ is a factor of $p - 1$ where $\bar{a}$ and $\bar{b}$ are two safe primes.
4. $\phi(N)$ is the Euler's phi function of N.
5. $\phi(\bar{N})$ is the Euler's phi function of $\bar{N}$.
6. g_1 is an element of $\mathbb{Z}_p^*$ of order N.
7. g_2 is an element of $\mathbb{Z}_p^*$ of order $\bar{N}$.

Let $U_s = (u_{s1}, u_{s2}, \ldots, u_{sn})$ be denoted as the n-signer group and $U_v = (u_{v1}, u_{v2}, \ldots, u_{vl})$ be denoted as l-verifier group. Any t members out of the n-signer group can represent to generate the authenticated-encryption signature, while any k members out of the l-verifier group are required to verify the signature and recover the message. First, the SA performs the following steps to generate the keys for both signer and verifier group.

Table 1 The secret and public keys of the scheme

		Secret keys	Public keys
Signer	Group	$d, f_s(0)$	e, Y_s
	Individual	$f_s(x_{si})$	y_{si}
Verifier	Group	$\beta, f_v(0)$	α, Y_v
	Individual	$f_v(x_{vj})$	y_{vj}

1. It picks randomly $e \in \mathbb{Z}_N^*$ such that $\gcd(e,\phi(N)) = 1$ and then calculates d such that $ed \equiv 1(\text{mod } \phi(N))$.
2. It picks another integer $\alpha \in \mathbb{Z}_{\bar{N}}^*$ such that $\gcd(\alpha,\phi(\bar{N})) = 1$ and then calculates β such that $\alpha\beta \equiv 1(\text{mod } \phi(\bar{N}))$.
3. It constructs two secret polynomials:
   i. $f_s(x) = a_0 + a_1x^1 + a_2x^2 + \ldots + a_{t-1}x^{t-1}(\text{mod } N)$
   ii. $f_v(x) = b_0 + b_1x^1 = b_2x^2 + \ldots + b_{k-1}x^{k-1}(\text{mod } N)$
   where $a_i, b_j \in [1, N-1]$, for $i = 0, 1, 2, \ldots, t-1$ and $j = 0, 1, 2, \ldots, k-1$.
4. It sets the group secret key for signer, $f_s(0) = a_0$ and the corresponding group public key $Y_s \equiv {g_1}^{f_s(0)}(\text{mod } p)$.
5. It sets the group secret key for verifier, $f_v(0) = b_0$ and the corresponding group public key $Y_v \equiv {g_1}^{f_v(0)}(\text{mod } p)$.
6. It sets the individual secret key for signer, $f_s(x_{si})$, where x_{si} is the public identity associated with u_{si}, for $i = 0, 1, 2, \ldots, n$, and the corresponding individual public key $y_{si} \equiv {g_1}^{f_s(x_{si})}(\text{mod } p)$.
7. It sets the individual secret key for verifier, $f_v(x_{vj})$, where x_{vj} is the public identity associated with u_{vj}, for $j = 0, 1, 2, \ldots, l$, and the corresponding individual public key $y_{vj} \equiv {g_1}^{f_v(x_{vj})}(\text{mod } p)$.

The summary of the secret and public keys for both signer and verifier is given in Table 1:

2.2 *Phase 2: Signature/Encryption*

Before signing/encrypting a message M, all signers jointly elect a clerk among themselves. The clerk validates all partial signatures and combines the partial signatures into a group signature. Then, each signer

1. Selects r_i, such that $0 < r_i < N$ and $\gcd(r_i, N) = 1$, then, computes $k_i \equiv {g_1}^{r_i}(\text{mod } p)$
2. By using the information of the public identity of other signers, calculates

$$L_{si} \equiv \prod_{j=1, j \neq i}^{t} \frac{-x_{sj}}{x_{si} - x_{sj}} (\text{mod} N)$$

3. Calculates $z_{si} \equiv k_i \cdot {Y_v}^{f_s(x_{si}) \cdot L_{si}}(\text{mod } p)$

4. Broadcasts k_i and z_{si} to other signer via a secure channel. After all k_i and z_{si} are received, calculates

$$K \equiv \prod_{i=1}^{t} k_i \pmod{p}$$

and

$$Z \equiv \prod_{i=1}^{t} z_{si} \pmod{p}$$

5. Calculates $s_i \equiv r_i \cdot Z + K \cdot f_s(x_{si}) \cdot L_{si} (\bmod N)$
6. Sends (k_i, s_i) to the clerk via a secure channel

Upon receiving (k_i, s_i) from all signers, the clerk checks the equality

$$g_1{}^{s_i} \equiv k_i{}^{Z} \cdot y_{si}{}^{K \cdot L_{si}} \pmod{p}.$$

If the equation holds for all t signers, then he calculates

$$S' \equiv \left(M \prod_{i=1}^{t} g_1{}^{-s_i} \right)^{d} \pmod{p}$$

and

$$S \equiv (S')^{\alpha} (\bmod \bar{N}).$$

Finally, he sends (K, S) as the group signature to the group of verifiers.

2.3 Phase 3: Verification/Decryption

After the verifier group receives the group signature (K, S), k out of l verifiers execute the following steps to recover the message:

1. By using the information of the public identity for each verifier, compute

$$L_{vj} \equiv \prod_{i=1, i \neq j}^{k} \frac{-x_{vi}}{x_{vj} - x_{vi}} \pmod{N}$$

2. Calculate

$$z_{vj} \equiv Y_s{}^{f_v(x_{vj}) \cdot L_{vj}} \pmod{p}$$

3. Send z_{vj} to other verifiers via a secure channel. After all z_{vj} are received, compute

$$Z \equiv K \cdot \prod_{j=1}^{k} z_{vj} \pmod{p}.$$

4. Recover the message M from the following equation:

$$M \equiv \left(S^e\right)^{\beta} \cdot Y_s{}^K \cdot K^Z \pmod{p}$$

Finally, the verifiers verify the validity of the group signature by checking the validity of the attached redundancy to the message M.

Theorem 1 *Following the applied protocol, then the verification of the signature and retrieval of the message phase is correct.*

Proof: The equation in signature verification and message retrieval is true for correct signature since,

$$\begin{aligned}
&\left(S^e\right)^{\beta} \cdot Y_s{}^K \cdot K^Z \\
&\equiv \left(S'^{e}\right)^{\alpha\beta} \cdot Y_s{}^K \cdot K^Z \pmod{p} \\
&\equiv S'^{e} \cdot Y_s{}^K \cdot K^Z \pmod{p} \\
&\equiv \left(M \prod_{i=1}^{t} g_1{}^{-s_i}\right)^{ed} \cdot Y_s{}^K \cdot K^Z \pmod{p} \\
&\equiv M \left(\prod_{i=1}^{t} g_1{}^{-s_i}\right) \cdot Y_s{}^K \cdot K^Z \pmod{p} \\
&\equiv M \left(\prod_{i=1}^{t} k_i{}^{-Z} \cdot y_{si}{}^{-K \cdot L_{si}}\right) \cdot Y_s{}^K \cdot K^Z \pmod{p} \\
&\equiv M \left(\prod_{i=1}^{t} k_i{}^{-Z}\right) \left(\prod_{i=1}^{t} y_{si}{}^{-K \cdot L_{si}}\right) \cdot Y_s{}^K \cdot K^Z \pmod{p} \\
&\equiv M \cdot K^{-Z} \cdot Y_s{}^{-K} \cdot Y_s{}^K \cdot K^Z \pmod{p} \\
&\equiv M \pmod{p}
\end{aligned}$$

3 Security Analysis

1. The attacker tries to obtain the group secret keys $d, \beta, f_s(0)$, and $f_v(0)$ from all information of the system. This means that he has to solve $ed \equiv 1 (\bmod\ \phi(N))$,

$\alpha\beta \equiv 1(\text{mod } \phi(\bar{N}))$, $Y_s \equiv {g_1}^{f_s(0)}(\text{mod } p)$, and $Y_v \equiv {g_1}^{f_v(0)}(\text{mod } p)$, which is clearly infeasible due to the difficulty of solving factoring and discrete logarithm problems, respectively.

2. Derivation of the individual secret keys $f_s(x_{si})$ and $f_v(x_{vj})$ from the individual public keys $y_{si} \equiv {g_1}^{f_s(x_{si})}(\text{mod } p)$ and $y_{vj} \equiv {g_1}^{f_v(x_{vj})}(\text{mod } p)$ are equivalent to solve discrete logarithms problems.
3. Assume that factoring problem is solvable.
   a. The attacker could find d and try to generate the signature S from the signatire/encryption phase. However, without knowing the value of s_is from all signers, which is only can be computed by using the individual secret key for the signer $f_s(x_{si})$, the attacker still cannot sign the message.
   b. In the verification/decryption phase, the attacker could find β. But he/she still cannot recover the message since he/she need to find the value of Z, which is can only be obtained if he/she knows the individual secret key for the verifier $f_v(x_{vj})$.
4. Suppose that discrete logarithm problem can be solved.
   a. In this case, the attacker could find Z and $f_s(x_{si})$, then can compute s_is. However, without knowing the value of d, he/she still cannot generate the signature in sign/encryption phase.
   b. In verify/decryption phase, he could find the individual secret key $f_v(x_{vj})$ and then can compute Z. However, he/she still cannot recover the message since he/she does not know the value of β.

4 Performance Evaluation

It is shown that the proposed authenticated encryption scheme is secure against some attacks. Next, the efficiency of this scheme is evaluated, in terms of number of secret and public keys, computational complexity, and communication cost. The following notations are used to analyze the efficiency of this scheme:

- SK and PK are the number of secret and public keys, respectively.
- T_{exp} is the time complexity for executing the modular exponentiation computation.
- T_{mul} is the time complexity for executing the modular multiplication computation.
- T_{inv} is the time complexity for executing the modular inverse computation.
- $|\mu|$ denotes the bit length of μ.
- t is the number of signers involved in sign/encryption phase.
- k is the number of recipients/verifiers involve in verify/decryption phase.

The performance of this scheme is shown in Table 2.

Table 2 The performance of the proposed authenticated encryption scheme

		Proposed threshold-authenticated encryption scheme
Number of keys	SK	$4 + t + k$
	PK	$2 + t + k$
Computational complexity	Sign/Encrypt	$(6t+2)T_{exp}+(4t^2+5t)T_{mul}+(t^2-t)T_{inv}$
	Verify/Decrypt	$(5k)T_{exp}+(3k^2+k)T_{mul}+(k^2-k)T_{inv}$
Size of parameters/ communication cost		$(t+k+1)\,\lvert N\rvert+\lvert\bar{N}\rvert\,(4t+3k)\,\lvert p\rvert$

5 Conclusion

In this paper, a new threshold-authenticated encryption scheme based on two common problems in cryptography: factoring and discrete logarithms, is proposed. Beside the scheme, it is also shown that the scheme that has been developed is secure against some cryptographic attacks and requires a reasonable time complexity in both the signing and verifying phases.

Acknowledgments The first author acknowledges the financial support received from Universiti Utara Malaysia under the LEADS grant code 12030.

References

1. Chen, T.S., Kuang, K.H., Chung, Y.F.: A practical authenticated encryption scheme based on the elliptic curve cryptosystem. Comp. Stand. Inter., **26**, 461–469 (2004)
2. Nyberg, K., Rueppel, R.A.: A new signature scheme based on the DSA giving message recovery. In: Denning, D., Pyle, R., Ganesan, R., Sandhu, R., & Ashby, V. (eds.) Proceedings of the First ACM Conference on Computer and Communications Security, pp. 58–61. ACM Press, New York (1993)
3. Hsu, C.L., Wu, T.C.: Authenticated encryption scheme with (*t, n*) shared verification. IEE Proc.-E, **145**(2), 117–120 (1998)
4. Desmedt, Y.: Society and group oriented cryptography: a new concept. In: Pomerance, C. (ed.) Advances in Cryptology, Proceedings of Crypto '87. Lecture Notes in Computer Science, vol. 293, pp. 120–127. Springer, Heidelberg (1988)
5. Desmedt, Y., Frankel, Y.: Shared generation of authenticators. In: Feigenbaum, J. (ed.) Advances in Cryptology, Proceedings of Crypto '91. Lecture Notes in Computer Science, vol. 576, pp. 457–469. Springer, Heidelberg (1992)
6. Shamir, A.: How to share a secret. Commun. ACM, **22**(11), 612–613 (1979)
7. Wang, C.T., Chang, C.C., Lin, C.H.: Generalization of threshold signature and authenticated encryption for group communications. IEICE T. Fund. Electr., **E83-A**(6), 1228–1237 (2000)
8. Hsu, C.L., Wu, T.S., Wu, T.C.: Improvements of generalization of threshold signature and authenticated encryption for group communications. Inform. Process. Lett., **81**, 41–45 (2002)
9. Rivest, R., Shamir, A., Adleman, L.: A method for obtaining digital signature and public-key cryptosystem. Commun. ACM, **21**(2), 120–126 (1978)
10. ElGamal, T.: A public key cryptosystem and a signature scheme based on discrete logarithms. IEEE T. Inform. Theory, **31**(4), 469–472 (1985)

On Differential Invariants of Some Classical Groups

Ural D. Bekbaev and Isamiddin S. Rakhimov

Abstract The paper deals with the description of the field of differential rational invariants under simultaneous changing of the differential operator and an action of subgroups of affine transformation group on system of finite points of finite-dimensional differential vector space. In particular, the results in terms of differential geometry mean that we describe the field of differential rational invariants of finite system of geometric curves with respect to various group motions.

1 Introduction

The roots of differential invariants go back to the nineteenth century when fruitful applications of Lie groups in the theory of differential equations were found. The first differential invariant discovered was the Schwarzian derivative $\frac{y'''}{y''} - \frac{3}{2}\left(\frac{y''}{y'}\right)^2$ of function $y(t)$ of one variable. It is invariant under a fractional linear change of an independent variable and, separately, under a fractional linear change of the dependent variable. We give several well known examples imitated from [7] (for more examples see annotated bibliography therein) to motivate our research.

Example 1 Let $\mathbf{x}' = A\mathbf{x}$ be a system of ordinary linear differential equations of the first order, where

$$\mathbf{x} = \begin{bmatrix} x_1(t) \\ \cdot \\ \cdot \\ \cdot \\ x_n(t) \end{bmatrix} \qquad A = \left(a_{ij}(t)\right)_{i,j=1,\dots,n},$$

the $x_i(t)$ and $a_{ij}(t)$ are functions of t from a given class $\mathfrak{F}$ and $\mathbf{x}' = \frac{d\mathbf{x}}{dt}$.

I. S. Rakhimov (✉)
Laboratory of Cryptography, Analysis and Structure,
(INSPEM) & Department of Mathematics, FS, UPM, Serdang, Malaysia
e-mail: risamiddin@gmail.com

U. D. Bekbaev
Kulliyyah of Engineering, IIUM, Selayang, Malaysia
e-mail: bekbaev@iium.edu.my

A. Kilicman et al. (eds.), *International Conference on Mathematical Sciences and Statistics 2013*, DOI 10.1007/978-981-4585-33-0_26,

It is easy to see the change $\mathbf{x} = B\mathbf{y}$ of dependent matrix $\mathbf{x}$, where $B = B(t)$ is an invertible $n \times n$ matrix, which gives rise to the system of differential equations:

$$\mathbf{y}' = (B^{-1}AB - B^{-1}B')\mathbf{y}.$$

Thus, this action is the same as the action of group $GL_n(\mathfrak{F})$ on $Mat_n(\mathfrak{F})$ given by

$$B(A) = B^{-1}AB - B^{-1}B' \text{ for } A \in Mat_n(\mathfrak{F}),\ B \in GL_n(\mathfrak{F}).$$

To describe the orbits of this action one needs to construct functions f of a_{ij} and their derivatives such that the values of f for A and $B(A)$ are the same.

Let $R = \mathfrak{F}\{a_{ij}\}_{i,j=1,2,\ldots,n}$ be the ring of differential polynomials of a_{ij}. If R is an integral domain then one can define its field of differential rational functions $K = \mathfrak{F} < a_{ij} >$. The action above induces an action of $GL_n(\mathfrak{F})$ on R and K. By $R^{GL_n(\mathfrak{F})}$ and $K^{GL_n(\mathfrak{F})}$ one denotes the subring of invariant differential polynomials and subfields of invariant differential rational functions, respectively. The study of $R^{GL_n(\mathfrak{F})}$ and $K^{GL_n(\mathfrak{F})}$ is of great interest. This problem has also a geometric interpretation (see [7]).

There is a result of Nichols and Weisfeiler [6] asserting that if $\mathfrak{F}$ is differentially closed, then $GL_n(\mathfrak{F})$ acts transitively on $Mat_n(\mathfrak{F})$, which means that there is only one orbit, e.g., $R^{GL_n(\mathfrak{F})} = \mathbb{C}$. However, results by Babbitt and Varadarajan [1], [2], [3] explored a difficult picture in a slightly general case, when the authors take the field of formal Laurent series in t or the field of convergent (outside 0) series as $\mathfrak{F}$.

Example 2 Consider a system of ordinary linear differential equations of the second order: $\mathbf{x}'' = 2A_1\mathbf{x}' + A_2\mathbf{x}$. The action $\mathbf{x} = B\mathbf{y}$ leads to the system

$$\mathbf{y}'' = 2(B^{-1}A_1B - B^{-1}B')\mathbf{y}' + (B^{-1}A_2B + 2B^{-1}A_1B' - B^{-1}B'').$$

Thus, we get an action $GL_n(\mathfrak{F})$ on $M = Mat_n(\mathfrak{F}) \bigoplus Mat_n(\mathfrak{F})$ given by

$$B\begin{bmatrix} A_1 \\ A_2 \end{bmatrix} = \begin{bmatrix} B^{-1}A_1B - B^{-1}B' \\ B^{-1}A_2B + 2B^{-1}A_1B' - B^{-1}B'' \end{bmatrix}.$$

The detailed study of this case has been given in [7].

Example 3 Consider the class of linear differential equations of the form

$$L(y) = y^{(n)} + a_1(t)y^{(n-1)} + \ldots + a_n(t)y = 0.$$

The change of variables (both dependent and independent) of the form

$$\begin{cases} y_1 = \lambda(t)y \\ t_1 = \mu(t) \end{cases}$$

form a group if $\lambda \neq 0$ and $\mu' \neq 0$. The class above is stable under the action of G. One is interested in finding differential invariants.

The most interesting action occurs in case $a_1 = a_2 = 0$ in $L(y)$ (see [7]).

First, one considers polynomial differential invariants, i.e. the polynomials in $a_1, a_2, \ldots, a_n$ and their derivatives which are invariant under this group. It turns out that such invariants are very few. It is not difficult to see that they are, in essence, differential polynomials P in a_i that the substitution $y_1 = \lambda(t)y$, $t_1 = \mu(t)$ gives P times some power of the Jacobian of the substitution.

2 Main Results

First, we prove a few propositions in the case of differential fields with commuting differentials $d_1, d_2, ..., d_m$.

Proposition 1 *Let $K \subset L$ be a differential extension, $g = [g_{ij}]_{i,j=1,...,m} \in GL(m, L)$ such that $d_k g_{ij} = d_i g_{kj}$ for $i, j, k = 1, ..., m$, $\delta = g^{-1}d$, where d (respectively, δ) is the column vector with "components" $d_1, d_2, ..., d_m$ (respectively, $\delta_1, \delta_2, ..., \delta_m$). If K is invariant with respect to $\delta_1, \delta_2, ..., \delta_m$, then*

$$\delta - tr.degL/K \geq d - tr.degL/K < (g_{ij}) > .$$

(Here, $\delta - tr.degL/K$ is the differential transcendental degree of δ-differential field L over δ-differential field K.)

Proof Let $L\{\mathbf{x}, \delta\}$ be the ring of δ-polynomials over L at differential variables $\mathbf{x} = (x_1, x_2, ..., x_n)$, and $L < \mathbf{x}, \delta >$ be its field of fractions. Consider the mapping

$$\Phi : L\{\mathbf{x}, \delta\} \longrightarrow L\{\mathbf{x}, d\}, \text{ defined by } \Phi(f^{\delta}\{\mathbf{x}\} = f^{g^{-1}d}\{\mathbf{x}\}, \text{ for } f^{\delta}\{\mathbf{x}\} \in L\{\mathbf{x}, \delta\}.$$

The mapping $\psi : L\{\mathbf{x}, d\} \longrightarrow L\{\mathbf{x}, \delta\}$, where $\psi(f^d\{\mathbf{x}\}) = f^{g\delta}\{\mathbf{x}\}$ is the inverse of Φ. Therefore, if $f^{\delta}\{\mathbf{x}\} \in K\{\mathbf{x}, \delta\}$ is a nonzero polynomial, then, being its image the polynomial $\Phi(f^{\delta}\{\mathbf{x}\}) \in K < (g_{ij})_{i,j=1}^m > \{\mathbf{x}, d\}$ also is not zero. This implies that if a set of elements $a_1, a_2, ..., a_n$ from L is δ-algebraically independent over K, then it is also d-algebraically independent over $K < (g_{ij})_{i,j=1}^m >$. Hence,

$$\delta - tr.degL/K \geq d - tr.degL/K < (g_{ij}) > .$$

Proposition 2 *Let*

$$G = \{[g_{ij}] \in GL(m, K < \mathbf{x}, d >) : \ d_k g_{ij} = d_i g_{kj} \ \ for \ \ i, j, k = 1, 2, ..., m\}$$

and $[\chi_{ij}^d < \mathbf{x} >] \in G$ such that for any $g \in G$ the identity $[\chi_{ij}^{g^{-1}d} < \mathbf{x} >] = g^{-1}[\chi_{ij}^d < \mathbf{x} >]$ holds true. Then, for $\delta = [\chi_{ij}^d < \mathbf{x} >]^{-1}d$ the $[\chi_{ij}^{\delta} < \mathbf{x} >]$ is the identity matrix of order m.

Proof Straightforward.

Proposition 3 *Let K be a differential field of characteristic zero. Then, each element $f < \mathbf{x} >$ of $K < \mathbf{x}, d > \backslash K$ is d-transcendental over K.*

Proof Let Θ be free commutative semigroup with generators $d_1, d_2, ..., d_m$. Consider linear ordering in Θ as follows:

$$d_1^{i_1} d_2^{i_2}, ..., d_m^{i_m} > d_1^{j_1} d_2^{j_2}, ..., d_m^{j_m}$$

if $\left(\sum_{k=1}^m i_k, i_2, ..., i_m \right) > \left(\sum_{k=1}^m j_k, j_2, ..., j_m \right)$ with respect to lexicographical order.

Now, represent $f < \mathbf{x} >$ in the form of irreducible fraction as follows: $f < \mathbf{x} > = \frac{P\{\mathbf{x}\}}{Q\{\mathbf{x}\}}$. Suppose that x_1 takes part in $\frac{P\{\mathbf{x}\}}{Q\{\mathbf{x}\}}$ and $\theta_1 = max\{d - deg_{x_1} P\{\mathbf{x}\}, d - deg_{x_1} Q\{\mathbf{x}\}$.

We show that the differential order of $\theta f < \mathbf{x} >$ with respect to x_1 is equal to $\theta\theta_1$. It is obvious if $d - deg_{x_1} P\{\mathbf{x}\} \neq d - deg_{x_1} Q\{\mathbf{x}\}$. Now, consider the case $\theta_1 = d - deg_{x_1} P\{\mathbf{x}\} = d - deg_{x_1} Q\{\mathbf{x}\}$. Then, for each d_i the coefficient of $d_i\theta_1 x_1$ in the denominator of $d_i \frac{P<\mathbf{x}>}{Q<\mathbf{x}>}$ will be $\frac{\partial P<\mathbf{x}>}{\partial\theta_1<\mathbf{x}>} Q < \mathbf{x} > - \frac{\partial Q<\mathbf{x}>}{\partial\theta_1<\mathbf{x}>} P < \mathbf{x} >$, which is nonzero due to the following two facts: the ring $K < \mathbf{x}, d >$ is unique factorization domain and $char K = 0$. Now since $\{x_1, x_2, ..., x_n\}$ is a d-algebraic independent over K, we conclude that the system elements $\{\theta f : \theta \in \Theta\}$ is an algebraic independent over the field K.

Further, let (K, d) be an ordinary differential field of characteristic zero with the property: if $f^d\{t\} \in K\{t, d\}$ is a differential polynomial at single variable t and $f^d\{g\} = 0$ for all $g \in K^*$ then $f^d\{t\}$ is a trivial differential polynomial. Let C stand for field of constant of K. The group of affine transformations $C^n \rtimes GL(n, C)$ operates naturally on the differential space $V = K^n$. In this paper the actin of a subgroup H of the group $C^n \rtimes GL(n, C)$ on V is understood the restriction of the action of $C^n \rtimes GL(n, C)$ to H.

Let H be a subgroup of $C^n \rtimes GL(n, C)$.

Definition 1 A differential rational function (d.r.f.) $f^d < \mathbf{x} > \in K < \mathbf{x}, d >$ is said to be (K^*, H) invariant (relatively invariant with a weight m) if for all $g \in K^*$ and $h \in H$ the following holds true:

$$f^{g^{-1}d} < h\mathbf{x} > = f^d < \mathbf{x} > \quad (f^{g^{-1}d} < h\mathbf{x} > = g^{-m} f^d < \mathbf{x} >, respectively).$$

Theorem 1 *Let $p^d < \mathbf{x} > \in K < \mathbf{x}, d >$ be a differential function such that $p^{g^{-1}d} < h\mathbf{x} > = g^{-1} p^d < \mathbf{x} >$ for all $g \in K^*$ and $h \in H$. Put $\delta = (p^d < \mathbf{x} >)^{-1} d$. Then, the field of all (K^*, H)-invariant functions is a differential field with respect to K-linear differential operator δ and as a such differential field it has the transcendental degree equal to $n - 1$ over K.*

Proof Due to a result of [4] it is known that the field (K^*, H)-invariant differential rational functions is a δ-differential field. According to Proposition 1

$$\delta - tr.deg L/K \geq d - tr.deg L/K < (g_{ij}) >$$

and in virtue of Proposition 3 the last is equal to $n - 1$.

On the other hand, due to Proposition 2 $p^\delta < \mathbf{x} > = 1$. Hence, $\delta - tr.deg K < \mathbf{x}, \delta > /K \leq n - 1$ and so

$$\delta - tr.deg K < \mathbf{x}, \delta > /K = n - 1.$$

In order to prove the theorem it is sufficient to show that each x_i, $i = 1, 2, ..., n$ is δ-algebraic over the field of (K^*, H)-invariant rational functions. To show that we consider the following differential equation with respect to y :

$$(W^\delta(\delta x_1, \delta x_2, ..., \delta x_n))^{-1} W^\delta(\delta x_1, \delta x_2, ..., \delta x_n, \delta y) = 0,$$

where W is a Wronskian. Obviously, x_i, $i = 1, 2, ..., n$ are solutions of this equation and the coefficients of the equation are (K^*, H)-invariant rational functions.

Further, $U_i^d\{\mathbf{x}\} = [\mathbf{x} d\mathbf{x} \cdot ... \cdot d^{i-1}\mathbf{x} d^n\mathbf{x} d^{i+1}\mathbf{x} \cdot ... \cdot d^{n-1}\mathbf{x}]$, that is, in the determinant $[\mathbf{x} d\mathbf{x} \cdot ... \cdot d^{n-1}\mathbf{x}]$ the column $d^i\mathbf{x}$ is replaced by $d^n\mathbf{x}$.

Theorem 2 *The field $K(V)^{(K^*,GL(n,C))}$ of $(K^*,GL(n,C))$-invariant functions of the space V is a differential field with respect to K-linear differential operator*

$$\delta=\left(\frac{n(n+1)}{2}\frac{dU_n^d\{\mathbf{x}\}}{U_n^d\{\mathbf{x}\}}-\frac{n(n-1)}{2}\frac{dU_0^d\{\mathbf{x}\}}{U_0^d\{\mathbf{x}\}}\right)^{-1}d$$

and as a such differential field, it is generated over K by the following δ-algebraic independent (over K) system of elements:

$$\frac{U_0^\delta\{x\}}{U_n^\delta\{x\}},\ \frac{U_1^\delta\{x\}}{U_n^\delta\{x\}},\ldots,\ \frac{U_{n-2}^\delta\{x\}}{U_n^\delta\{x\}}.$$

Proof It is not difficult to see that the function

$$p^d<\mathbf{x}>=\frac{n(n+1)}{2}\frac{dU_n^d\{\mathbf{x}\}}{U_n^d\{\mathbf{x}\}}-\frac{n(n-1)}{2}\frac{dU_0^d\{\mathbf{x}\}}{U_0^d\{\mathbf{x}\}}$$

possesses the property $p^{g^{-1}d}<h\mathbf{x}>=g^{-1}p^d<\mathbf{x}>$ for $g\in K^*$ and $h\in GL(n,C)$. Therefore, according to Theorem 1, the field $K(V)^{(K^*,GL(n,C))}$ is δ-differential field. The field $(K^*,1)$-invariant functions coincides with $K<\mathbf{x},\delta>$. We specify $(K^*,GL(n,C))$-invariant elements of $K<\mathbf{x},\delta>$. To do this, we note that the K-linear operator δ is invariant under the action of $GL(n,C)$. Therefore, we can follow the proof given in [5] (Theorem 8.7) and show that the system

$$\frac{U_0^\delta\{\mathbf{x}\}}{U_n^\delta\{\mathbf{x}\}},\ \frac{U_1^\delta\{\mathbf{x}\}}{U_n^\delta\{\mathbf{x}\}},\ldots,\ \frac{U_{n-2}^\delta\{\mathbf{x}\}}{U_n^\delta\{\mathbf{x}\}}$$

is the set of δ-generators of $K(V)^{(K^*,GL(n,C))}$.

On the other hand, if we consider

$$p^d<\mathbf{x}>=n\frac{U_{n-1}^d\{\mathbf{x}\}}{U_n^d\{\mathbf{x}\}}-\frac{n(n-1)}{2}\left(\frac{U_0^d\{\mathbf{x}\}}{U_n^d\{\mathbf{x}\}}\right)^{-1}d\frac{U_0^d\{\mathbf{x}\}}{U_n^d\{\mathbf{x}\}}$$

then in view of Proposition 2 $p^\delta<\mathbf{x}>=1$, this implies that $\frac{U_{n-1}^\delta\{x\}}{U_n^\delta\{x\}}$ is δ-rationally expressed via $\frac{U_0^\delta\{x\}}{U_n^\delta\{x\}}$. Therefore, the δ-algebraically independence of the generators given in the theorem follows from Theorem 1.

The general scheme of proving the theorems given below is similar to that of Theorem 2, however, in each case we choose as much as possible, a simple polynomial for $p^d<\mathbf{x}>$.

Consider the natural action of the groups $GL(n,C)$ and $SL(n,C)$ on $V^p\times(V^*)^q$, where $p,q\in\mathbb{N}$.

Let $\mathbf{x}^1,\mathbf{x}^2,\ldots,\mathbf{x}^p,\mathbf{y}^1,\mathbf{y}^2,\ldots,\mathbf{y}^q$ be n-dimensional differential variables. By $U_{ij}^d\{\mathbf{x}^1,\mathbf{x}^j\}$ we denote the determinant $[\mathbf{x}^1 d\mathbf{x}^1\ldots d^{i-1}\mathbf{x}\ \mathbf{x}^j\ d^{i+1}\mathbf{x}\ldots d^{n-1}\mathbf{x}^1]$. The Euclidian inner product is denoted by $(\cdot,\cdot)$.

In the next two theorems $\delta = (p^d < \mathbf{x}^1 >)^{-1}d$, where

$$p^d < \mathbf{x}^1 >= \left(\frac{n(n+1)}{2} \frac{dU_n^d\{\mathbf{x}^1\}}{U_n^d\{\mathbf{x}^1\}} - \frac{n(n-1)}{2} \frac{dU_0^d\{\mathbf{x}^1\}}{U_0^d\{\mathbf{x}^1\}} \right).$$

Theorem 3 *The field $K < V^p \times (V^*)^q >^{(K^*, GL(n,C))}$ is a differential field with respect to K-linear differential operator δ and as a such differential field it is generated by the following δ-algebraic independent system of elements:*

$$\frac{U_{ij}^\delta\{\boldsymbol{x}^1, \boldsymbol{x}^j\}}{U_n^\delta\{\boldsymbol{x}^1\}}, \text{ where } i = 0, 1, ..., n-1, j = 2, 3, ..., p;$$

$$\frac{U_i^\delta\{\boldsymbol{x}^1\}}{U_n^\delta\{\boldsymbol{x}^1\}}, \text{ where } i = 0, 1, ..., n-2;$$

$$\left(\delta^i \boldsymbol{x}^1, \boldsymbol{y}^j\right), \text{ where } i = 0, 1, ..., n-1, \ j = 1, 2, ..., q.$$

Theorem 4 *The field $K < V^p \times (V^*)^q >^{(K^*, SL(n,C))}$ is a differential field with respect to K-linear differential operator δ and as a such differential field it has the differential transcendental degree equal to $n(p+q) - 1$ and is generated by the following δ-algebraic independent system of elements:*

$$U_i^\delta\{\boldsymbol{x}^1\}, U_{ij}^\delta\{\boldsymbol{x}^1, \boldsymbol{x}^j\}, \ \left(\delta^i \boldsymbol{x}^1, \boldsymbol{y}^k\right), \text{ where } i = 0, 1, ..., n-1, \ j = 2, ..., q, \ k = 1, 2, ..., q.$$

In the following two theorems, the field of differential rational functions under natural action of the groups $O(n, C)$ and $SO(n, C)$ on V^p is described. In the theorems $\delta = (\mathbf{x}^1, d\mathbf{x}^1)^{-1}d$.

Theorem 5 *The field $K < V^p >^{(K^*, O(n,C))}$ is a differential field with respect to K-linear differential operator δ and as a such differential field it has the differential transcendental degree equal to $np - 1$ and is generated by the following system of elements:*

$$\left(\delta^i \boldsymbol{x}^1, \delta^i \boldsymbol{x}^1\right), \ \left(\delta^i \boldsymbol{x}^1, \boldsymbol{x}^j\right), \text{ where } i = 0, 1, ..., n-1, \ j = 2, ..., p.$$

Theorem 6 *The field $K < V^p >^{(K^*, SO(n,C))}$ is a differential field with respect to K-linear differential operator δ and as a such differential field it has the differential transcendental degree equal to $np - 1$ and is generated by the following system of elements:*

$$U_n^\delta\{\boldsymbol{x}^1\}, \left(\delta^i \boldsymbol{x}^1, \delta^i \boldsymbol{x}^1\right), \ \left(\delta^k \boldsymbol{x}^1, \boldsymbol{x}^j\right), \text{ where } i = 0, 1, ..., n-2, \ k = 0, 1, ..., n-1,$$
$$j = 2, ..., p.$$

In the next theorem, n is an even number and $\delta = [\mathbf{x}^1 d\mathbf{x}^1]^{-1}d$, where

$$[\mathbf{x}^1 \mathbf{x}^j] = (x_1^1 x_2^j - x_2^1 x_1^j) + (x_3^1 x_4^j - x_4^1 x_3^j) + ... + (x_{n-1}^1 x_n^j - x_n^1 x_{n-1}^j).$$

Theorem 7 *The field* $K < V^p >^{(K^*, Sp(n,C))}$ *is a differential field with respect to* K*-linear differential operator* δ *and as a such differential field, it has the differential transcendental degree equal to* $np-1$ *and is generated by the following* δ*-algebraic independent system of elements:*

$[\delta^i \boldsymbol{x}^1, \delta^{i+1}\boldsymbol{x}^1]$, $i = 1, ..., n-1$, $[\delta^k \boldsymbol{x}^1, \boldsymbol{x}^j]$, where $k = 0, 1, ..., n-1$, $j = 2, ..., p$.

From geometrical point of view, it naturally makes sense to consider analogous invariants for the groups $C^n \rtimes O(n,C)$, $C^n \rtimes SO(n,C)$, $C^2 n \rtimes Sp(2n,C)$, $C^n \rtimes GL(n,C)$, $C^n \rtimes SL(n,C)$ as well.

In the following theorems, $\delta = \left(d \frac{[d\mathbf{x}^1 d^2\mathbf{x}^1 ... d^n\mathbf{x}^1]^2}{(d\mathbf{x}^1, d\mathbf{x}^1)^{\frac{n(n+1)}{2}}} \right)^{-1} d.$

Theorem 8 *The field* $K < V^p >^{(K^*, C^n \rtimes O(n,C))}$ *is a differential field with respect to* K*-linear differential operator* δ *and as a such differential field it has the differential transcendental degree equal to* $np-1$ *and is generated by the following system of elements:*

$$\left(\delta^i \boldsymbol{x}^1, \delta^i \boldsymbol{x}^1\right), \ \left(\delta^i \boldsymbol{x}^1, \boldsymbol{x}^j - \boldsymbol{x}^1\right), \text{ where } i = 0, 1, ..., n, \ j = 2, ..., p.$$

Theorem 9 *The field* $K < V^p >^{(K^*, C^n \rtimes SO(n,C))}$ *is a differential field with respect to* K*-linear differential operator* δ *and as a such differential field it has the differential transcendental degree equal to* $np-1$ *and is generated by the following system of elements:*

$$U_n^\delta\{\delta \boldsymbol{x}^1\}, \ \left(\delta^i \boldsymbol{x}^1, \delta^i \boldsymbol{x}^1\right), \text{ where } i = 1, ..., n-1, \ \left(\delta^k \boldsymbol{x}^1, \boldsymbol{x}^j - \boldsymbol{x}^1\right), \text{ where } i = 1, ..., n, \ k = 2, ..., p.$$

Theorem 10 *The field* $K < V^p >^{(K^*, C^n \rtimes Sp(n,C))}$ *is a differential field with respect to* K*-linear differential operator* δ *and as a such differential field it has the differential transcendental degree equal to* $np-1$ *and is generated by the following* δ*-algebraic independent system of elements:*

$$[\delta^i \boldsymbol{x}^1, \delta^{i+1}\boldsymbol{x}^1], \ [\delta^k \boldsymbol{x}^1(\boldsymbol{x}^j - \boldsymbol{x}^1)], \text{ where } i = 1, ..., n, \ j = 2, ..., p.$$

Let

$$p_1^d < \mathbf{x} >= \frac{U_{n-2}^d\{d\mathbf{x}\}}{U_n^d\{d\mathbf{x}\}} + \frac{n-1}{3} d \frac{U_{n-2}^d\{d\mathbf{x}\}}{U_{n-1}^d\{d\mathbf{x}\}} - \frac{(n-1)(3n+2)}{6n(n+1)} \left(\frac{U_{n-1}^d\{d\mathbf{x}\}}{U_n^d\{d\mathbf{x}\}} \right)^2$$

and

$$p^d < \mathbf{x} >= \frac{n(n+1)}{2} \frac{dp_1^d < \mathbf{x} >}{p_1^d < \mathbf{x} >} - 2 \frac{dU_n^d\{d\mathbf{x}\}}{U_n^d\{d\mathbf{x}\}}.$$

It is easy to show that $p^{g^{-1}d} < h\mathbf{x} >= g^{-1} p^d < \mathbf{x} >$ for all $g \in K^*$ and $h \in C^n \rtimes GL(n,C)$. That is, $p^d < \mathbf{x} >$ is an invariant with the weight one for $C^n \rtimes GL(n,C)$. To the best knowledge of the authors, this is first time when a differential rational invariant with weight one for the group $C^n \rtimes GL(n,C)$ is given.

In the next two theorems, we take $\delta = (p^d < \mathbf{x}^1 >)^{-1} d$.

Theorem 11 *The field $K < V^p >^{(C^n \rtimes GL(n,C))}$ is a differential field with respect to K-linear differential operator δ and as a such differential field it has δ-differential transcendental degree equal to $np - 1$ and is generated by the following system of elements:*

$$\frac{U_i^\delta\{\delta \boldsymbol{x}^1\}}{U_n^\delta\{\delta \boldsymbol{x}^1\}}, \quad \frac{U_{ij}^\delta\{\delta \boldsymbol{x}^1, \boldsymbol{x}^j - \boldsymbol{x}^1\}}{U_n^\delta\{\delta \boldsymbol{x}^1\}}, \text{ where } i = 0, 1, ..., n-1, j = 2, 3, ..., p.$$

Theorem 12 *The field $K < V^p \times (V^*)^q >^{(K^*, C^n \rtimes SL(n,C))}$ is a differential field with respect to K-linear differential operator δ and as a such differential field it has the differential transcendental degree equal to $np - 1$ and is generated by the following system of elements:*

$$U_i^\delta\{\delta \boldsymbol{x}^1\}, \ U_{ij}^\delta\{\delta \boldsymbol{x}^1, \boldsymbol{x}^j - \boldsymbol{x}^1\}, \ \text{where } i = 0, 1, ..., n-1, \ j = 2, ..., p.$$

The following proposition establishes a link between invariant and relatively invariant differential rational functions. Here H stands for any subgroup of $GL(n, K)$.

Proposition 4 *If $p^d < \mathbf{x} >$ is nonzero (K^*, H)-relatively invariant differential rational function with a weight one, then any (K^*, H)-relatively invariant differential rational function has the form $(p^d < \mathbf{x} >)^m f^d < \mathbf{x} >$, where $f^d < \mathbf{x} >$ is a (K^*, H)-invariant differential function.*

Proof Straightforward.

Remark 1

i. In differential geometry, one usually considers (K^*, H)-invariant differential rational functions in sense of the paper with constant coefficients. In Theorems 3-12 the generators and the respective functions $p^d < \boldsymbol{x} >$ given are the same as in classical case. Therefore, all the theorems stay to be valid after their natural reformulations, if we restrict ourselves to the field of (K^*, H)-differential rational invariant functions with coefficients from C.
ii. If in Theorems 5 and 6 we take $\delta = \frac{(x^1, dx^1)}{(dx^1, dx^1)}$, then the generator $(\delta \boldsymbol{x}^1, \delta \boldsymbol{x}^1)$ can be removed from the list of generators.

Acknowledgement This research was supported by the Universiti Putra Malaysia (UPM) through the research grant 05-02-12-2188RU.

References

1. Babbitt, D.G., Varadarajan, V.S.: 'Formal reduction theory of memomorphic differential equations'. Pac. J. Math. **109**, 1–80 (1983)
2. Babbitt, D.G., Varadarajan, V.S.: 'Local moduli for meromorphic differential equations'. Bull. Am. Math. Soc. (New Series) **12**(1), 95–98 (1985)
3. Babbitt, D.G., Varadarajan, V.S.: 'Deformation of nilpotent matrices over rings and reduction of analytic families of memomorphic differential equations'. Mem. Am. Math. Soc. **55**(325), 1–153 (1985)

4. Bekbaev U.D.: ‘On differential rational invariants of differential equations’. Priprint VINITI(1989), n 4975–V89 (in russia)
5. Hadjiev Dj: Applications of theory invariants to differential geometry of curves, FAN, (1988), 135 pp. (in russian)
6. Nichols, W., Weisfeiler, B.: Differential formal groups of J.F. Ritt. Amer. J. Math. **104**, 943–1003 (1982)
7. Weisfeiler, B.: ‘Comments on differential invariants’, infinite-dimensional groups with applications. Math. Sci. Res. Inst. Publ. **4**, 355–370 (1985)

Structure of Relative Relation Modules of Finite Groups

Mohammad Yamin and P. K. Sharma

Abstract Decomposability properties of modules of finite groups are known to be useful in describing the structure of certain groups. Let E be a free product of a finite number of cyclic groups, and S a normal subgroup of E such that $E/S \cong G$ is finite. For a prime p, $\hat{S} = S/S^{'}S^{p}$ may be regarded as F_pG-module. Whenever E is a free group, $\hat{S}$ is called relation module (modulo p) of G; in general $\hat{S}$ is called relative relation module (modulo p). Many researchers have studied relation and relative relation modules. In this paper, we describe the structure of relative relation modules of finite groups, and in particular, those of abelian p-groups. We also describe the decomposition of relative relation modules of SL(2,p) and PSL(2,p). Decomposition of $\hat{S}$ described in this paper may be useful in studying factor group of PSL(2, Z), which is an interesting problem in its own right.

1 Introduction

Let G be a fixed finite group of order n generated by $X = \{g_i : 1 \leq i \leq d\}$. Let G_i be the cyclic subgroup of G generated by g_i of order n_i and E_i be cyclic group generated by e_i of order m_i, where $m_i = k_i n_i, 1 \leq i \leq d$. Let p be fixed prime and F_p the field of p-elements. Let $k_i < \infty$ and $p \not| k_i$ if $i \leq \delta$ and $\delta \leq d$, and $k_i = \infty$ or $p|k_i$ if $\delta + 1 \leq i \leq d$. Let E be the free product of $E_i, 1 \leq i \leq d$. Let S be the kernel of the natural epimorphism, defined by $e_i \to g_i$, $1 \leq i \leq d$. Then we get a short, exact sequence of groups.

$$1 \to S \to E \overset{\psi}{\to} G \to 1 \tag{1}$$

The elementary abelian p-group $\hat{S} = S/S^{'}S^{p}$, regarded as an F_pG module, is called relative relation module (modulo p). When E is a free group, that is when $\delta = 0$, $\hat{S}$

M. Yamin (✉)
King Abdulaziz University, Jeddah, Saudi Arabia
e-mail: myamin@kau.edu.sa

P. K. Sharma
D.A.V. College, Jalandhar, Punjab, India
e-mail: pksharma@davjalandhar.com

A. Kilicman et al. (eds.), *International Conference on Mathematical Sciences and Statistics 2013*, DOI 10.1007/978-981-4585-33-0_27,

is called relation module (modulo p). Gaschutz [3], Gruenberg [5], [6] and Mittal and Passi [10] have studied relation modules. Relative relation modules have been studied by Yamin [13]–[15], Kimmerle [9] and Sharma [11], [12]. The aim of this paper is to describe certain properties of relative relation modules in non-trivial cases. Let us recall an important result that we shall need later.

Proposition 1 Given (1), the sequence, which we call the relative relation sequence

$$1 \to \hat{S} \to \oplus_{i=1}^{\delta} g_i^G \oplus_{i=\delta+1}^{d} b_i F_p G \stackrel{\hat{\psi}}{\to} g \to 1 \tag{2}$$

is F_pG-exact, where g_i^G denotes the augmentation ideal of $b_i F_p G$ induced to G and $\hat{\psi}$ is determined by

$$b_i(g_i - 1) \to g_i - 1 \text{ where } 1 \leq i \leq \delta, \text{ and}$$
$$b_i \to g_i - 1 \text{ where } 1 + \delta \leq i \leq d$$

For a proof, see [14].

Remark 1 When $\delta = 0$, the sequence (2) reverts to what is known as the relation sequence described and used in [5], [6].

2 Comparison of Relative Relation modules

Here we consider two different generating sets for a group and show that the resulting relative relation modules are isomorphic. To do that, let $X_1 = \{g_i : 1 \leq i \leq d_1\}$ and $X_2 = \{h_i : 1 \leq i \leq d_2\}$ be two generating sets of the group G, then the corresponding relative relation sequences will be

$$1 \to \hat{S_1} \to \oplus_{i=1}^{\delta_1} g_i^G \oplus_{i=\delta_1+1}^{d_1} b_i F_p G \stackrel{\hat{\psi_1}}{\to} g \to 1 \tag{3}$$

$$1 \to \hat{S_2} \to \oplus_{i=1}^{\delta_2} g_i^G \oplus_{i=\delta_2+1}^{d_2} b_i F_p G \stackrel{\hat{\psi_2}}{\to} g \to 1 \tag{4}$$

where $\hat{S_i} = S_i/S_i^{'}S_i^p$. Whenever $\delta_i = 0$, we identify $\hat{S_i}$ with $\hat{R_i}$. When $\delta_1 = \delta_2 = 0$ and $d_1 \leq d_2$, then by the application of Schanuel's lemma together with the Krull–Schmidt theorem, we have

$$\hat{R_2} \cong \hat{R_1} \oplus L, \text{ where } L \text{ is a free module of rank } d_2 - d_1. \tag{5}$$

The analogous result for relative relation modules is far from being true (see [14], Ex. 4.4). However, under certain conditions, we have the following result.

Proposition 2 ([14]) Suppose $d_1 = d_2 = d$ and $\delta_1 = \delta_2 = \delta$. Let x be a fixed element of G such that $< h_i > = < g_i >^x$.

The above proposition readily gives us the following corollary.

Corollary 1 Let p be semi-coprime to the order of G and suppose $d_1 = d_2 = d$ and $\delta_1 = \delta_2 = \delta$ and $h_i = g_i^{x_i}$, $1 \le i \le \delta$. Then $\hat{S}_1 \cong \hat{S}_2$.

This follows by Schanuel's lemma and the fact that in this case, g_i^G and h_i^G are projective.

Remark 2 For an isomorphism between $\hat{S}_1$ and $\hat{S}_2$, in the semi-coprime case, it is only necessary to have an isomorphism between the middle terms of the relative relation sequences (3) and (4).

3 Decomposability of Relative Relation Modules

In this section, we shall describe some important decomposition properties of relative relation modules in terms of projective and non-projective direct summands. Consider a decomposition

$$\hat{S} \cong P(\hat{S}) \oplus N(\hat{S}),$$

where $P(\hat{S})$ is F_pG-projective and $N(\hat{S})$ contains no projective summands. $P(\hat{S})$ is called the projective part and $N(\hat{S})$ the non-projective part of $\hat{S}$.

When p is coprime to $|G|$, the decomposition of $\hat{S}$ is easily described. When p is semi-coprime to $|G|$ i.e. when $p||G|$ and $p \not| |G_i|$, $1 \le i \le \delta$, it has been proved by Yamin [14] that $N(\hat{S})$ is non-zero and indecomposible (and also isomorphic to $N(\hat{S})$ in case when $\delta = 0$). In the non-semi coprime case, $N(\hat{S})$ may be zero or decomposable (and may not be a homomorphic image of $N(\hat{S})$ in the case when $\delta = 0$). It has also been proved by Yamin [15] that $\hat{S}$ is non-projective and indecomposable in the case when G is a p-group and $\delta = d$. In the case when $1 < \delta < d, d > 2$, it has been proved by the authors, to be reported elsewhere, that $\hat{S}$ is non-projective and decomposable into two indecomposable submodules in the case when G is an elementary abelian p-group.

The following Theorem from [14] is very useful in describing the structure of the relative relation module $\hat{S}$ in the semi-coprime case.

Theorem 1 Suppose p divides $|G|$ but $p \not| |G_i|$, $1 \le i \le \delta$. Let $1 \to J \to V \xrightarrow{f} W \xrightarrow{g} F_p \to 1$ is an F_pG-exact sequence, where V and W are projective modules, $\ker g = \phi W$ and $J \le \phi W$. Then

$$\hat{S} \oplus V \oplus F_pG \cong J \oplus_{i=1}^{\delta} g_i^G \oplus_{i=\delta+1}^{d} b_i F_pG \oplus W, \tag{6}$$

and J is non-zero, non-projective and indecomposable module.

Next, suppose that $X_2 = X_1 \cup \{g_0\}, \delta_2 = \delta_1 + 1$. Then we prove:

Lemma 1 $\hat{S}_1$ is a submodule of $\hat{S}_2/\hat{S}_1 \cong g_0^G$

Proof Let $V = \oplus_{i=1}^{\delta_1} g_i^G \oplus_{i=\delta_1+1}^{d} b_i F_pG$ and $W = g_0^G$. Clearly $\hat{\psi}_1$ and the restriction of $\hat{\psi}_2$ to V coincide. Therefore, $\hat{S}_2 \cap V = \hat{S}_1$. Moreover, $(\hat{S}_2/\hat{S}_1) \cap (V/\hat{S}_1) =$

$(V \oplus W)/\hat{S_1} = \{0\}$; therefore the sum of $\hat{S_2}/\hat{S_1}$ and $V/\hat{S_1}$ is a direct sum. Since $dim(\hat{S_2}/\hat{S_1}) + dim(V/\hat{S_1}) = dim(V \oplus W)/\hat{S_1}$,

$$(\hat{S_2}/\hat{S_1}) \oplus (V/\hat{S_1}) = (V \oplus W)/\hat{S_1} \cong V/\hat{S_1} \oplus W$$

An application of the Krull–Schmidt theorem now gives the required result. Next, we prove the following:

Proposition 3 Suppose either $g_0 \in X_1$ or p does not divide the order of $< g_0 >$. Then $\hat{S_2} = \hat{S_1} \oplus g_0^G$.

Proof Suppose $g \in X_1$. We now suppose without loss of generality that $g_0 = g_1$. Let $V = \oplus_{i=1}^{\delta_1} g_i^G \oplus_{i=\delta_1+1}^{d} b_i F_p G$. Let η_1 be the identity homomorphism of V, and η_2 the epimorphism of $V \oplus g_0^G$ to V such that the restriction of η_2 to V, and g_0^G is the identity isomorphism. Let $\hat{\psi}_1 = \eta_1 \hat{\psi}_2$ and $\hat{\psi}_2 = \eta_2 \hat{\psi}_1$. Now we apply the same argument as in the proof of Schanuel's lemma ([5], Lemma 11, p. 162) in conjunction with the Krull–Schmidt theorem to deduce the the desired result.

When $p \not| | < g_0 > |$, the result is obvious because g_0^G is projective.

It is not known whether $\hat{S_2}$ splits over $\hat{S_1}$ and g_0^G if is it only suppose that g_0 is a conjugate of an element in X_1 in G.

Remark 3 If in the Proposition 3 we suppose that G is a p-group and $g_0 \in X_1$, then $N(\hat{S_2}) \cong N(\hat{S_1}) \oplus g_0^G$. This follows from the fact that, g_0^G, being embedded in $F_p G$, is non-projective. This substantiates our earlier claim that the non-projecitive part of a relative relation module can be decomposable.

4 A Complete Decompsition of Relative Relation Modules of SL(2, p) and PSL(2, p)

In this section, we shall discuss the structure of relative relation modules in special cases. To be very specific, we shall discuss the structure of relative relation modules of special linear groups SL(2, p) and projective special linear groups PSL(2, p), and provide a complete decomposition of $\hat{S}$ for these groups. The description of $\hat{S}$ of PSL(2, p) may provide some important information about the factor groups of PSL(2, Z), which is an interesting problem in its own right. Some part of this section has also been included in [12], but we still present it here for the sake of notation in proving other results or for making contextual observations. For the discussion in this section, we recall some well-known results from [2], often without a specific reference for them.

For an $F_p G$ module V, define φV to be the smallest submodule of V such that $V/\varphi V$ is completely reducible. Set $\varphi^k V = \varphi(\varphi^{k-1} V)$. This is known as the upper Loewy series of V. Similarly, we define the lower Loewy series of V as $\sigma^0 V = \{0\}$ and $\sigma^k V$ the largest submodule of $\sigma^{k-1} V$ such that $\sigma^{k-1} V/\sigma^k V$ is completely reducible. Let g denote the augmentation ideal of the group algebra $F_p G$, g_i the augmentation ideal of $F_p G_i$, T_i the trivial irreducible submodule of $F_p G_i$.

Now, the multiplicity of an indecomposable module A in V is defined to be the number of summands in an unrefinable direct decomposition of V isomorphic to A. If W is another module, dimension of the space $Hom_{F_pG}(V, W)$ is called the intertwining number of W in V, and is denoted by $I(V,W)$. It is well known ([2], Cor. 43.16, 320 Ex. 43.6, p. 322) that $I(\oplus V_i, W) = \sum I(V_i, W)$. If F_pG is semi-simple, then $I(V, W) = I(W, V)$. For a subgroup H of G, we call $\{v \in V | vh = v \ for\ all\ h \in H\}$, which is known as the H-fixed point space of V. It is well known that the dimension of the fixed point space does not change with an extension of the ground field. We use these facts in proving the following well-known result.

Lemma 2 Let F_p be the splitting field for G. Let P be a principal indecomposable module with $P/\varphi P \cong U$, r the dimension of U and s the dimension of G_i, the fixed point space of U. Then the multiplicity of P in T_i^G is s, and the multiplicity of P in g_i^G is $r - s$.

Proof As $p \not| |G_i|$, both T_i^G and g_i^G are F_pG-projective , and their direct sum is isomorphic to the right regular module F_pG, i.e. $F_pG \cong T_i^G \oplus g_i^G$. Therefore, in view of ([2], Theorem 61.16, p. 419), we only need to prove the first assertion. T_i^G, being projective, is isomorphic to a direct sum of a (finite) number of principal indecomposable modules of F_pG, and for each principal indecomposable module P_j, we have

$$I(P_j, U) = (P_j/\varphi P_j, U) = \begin{cases} I(U, V), & \text{if } P_j \cong P; \\ 0, & \text{otherwise.} \end{cases}$$

Let m be the multiplicity of P in T_i^G. Then , as F_p is a splitting field for G, Schur's lemma ([2], 27.3, p. 181) gives that $I(T_i^G, U) = m$.

On the other hand, let T be the direct sum of the trivial irreducible submodules of the completely reducible module U_{G_i}. Clearly, T is the G_i-fixed point space of U, and so its dimension must be s. By Nakayama's lemma ([7] Chapter 5, Sentence 16.6, p. 556), we have

$$I(T_i^G, U) = I(T_i, U_{G_i}) = I(T_i, T) = 0,$$

and so the multiplicity of P in T_i^G must be equal to s.

Let $G = \mathrm{SL}(2, p)$, the group of two by two matrices of determinant one over the field F_p, which may be generated by $X = \{g_1, g_2\}$, where

$$g_1 = \begin{pmatrix} 0 & 1 \\ -1 & 0 \end{pmatrix}, \quad g_2 = \begin{pmatrix} 0 & 1 \\ -1 & 1 \end{pmatrix}, \quad \text{where} \quad g_1^4 = g_2^6 = \begin{pmatrix} 1 & 0 \\ 0 & 1 \end{pmatrix}. \tag{7}$$

Let p be a prime other than 2 and 3. We wish to determine all the indecomposable summands of $\hat{S}$. To do that, we first give a description of irreducible F_pG-modules. This description is due to Brauer and Nesbitt [1].

For an integer $m \geq 1$, let V_m be the set of all homogeneous polynomials in two indeterminate, x and y say, over F_p of degree $m - 1$. Then V_m can be regarded as a vector space over F_p and $\{x^j y^{m-1-j}, 0 \leq j \leq m-1\}$ is an F_p basis. In fact, $\{V_m;\ 1 \leq m \leq p\}$ is a complete set of distinct irreducible G modules, where the G action is defined by

$$(x^j y^{m-1-j})g = (a_{11}x + a_{12}y)^j (a_{21}x + a_{22}y)^{m-1-j}, \text{ where} \tag{8}$$

$$g = \begin{pmatrix} a_{11} & a_{12} \\ a_{21} & a_{22} \end{pmatrix} \in G.$$

Each V_m is also an absolutely irreducible module; and so, F_p is a splitting field for G. For details, see ([1], Sect. 30); also ([4], (3.1)).

We suppose that $\{P_m, 1 \leq m \leq p\}$ is a complete set of all non-isomorphic principal indecomposable modules with $P_m/\varphi P_m \cong V_m$. Then the Loewy structure of each P_m is explicitly known. For details, see ([4], Sect. 6). From there, we conclude that there exists an exact sequence

$$0 \to Kerf \to P_{p-2} \xrightarrow{f} P_1 \xrightarrow{g} F_p \to 0, \tag{9}$$

which satisfies all the conditions of Theorem 1. Therefore, we have

$$\hat{S} \oplus P_{p-2} \oplus T_2^G \cong (kerf) \oplus g_1^G \oplus P_1, \text{ and also} \tag{10}$$

$$N(\hat{S}) = (kerf) \text{ and } dim(N(\hat{S})) = p + 1. \tag{11}$$

Now, we need only to find the multiplicity of V_m in g_1^G and in T_2^G. To do that, let $f_i(V_m)$ denote the dimension of the G_i-fixed point space of V_m and set $\alpha_m = m - f_1(V_m) - f_2(V_m)$. Let l_m be the multiplicity of P_m in $\hat{S}$. Then from an application of Lemma 2 and that of Eq. (10), we have

$$l_m = \begin{cases} \alpha_m + 1 & \text{if } m = 1; \\ \alpha_m - 1 & \text{if } m = p-2; \\ \alpha_m & \text{otherwise.} \end{cases} \tag{12}$$

Thus we need to find $f_1(V_m)$ and $f_2(V_m)$ for $m = 1, 2, \ldots, p$.

As F_p need not be a splitting field for G_i, let F be an extension of F_p so that F is a splitting field for both the G_i. Thus, $f_i(V_m)$ is the same as the dimension of the G_i-fixed point of $V_m^F = V_m \otimes_{F_pG} F_pG$, which is the same as the dimension of the subspace of V_m^F fixed by g_i. We may regard g_i as a matrix over F, which we can find by Maschke's theorem.

$$\widetilde{g_i} = \begin{pmatrix} \alpha_i & 0 \\ 0 & \beta_i \end{pmatrix}, \quad i = 1, 2, \alpha_i, \beta_i \in F$$

such that g_i is conjugate to $\widetilde{g_i}$ in GL (2, F). In fact,

$$\widetilde{g_1} = \begin{pmatrix} \omega & 0 \\ 0 & \omega^3 \end{pmatrix}, \ \widetilde{g_2} = \begin{pmatrix} \mu & 0 \\ 0 & \mu^5 \end{pmatrix},$$

where ω and μ are fourth and sixth roots of unity because

$$det(\widetilde{g}_i) = det(g_i) = 1, \text{ and } g_1^4 = g_2^6 = \begin{pmatrix} 1 & 0 \\ 0 & 1 \end{pmatrix}.$$

If we regard $\{x^j y^{m-1-j},\ 0 \leq j \leq m-1\}$ as an F_p basis of V_m^F, then from (8) we get the following results.

$$(x^j y^{m-1-j})g_1 = (\omega x)^j (\omega^3 y)^{m-1-j} = \omega^{3(m-1)-2j} x^j y^{m-1-j}, \text{ and} \tag{13}$$

$$(x^j y^{m-1-j})g_2 = (\mu x)^j (\mu^5 y)^{m-1-j} = \mu^{5(m-1)-4j} x^j y^{m-1-j}. \tag{14}$$

Thus, $f_1(V_m)$ is the number of *j*'s in which $3(m-1)-2j$ is divisible by 4 and $f_2(V_m)$ is the number of *j*'s for which $5(m-1)-4j$ is divisible by 6. First, we have

$$f_i(V_1) = dim V_1 = 1. \tag{15}$$

Next, we claim that if m is even,

$$f_i(V_m) = 0,\ i = 0, 1, 2. \tag{16}$$

If m is odd, we have

$$f_1(V_m) = \begin{cases} \frac{m+1}{2} & \text{if } 4|(m-1); \\ \frac{m-1}{2} & \text{if } 4 \not|(m-1) \end{cases} \quad \text{and}$$

$$f_2(V_m) = \begin{cases} \frac{m+2}{3} & \text{if } 3|(m-1); \\ \frac{m}{3} & \text{if } 3|m; \\ \frac{m-2}{3} & \text{if } 3|(m+1). \end{cases} \tag{17}$$

Using (13)–(17), our discussion may be summarized as follows:

Proposition 4 Let l_m be the multiplicity of P_m in $\hat{S}$, and $l_m P_m$ the direct sum of l_m copies of P_m, then we have $\hat{S} = N(\hat{S}) \oplus_{m=1}^{p} l_m P_m$, where $N(\hat{S})$ is given by (11) and

$$l_m = \begin{cases} 0 & \text{if } m = 1; \\ m & \text{if } m \text{ is even;} \\ \alpha_m - 1 & \text{if } m = p-2; \\ \alpha_m & \text{otherwise;} \end{cases} \quad \text{where } \alpha_m = \begin{cases} \frac{m-7}{6} & \text{if } 12|(m-1); \\ \frac{m+1}{6} & \text{if } 3|m-1, 4|m-1; \\ \frac{m-3}{6} & \text{if } 3|m, 4|m-1; \\ \frac{m-1}{6} & \text{if } 3|m-1, 4 \not|m-1; \\ \frac{m+1}{3} & \text{if } 3|m, 4 \not|m-1; \\ \frac{m+1}{6} & \text{if } 3|m+1, 4 \not|m-1. \end{cases} \tag{18}$$

Now let H = PSL (2, p), the factor group $G/Z(G)$, where $Z(G)$ is the centre of G, is a cyclic subgroup of G; in fact $Z(G) = < z >$ is a cyclic group of order 2. Then corresponding to $h_i \equiv g_i \bmod Z(G)$, i = 1, 2. Let $\hat{K}$ be the relative relation module of H. We shall now describe $\hat{K}$. It is well known that $\{P_1, P_3, \ldots, P_p\}$ is a complete set of distinct principal indecomposable modules of F_pH (See [8], Sec. 3). Clearly Proposition 2 is still valid if we replace $\hat{S}$ with $\hat{K}$. Moreover, $V = 2P_2 \oplus 4P_4 \oplus \cdots \oplus (p-1)P_{p-1}$ is a direct summand of $\hat{S}$. In fact, $\hat{S} = \hat{K} \oplus V$ because $dim\hat{S} - dim\hat{K} = dimV$. Thus, the projective summand of $\hat{S}$ consists of all the $P_m's$ of $\hat{S}$ for which m is odd.

Now we are in a position to describe the structure of $\hat{S}$ for the group SL(2, p) and PSL(2, p), for $p \geq 5$.

Theorem 2 When $\delta = d = 2$, then the relative relation module does not contain all the distinct principal indecomposable modules of F_pG, when G is SL(2, p) or PSL(2, p) , for $p \geq 5$.

Proof From (10), we have $\hat{S} \oplus P_{p-2} \oplus T_2^G = N(\hat{S}) \oplus g_1^G \oplus P_1$. Now, using (13)–(17), we can calculate T_2^G and g_1^G as follows:

$$T_2^G = 1P_1 \oplus \cdots \oplus t_2 P_{p-2} \oplus m_2 P_p, \text{ and} \tag{19}$$

$$g_1^G = 2P_2 \oplus \cdots \oplus t_1 P_{p-2} \oplus (p-1)P_{p-1} \oplus m_1 P_p \tag{20}$$

$$t_2 = \begin{cases} \frac{p-2}{3} & \text{if } 3|(p-2); \\ \frac{p-4}{3} & \text{if } 3|(p-1) \end{cases} \quad \text{and} \quad m_2 = \begin{cases} \frac{p+2}{3} & \text{if } 3|(p-1); \\ \frac{p-2}{3} & \text{if } 3|(p+1) \end{cases} \tag{21}$$

$$t_1 = \begin{cases} \frac{p-3}{3} & \text{if } 4|(p-3); \\ \frac{p-1}{2} & \text{if } 4 \not|(p-3) \end{cases} \quad \text{and}\, m_1 = \begin{cases} \frac{p-1}{2} & \text{if } 4|(p-1); \\ \frac{p+1}{2} & \text{if } 4 \not|(p-1) \end{cases} \tag{22}$$

Therefore, by Krull–Schmidt theorem, we have

$$\hat{S} = N(\hat{S}) \oplus 2P_2 \oplus \cdots \oplus (t_1 - t_2 - 1)P_{p-2} \oplus (p-1)P_{p-1} \oplus (m_1 - m_2)P_p, \tag{23}$$

where either $N(\hat{S}) = P_1 \oplus F_p$ or $N(\hat{S}) = P_p \oplus F_p$. We also find that $t_1 - t_2 - 1 = 0$ for $p = 5$ which means that at least P_3 is not a direct summand of $\hat{S}$, and so $\hat{S}$ does not contain all the distinct principal indecomposable modules of F_pG, when G is SL(2, p). In addition, if we suppose that $\hat{K}$ is the relative relation module modulo p of PSL(2, p), then as shown earlier, we conclude

$$\hat{S} = \hat{K} \oplus V, \text{ where } V = 2P_2 \oplus 4P_2 \cdots \oplus (p-1)P_{p-1}. \tag{24}$$

Thus, from (22), (23) and by Krull–Schmidt theorem, and the fact that $N(\hat{S}) = N(\hat{K})$, we have $\hat{K} = N(\hat{K}) \oplus \cdots \oplus (t_1 - t_2 - 1)P_{p-2} \oplus (p-1)P_{p-1} \oplus (m_1 - m_2)P_p$ where t_1, t_2 and m_1, m_2 are given by (20) and (21).

Now we prove the following:

Theorem 3 When $\delta = 1$, then $\hat{S}$ contains all the distinct principal indecomposable modules of F_pG for the groups SL(2, p) and PSL(2, p), for $p \geq 5$.

Proof We know that $F_pG \cong T_i^G \oplus g_i^G$, for all i, $1 \leq i \leq \delta$. For $\delta = 1$, $g_2^G \cong F_pG$, and so $T_i^G = 0$. Therefore, from (10) , we have $\hat{S} \oplus P_{p-2} = N(\hat{S}) \oplus g_i^G \oplus P_1$.

Additionally, from (19), we have

$$g_1^G = 2P_1 \oplus \cdots \oplus t_1 P_{p-2} \oplus (p-1)P_{p-1} \oplus m_1 P_p$$

where t_1 and m_1 are given by (21). By Krull–Schmidt theorem, we have

$$\hat{S} = N(\hat{S}) \oplus P_1 \oplus 2P_2 \oplus \cdots \oplus (t_1 - 1)P_{p-2} \oplus (p-1)P_{p-1} \oplus m_1 P_p.$$

We may also check that $t_1 - 1 \geq 1$ and $m_1 \geq 2$ for all $p \geq 5$. Thus, $\hat{S}$ contains all the distinct principal indecomposable modules of F_pG.

Similarly, we can also prove that the corresponding relative relation modules for the group PSL(2, p) also contains all the distinct principal indecomposable modules of F_pG.

Finally, we make the following remark.

Remark 4 We know that PSL(2, Z) may be regarded as $Z_2 * Z_3$, free product of cyclic group of order 2 and 3 . The above description of $\hat{K}$ may lead to some useful information about the factor group of PSL(2, Z).

References

1. Brauer, R., Nesbitt, C.: On the modular characters of groups. Ann. Math. **42**(2), 556–590 (1941)
2. Curtis, C.W., Reiner, I.: Representation Theory of Finite Groups and Associative Algebras, Wiley, New York (1962)
3. Gaschutz, W.: Uber modulare Darstellungen endlicher Gruppen, die-von freien Gruppen induziert warden. Math. Z. **60**, 274–286 (1954)
4. Glover, D.L.: A study of certain modular representations. J. Algebra. **51**, 425–475 (1978)
5. Gruenberg, K.W.: Cohomological Topics in Group Theory (Lecture notes in Mathematics 143). Springer-Verlag, Berlin (1970)
6. Gruenberg, K.W.: Relation modules of finite groups, Regional Conference Series in Math, Number 25, A.M.S. Providence, R. I., (1976)
7. Hupert, B, Endliche Gruppen I, Die Grundlen-hern-der Berlin and New York and Sydney, (1967)
8. Janusz, G.J.: Indecomposable representation of groups with cyclic sylow subgroup. Trans. A.M.S. **125**, 288–295 (1966)
9. Kimmerle, W.: Relative relation modules on generators for integral group rings of finite groups. Math. Z. **172**, 143–156 (1980)
10. Mittal, J. N., Passi, I.B.S.: Annihilators of relation modules. J. Austral. Math. Soc. **16**(2), 228–233 (1973)
11. Sharma, P.K.: Some relative relation modules of finite groups. Ph.D. Thesis, Jamia Millia Islamia University, New Delhi (1992)
12. Sharma, P.K., Yamin, M.: Relative relation modules of SL(2, p) and PSL(2, p) groups. J. Indian Math. Soc. **80**(3–4), 341–348 (2013)

13. Yamin, M: Relative relation modules of finite groups, Ph.D. Thesis, Australian National University, Canberra (1983)
14. Yamin, M.: Relative relation modules of finite groups. Proc. Edinburgh Math. Soc. **34**, 433–442 (1991)
15. Yamin, M.: Minimal relative relation modules of finite p-groups. Proc. Amer. Math. Soc. **118**(1), 1–3 (1993)

Block Methods with Off-Steps Points for Solving First Order Ordinary Differential Equations

Lee Ken Yap and Fudziah Ismail

Abstract The block methods with off-step points are proposed for the solution of first-order ordinary differential equations. These methods are applied concurrently in block form to provide the approximation for both the main and off-step points. The stability properties are discussed. Some numerical examples are given to illustrate the efficiency of the methods.

1 Introduction

We consider the first-order ordinary differential equations (ODEs)

$$y'(x) = f(x, y), \; y'(a) = y_0 \tag{1}$$

with $x \in [a, b]$. In particular, the first-order differential equations arise in many real-world systems such as population growth, modeling phenomena of cooling, and radioactive decay. Therefore, first-order ODEs have attracted considerable attention. Many theoretical and numerical studies dealing with such equations have appeared in literature.

Lee [8] proposed block methods based on backward difference formula for solving first-order ODEs. These methods estimate the solution at the main points simultaneously. Ibrahim et al. [6] developed the block backward differentiation formula to solve first-order stiff ODEs while Majid and Suleiman [9] applied fully implicit block methods to solve first-order ODEs with variable step size. The studies on block methods in solving first-order ODEs demonstrated speed and accuracy advantages.

Enright and Higham [4] stated that the precise location of the off-step points can be used for minimizing interpolation errors and for developing various cheap

L. K. Yap (✉)
Department of Mathematical and Actuarial Sciences, Faculty of Engineering and Science, Universiti Tunku Abdul Rahman, Jalan Genting Klang, Setapak, Kuala Lumpur 53300, Malaysia
e-mail: lkyap@utar.edu.my

F. Ismail
Department of Mathematics, Faculty of Science, Universiti Putra Malaysia, Serdang, Selangor 43400, Malaysia
e-mail: fudziah@science.upm.edu.my

A. Kilicman et al. (eds.), *International Conference on Mathematical Sciences and Statistics 2013,* DOI 10.1007/978-981-4585-33-0_28,

strategies for stepsize control. Ebadi and Gokhale [3] demonstrated that the methods with off-step points have the improved stability region. Jator [7] and Akinfenwa et al. [1] proposed a class of self-starting hybrid block methods with off-step points. The derivation involved the interpolation and collocation of basic polynomials. The drawback of these approaches is it required to solve the large system of algebraic equations that leads to high computational cost.

The block method in this paper is extended from the line proposed by Lee [8]. The advantage of this new method is obtaining the approximation of y at the main and off-step points concurrently without reduction to finer step size.

2 Derivation of Explicit Block Method with Off-Step Points

In this section, we present the mathematical formulation of the explicit block method with off-step points. The derivation involves the interpolation polynomial, generating function, and Maclaurin Series to solve for the constant coefficients.

2.1 Derivation of Off-Step Point Method

Let the off-step point be defined as $x_{n+\frac{d}{2}} = x_n + \frac{d}{2}h$ for $d = \{1, 3, 5, \dots\}$. We integrate the first order differential equation (1) from x_n to $x_{n+\frac{d}{2}}$ on both sides, it then leads to

$$\int_{x_n}^{x_{n+\frac{d}{2}}} y'(x)\,dx = \int_{x_n}^{x_{n+\frac{d}{2}}} f(x, y)\,dx. \tag{2}$$

Next, we define the interpolation polynomial $P_{k,n}(x)$ to interpolate $f(x, y)$ at the k back values as follows

$$P_{k,n}(x_n + sh) = \sum_{m=0}^{k-1} (-1)^m \binom{-s}{m} \nabla^m f_n \tag{3}$$

with $s = \dfrac{x - x_n}{h}$. The interpolation polynomial (3) is substituted in (2).

$$y\left(x_{n+\frac{d}{2}}\right) - y(x_n) = \int_{x_n}^{x_{n+\frac{d}{2}}} \sum_{m=0}^{k-1} (-1)^m \binom{-s}{m} \nabla^m f_n\,dx$$

Change the limit of integration and replace $dx = h\,ds$ gives

$$y\left(x_{n+\frac{d}{2}}\right) = y(x_n) + h \int_0^{\frac{d}{2}} \sum_{m=0}^{k-1} (-1)^m \binom{-s}{m} \nabla^m f_n\,ds.$$

Table 1 Coefficients γ_m for first off-step point method

m	0	1	2	3	4	5
γ_m	$\frac{1}{2}$	$\frac{1}{8}$	$\frac{1}{12}$	$\frac{25}{384}$	$\frac{157}{2880}$	$\frac{243}{5120}$

Table 2 Coefficients γ_m for second off-step point method

m	0	1	2	3	4	5
γ_m	$\frac{3}{2}$	$\frac{9}{8}$	$\frac{9}{8}$	$\frac{147}{128}$	$\frac{753}{640}$	$\frac{6171}{5120}$

Table 3 Coefficients γ_m for third off-step point method

m	0	1	2	3	4	5
γ_m	$\frac{5}{2}$	$\frac{25}{8}$	$\frac{25}{6}$	$\frac{675}{128}$	$\frac{925}{144}$	$\frac{70135}{9216}$

Consequently,

$$y(x_{n+\frac{d}{2}}) = y(x_n) + h \sum_{m=0}^{k-1} \gamma_m \nabla^m f_n$$

with $\gamma_m = (-1)^m \int_0^{\frac{d}{2}} \binom{-s}{m} ds.$

In order to obtain a useful recurrence relation for the coefficients γ_m, the method of generating function is applied. Let the generating function $V(t)$ be defined as follows

$$\begin{aligned} V(t) &= \sum_{m=0}^{\infty} \gamma_m t^m = \int_0^{\frac{d}{2}} \sum_{m=0}^{\infty} (-t)^m \binom{-s}{m} ds \\ &= \int_0^{\frac{d}{2}} (1-t)^{-s} \, ds = \frac{1-(1-t)^{-\frac{d}{2}}}{\ln(1-t)}. \end{aligned}$$

Thus,

$$\sum_{m=0}^{\infty} \gamma_m t^m = \frac{1-(1-t)^{-\frac{d}{2}}}{\ln(1-t)}. \tag{4}$$

To obtain the coefficient γ_m, we apply the Maclaurin Series in (4). Recall that the Maclaurin Series is the Taylor Series expansion of a function about 0. The Maclaurin Series for a function $f(x)$ of order n can be easily obtained in *Mathematica* using statement `Series[f,{x, 0, n}]`. Hence,

$$\gamma_0 + \gamma_1 t + \gamma_2 t^2 + \ldots = \frac{d}{2} + \frac{d^2}{8} t + \frac{1}{48}\left(3d^2 + d^3\right) t^2 + \ldots$$

We compare the corresponding coefficients for the value of γ_m. Tables 1–3 show the values of γ_m for first, second, and third off-step points with $d = 1, 3, 5$, respectively.

2.2 Derivation of the Main Point Method

Similarly, let the main point be defined as $x_{n+d} = x_n + dh$ for $d = 1, 2, 3, \ldots$. We integrate (1) from x_n to x_{n+d} and replace $f(x, y)$ with the interpolation polynomial, which gives us

$$y(x_{n+d}) - y(x_n) = \int_{x_n}^{x_{n+d}} \sum_{m=0}^{k-1} (-1)^m \binom{-s}{m} \nabla^m f_n \, dx$$

which can be written as

$$y(x_{n+d}) = y(x_n) + h \int_0^d \sum_{m=0}^{k-1} (-1)^m \binom{-s}{m} \nabla^m f_n \, ds$$

with $dx = h\, ds$.

Here, we consider the multistep method in the form of

$$y(x_{n+d}) = y(x_n) + h \sum_{m=0}^{k-1} \beta_m \nabla^m f_n$$

with $\beta_m = (-1)^m \int_0^d \binom{-s}{m} ds$. With the generating function, we obtain the recursion relation as follows

$$B(t) = \sum_{m=0}^{\infty} \beta_m t^m = \int_0^d \sum_{m=0}^{\infty} (-t)^m \binom{-s}{m} ds = \int_0^d (1-t)^{-s} \, ds = \frac{1-(1-t)^{-d}}{\ln(1-t)}.$$

Consequently,

$$\sum_{m=0}^{\infty} \beta_m t^m = \frac{1-(1-t)^{-d}}{\ln(1-t)}. \qquad (5)$$

By applying Maclaurin Series for the function $\frac{1-(1-t)^{-d}}{\ln(1-t)}$ in (5) and comparing the corresponding coefficients, we manage to obtain the value of β_m. Tables 4–6 show the values of β_m for first, second, and third main point methods with $d = 1, 2, 3$, respectively.

By using integration coefficients and backward difference formula, we derive the explicit block methods with off-step points.

$$y_{n+\frac{1}{2}} = y_n + h\left(\frac{17}{24} f_n - \frac{7}{24} f_{n-1} + \frac{1}{12} f_{n-2}\right)$$

$$y_{n+1} = y_n + h\left(\frac{23}{12} f_n - \frac{4}{3} f_{n-1} + \frac{5}{12} f_{n-2}\right)$$

Table 4 Coefficients β_m for first main point method

m	0	1	2	3	4	5
β_m	1	$\frac{1}{2}$	$\frac{5}{12}$	$\frac{3}{8}$	$\frac{251}{720}$	$\frac{95}{288}$

Table 5 Coefficients β_m for second main point method

m	0	1	2	3	4	5
β_m	2	2	$\frac{7}{3}$	$\frac{8}{3}$	$\frac{269}{90}$	$\frac{33}{10}$

Table 6 Coefficients β_m for third main point method

m	0	1	2	3	4	5
β_m	3	$\frac{9}{2}$	$\frac{27}{4}$	$\frac{75}{8}$	$\frac{987}{80}$	$\frac{2499}{160}$

$$y_{n+\frac{3}{2}} = y_n + h\left(\frac{15}{4}f_n - \frac{27}{8}f_{n-1} + \frac{9}{8}f_{n-2}\right)$$

$$y_{n+2} = y_n + h\left(\frac{19}{3}f_n - \frac{20}{3}f_{n-1} + \frac{7}{3}f_{n-2}\right)$$

$$y_{n+\frac{5}{2}} = y_n + h\left(\frac{235}{24}f_n - \frac{275}{24}f_{n-1} + \frac{25}{6}f_{n-2}\right)$$

$$y_{n+3} = y_n + h\left(\frac{57}{4}f_n - 18f_{n-1} + \frac{27}{4}f_{n-2}\right)$$

3 Stability Properties

One of the practical criteria for a good method to be useful is that it must have a region of absolute stability. The most convenient method for finding the region of absolute stability is the boundary locus technique.

We apply test equation $y' = \lambda y$ in the explicit 1-point 1 off-step point block method.

$$y_{n+\frac{1}{2}} = y_n + h\left(\frac{17}{24}\lambda y_n - \frac{7}{24}\lambda y_{n-1} + \frac{1}{12}\lambda y_{n-2}\right)$$

$$y_{n+1} = y_n + h\left(\frac{23}{12}\lambda y_n - \frac{4}{3}\lambda y_{n-1} + \frac{5}{12}\lambda y_{n-2}\right)$$

These equations can be presented as follows

$$IY_{m+1} = (A + Bh\lambda)Y_m + Ch\lambda Y_{m-1} \tag{6}$$

with

$$I = \begin{bmatrix} 1 & 0 \\ 0 & 1 \end{bmatrix},\ A = \begin{bmatrix} 0 & 1 \\ 0 & 1 \end{bmatrix},\ B = \begin{bmatrix} -\frac{7}{24} & \frac{17}{24} \\ -\frac{4}{3} & \frac{23}{12} \end{bmatrix},\ C = \begin{bmatrix} 0 & \frac{1}{12} \\ 0 & \frac{5}{12} \end{bmatrix},$$

$$Y_{m+1} = \begin{bmatrix} y_{n+\frac{1}{2}} \\ y_{n+1} \end{bmatrix}, Y_m = \begin{bmatrix} y_{n-1} \\ y_n \end{bmatrix} \text{ and } Y_{m-1} = \begin{bmatrix} y_{n-3} \\ y_{n-2} \end{bmatrix}.$$

The stability polynomial is obtained by taking

$$Det[It^2 - (A + B\overline{h})t - C\overline{h}] = 0$$

where $\overline{h} = h\lambda$. It leads to

$$t^4 - \left(1 + \frac{13}{8}\overline{h}\right)t^3 + \left(\frac{5}{8}\overline{h} + \frac{37}{96}\overline{h}^2\right)t^2 - \frac{1}{96}\overline{h}^2 t = 0. \tag{7}$$

The stability polynomial is solved for $\overline{h}$, which gives $|t| \leq 1$, whereby the stability region is obtained by tracing the values of $\overline{h}$. Similar strategy is applied to obtain the stability region for the 2-point 2 off-step points block method and 3-point 3 off-step points block method.

The stability polynomial for the 2-point 2 off-step points block method and 3-point 3 off-step points block method are defined in (8) and (9), respectively.

$$t^4 - \left(1 + \frac{27}{8}\overline{h}\right)t^3 + \left(\frac{11}{8}\overline{h} + \frac{181}{96}\overline{h}^2\right)t^2 - \left(\frac{13}{96}\overline{h}^2 + \frac{1}{16}\overline{h}^3\right)t = 0 \tag{8}$$

$$t^6 - \left(1 + \frac{41}{8}\overline{h}\right)t^5 + \left(\frac{17}{8}\overline{h} + \frac{433}{96}\overline{h}^2\right)t^4 - \left(\frac{37}{96}\overline{h}^2 + \frac{5}{16}\overline{h}^3\right)t^3 = 0 \tag{9}$$

Figures 1–3 show the stability regions for all the three block methods. The region of absolute stability lies inside the boundary.

3.1 Zero Stability

To analyze the zero stability of all the three block methods, we apply $\overline{h} = 0$ in the stability polynomials (7), (8), and (9) to obtain the first characteristic polynomials. Since the roots of the first characteristic polynomials have modulus less than or equal to one, this imply the zero-stability of the three block methods.

3.2 Consistency

According to Henrici [5], the linear multistep method and the linear difference operator are said to be of order p if $C_0 = C_1 = C_2 = \ldots = C_p = 0$ and $C_{p+1} \neq 0$.

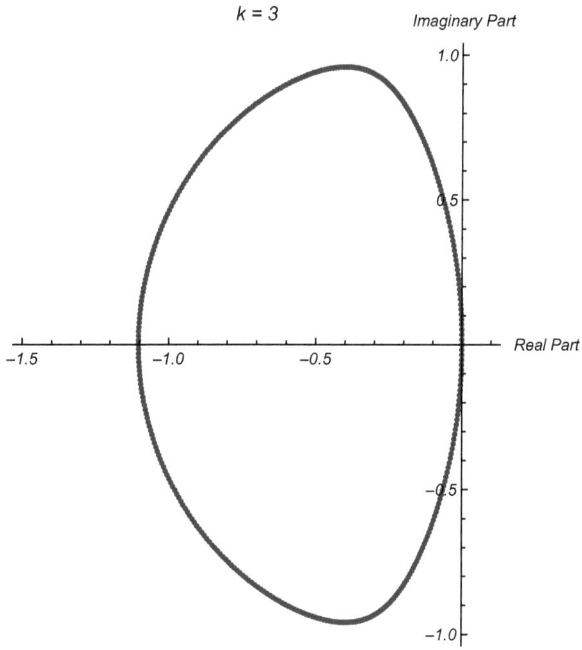

Fig. 1 Stability region for explicit 1-point with 1 off-step point block method

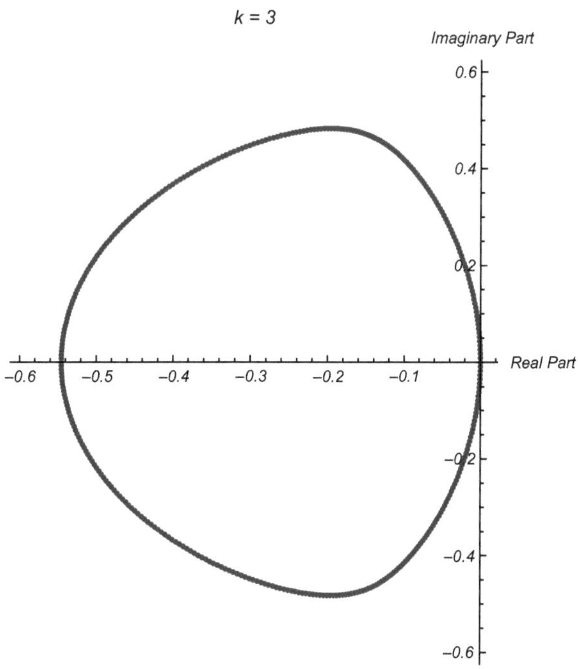

Fig. 2 Stability region for explicit 2-point with 2 off-step points block method

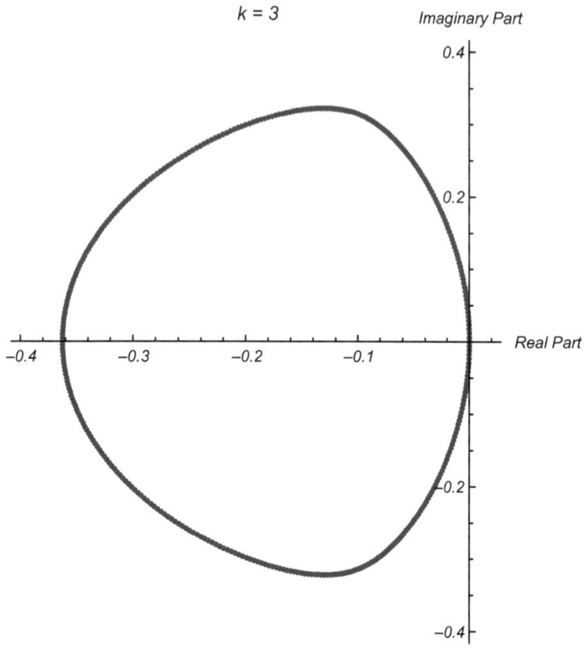

Fig. 3 Stability region for explicit 3-point with 3 off-step points block method

To check the order of the block methods, we rewrite (6) in standard form to define the coefficients α and β. Then, the difference operator is applied to obtain

$$C_0 = C_1 = C_2 = C_3 = \begin{bmatrix} 0 \\ 0 \end{bmatrix} \text{ and } C_4 = \begin{bmatrix} \frac{25}{384} \\ \frac{3}{8} \end{bmatrix}.$$

For 2-point with two off-step points block method and 3-point with three off-step points block method, we obtain the error constants, $C_4 = \left(\frac{25}{384}, \frac{3}{8}, \frac{147}{128}, \frac{8}{3}\right)^T$ and $C_4 = \left(\frac{25}{384}, \frac{3}{8}, \frac{147}{128}, \frac{8}{3}, \frac{675}{128}, \frac{75}{8}\right)^T$, respectively. As a whole, we can conclude that the new block multistep methods with off-step points are of order three. Since the block methods with off-step points have order $p > 1$, they are consistent.

4 Numerical Examples

To illustrate the efficiency of the new methods, the following problems are solved numerically.

Problem 1 See [2]

$$y' = \varepsilon(y - F) + F' \text{ with } F = xe^{-2x} \text{ and } \varepsilon = -10, \;\; y(1) = \frac{1}{e^2}, \;\; x \in [1, 5].$$

Exact solution: $y(x) = xe^{-2x}$

Table 7 Numerical results for Problem 1

h	Method	Step	Max error
10^{-2}	E1P	100	4.6771×10^{-8}
	E2P	52	1.3482×10^{-7}
	E3P	36	2.5809×10^{-7}
10^{-3}	E1P	1000	4.7996×10^{-11}
	E2P	502	1.6681×10^{-10}
	E3P	336	3.8177×10^{-10}
10^{-4}	E1P	10000	4.8156×10^{-14}
	E2P	5002	1.7074×10^{-13}
	E3P	3336	3.9924×10^{-13}
10^{-5}	E1P	100000	6.6613×10^{-16}
	E2P	50002	4.7185×10^{-16}
	E3P	33336	6.6613×10^{-16}

Table 8 Numerical results for Problem 2

h	Method	Step	Max error
10^{-2}	E1P	100	5.7524×10^{-8}
	E2P	52	2.0228×10^{-7}
	E3P	36	4.6563×10^{-7}
10^{-3}	E1P	1000	5.8301×10^{-11}
	E2P	502	2.0706×10^{-10}
	E3P	336	4.8444×10^{-10}
10^{-4}	E1P	10000	5.8537×10^{-14}
	E2P	5002	2.0670×10^{-13}
	E3P	3336	4.8694×10^{-13}
10^{-5}	E1P	100000	2.4980×10^{-15}
	E2P	50002	2.2760×10^{-15}
	E3P	33336	2.6645×10^{-15}

Problem 2 See [2]

$$y_1' = -y_1 + y_2(1 - y_1 - y_2), \quad y_2' = y_1 - y_2(1 - y_1) - e^{-x},$$

$$y_1(1) = \frac{1}{e}, \; y_2(1) = 0, \; x \in [1,2].$$

Exact solution: $y_1(x) = e^{-x}, \; y_2(x) = 0$

Tables 7 and 8 give the performance comparison of new block methods in terms of the total number of steps taken and maximum error. The notations used in the tables are as follows:

h	Step size used
Method	Method employed
Step	Total number of steps taken to obtain the solution
E1P	Implementation of explicit 1-point with 1 off step point block method
E2P	Implementation of explicit 2-point with 2 off-step points block method
E3P	Implementation of explicit 3-point with 3 off-step points block method

The maximum error is defined as

$$\text{Max Error} = \max_{1 \le i \le \text{Step}} \left(\left| y\left(x_i\right) - y_i \right| \right)$$

5 Discussion and Conclusion

The numerical results demonstrate the accuracy of the method. It performs better as the step size decreases. It is observed that as the step size gets smaller, both E2P and E3P methods reduce the total steps to almost one half and one third compared to E1P method. These results are expected since the m-point block methods calculate the values of y at m main and off-step points simultaneously.

Since the new block methods with off-step points include the approximations at main and additional off-step points, it is believed to provide better estimation in interpolating the delay term of delay differential equations.

References

1. Akinfenwa, O.A., Yao, N.M., Jator, S.N.: Implicit two step continuous hybrid block methods with four off-steps points for solving stiff ordinary differential equation. World Acad. Sci. Eng. Techno. **75**, 425–428 (2011)
2. D'Ambrosio, R., Esposito, E., Paternoster, B.: Exponentially fitted two-step runge-kutta methods: construction and parameter selection. Appl. Math. Comput. **218**, 7468–7480 (2012)
3. Ebadi, M., Gokhale, M.Y.: Class 2+1 hybrid BDF-like methods for the numerical solutions of ordinary differential equations. Calcolo. **48**, 273–291 (2011)
4. Enright, W.H., Higham, D.J.: Parallel Defect Control. BIT. **31**, 647–663 (1991)
5. Henrici, P.: Discrete Variable Methods in Ordinary Differential Equations. Wiley, New York (1962)
6. Ibrahim, Z.B., Othman, K.I., Suleiman, M.: Implicit r-point block backward differentiation formula for solving first-order stiff ODEs. Appl. Math. Comput. **186**(1), 558–565 (2007)
7. Jator, S.N.: Solving second order initial value problems by a hybrid multistep method without predictors. Appl. Math. Comput. **217**, 4036–4046 (2010)
8. Lee, L.S.: Two and Three-Point Block Methods for Solving First Order Ordinary Differential Equations in Parallel. Dissertation for Master of Science, Universiti Putra Malaysia (2010)
9. Majid, Z.A., Suleiman, M.: Implementation of four-point fully implicit block method for solving ordinary differential equations. Appl. Math. Comput. **184**(2), 514–522 (2007)

Implicit Finite Difference Solutions of One-Dimensional Burgers' Equation Using Newton–HSSOR Method

J. Sulaiman, M. K. Hasan, M. Othman and S. A. A. Karim

Abstract In this paper, we present the application of half-sweep successive over-relaxation (HSSOR) iterative methods together with Newton scheme, collectively Newton–HSSOR, in solving the nonlinear systems generated from the half-sweep Crank–Nicolson finite difference discretization scheme for a one-dimensional Burgers' equation. To linearize nonlinear systems, the Newton scheme is proposed to transform the nonlinear system into the form of linear system. In addition to that, the basic formulation and implementation of Newton–HSSOR iterative method are also shown. For comparison purpose, we also consider combinations between the full-sweep Gauss–Seidel (FSGS) and full-sweep successive over-relaxation (FSSOR) iterative methods with Newton scheme, which are indicated as Newton–FSGS and Newton–FSSOR methods respectively. Consequently, two illustrative examples are included to demonstrate the validity and applicability of tested methods. Finally, it can be concluded that the Newton–HSSOR method shows superiority over other tested methods.

J. Sulaiman (✉)
Mathematics with Economics Programme, Universiti Malaysia Sabah,
88400 Kota Kinabalu, Sabah, Malaysia
e-mail: jumat@ums.edu.my

M. K. Hasan
School of Information Technology, Universiti Kebangsaan Malaysia,
43600 UKM Bangi, Selangor, Malaysia
e-mail: khatim@ftsm.ukm.my

M. Othman
Department of Communication Technology and Network, Universiti Putra Malaysia,
43400 UPM Serdang, Selangor, Malaysia
e-mail: mothman@fsktm.upm.edu.my

S. A. A. Karim
Fundamental and Applied Sciences Department, Universiti Teknologi Petronas,
31750 Tronoh, Perak Darul Ridzuan, Malaysia
e-mail: samsul_ariffin@petronas.com.my

A. Kilicman et al. (eds.), *International Conference on Mathematical Sciences and Statistics 2013*, DOI 10.1007/978-981-4585-33-0_29,

1 Introduction

A study of Burgers' equation is of great importance since this mathematical model is encountered in various problems in science and engineering. These problems have been classified by nonlinear differential equations. To solve these problems, it is still very difficult to gain the solution of nonlinear equations either analytically or numerically. In the past few years, many efforts have focused on getting solutions of Burgers' equation by using several numerical methods: finite difference method [1]–[3], finite element method (FEM) [4]–[6], Galerkin method [7], [8], and spectral least-squares method [9]–[11]. In addition to these methods, many researchers have also proposed several spline functions with some numerical techniques in getting numerical solutions of Burgers' equation, e.g., spline method [12]–[16] and B-spline method [17]–[19]. Based on the numerical analytic methods, Abbasbandy and Darvishi [20], [21] have discussed the Adomian method to find a numerical solution to the problem. Besides this method, a numerical solution of variational iteration method was discussed by Abdoua and Soliman [22]. They pointed out that the results of the variational iteration method are more effective than the Adomian decomposition method.

In this paper, however, the finite difference method will be considered to develop a reliable algorithm in solving the nonlinear Burgers' equation. Since the nonlinear Burgers' equation can be classified as a family of time-dependent problems, the construction of the approximation equations based on finite difference discretization schemes can fall into three categories: explicit, implicit, and semi-implicit. Due to the stability of computation, an implicit finite difference approximation equation is preferred to solve the Burgers' equation. Then this implicit approximation equation leads to a nonlinear system at each time level. It is well known that to linearize nonlinear systems at each time level, the Newton method has been used to transform the original system into the form of a linear system. In addition to solve any large and sparse linear system at each time level, iterative methods are natural options for efficient solutions.

From literature survey, we can see in detail that a lot of work has been done on various iterative methods to yield a fast numerical solution of linear systems (see Young [23], [24]; Hackbusch [25]; Saad [26]). Apart from those iterative methods, the finding on the half-sweep iteration concept has been firstly introduced by Abdullah [27] via explicit decoupled group (EDG) iterative method in solving two-dimensional Poisson equations. Actually, the crucial characteristic of this concept is to reduce the computational complexity of linear systems generated from the corresponding approximation equations. Due to the advantages of this concept, further investigations have been extensively conducted in [27]–[37] for demonstrating the capability of the half-sweep iteration. Apart from these one-stage iteration concepts, combinations between half-sweep iteration concept with two-stage iterative methods, namely, half-sweep iterative alternating decomposition explicit (HSIADE) [38], half-sweep arithmetic mean (HSAM) [39], [40], and half-sweep arithmetic mean (HSGM) [41] have also been developed for solving linear systems. Based on the numerical results, it can be pointed out that these their proposed two-stage iterative methods are one of most efficient iterative methods in solving any system of linear equations. Since this half-sweep iteration concept has low computational complexity, this concept has been extended to derive a family of multigrid methods [42], [43]. Consequently, Hassan et al. [44] have established a family of finite-difference time-domain (FDTD) methods using this concept in solving wave propagation problems. Meanwhile, Saudi and Sulaiman [45]–[47] applied to solve the robotic path planning.

In this paper, however, we are concerned with the application of family of SOR iterative methods such as full-sweep SOR (FSSOR) and half-sweep SOR (HSSOR) by applying Newton scheme in solving nonlinear Burgers' equation. For simplicity, the combination of both proposed iterative methods with Newton scheme is identified as Newton–FSSOR and Newton–HSSOR methods respectively.

To investigate the capability of Newton–FSSOR and Newton–HSSOR, let us consider the one-dimensional nonlinear Burgers' equation in the form

$$\frac{\partial U}{\partial t} + U\frac{\partial U}{\partial x} - \nu\frac{\partial^2 U}{\partial y^2} = 0, \quad a \le x \le b, \ t \ge 0 \tag{1}$$

under the following boundary conditions

$$\begin{aligned} U(a,t) &= g_1(t), \quad t > 0 \\ U(b,t) &= g_2(t), \quad t > 0 \end{aligned}$$

and the initial condition

$$U(x,0) = g_3(x), \quad a \le x \le b$$

where $g_0(t)$, $g_1(t)$, and $g_2(x)$ are the prescribed functions of the variables. Actually, the problem (1) governs the mathematical model of turbulence, which contains the velocity for space x and time t, and the kinematic viscosity parameter related to the Reynolds number (*RE*), defined as $\nu = \frac{1}{RE}$. For simplicity, the next discussion will be restricted to uniform solid node points only as shown in Eq. (2). Let us assume that the solution domain Eq. (1) can be uniformly divided into an $m = 2^p, p \ge 2$ subinterval in which its distance, Δx is defined in Eq. (2).

$$\Delta x = \frac{(b-a)}{m} = h, \qquad n = m - 1 \tag{2}$$

Based on Eq. (2), it can be shown that the finite grid networks can facilitate how the implementation of proposed computational algorithms can be executed. For that reason, the implementation of the point iterative algorithms will be applied onto the interior node points until the iterative convergence fixed is achieved (see in [27]–[37]).

The paper is organized as follows: In Sect. 2, the use of the half-sweep reduction technique is used to derive a nonlinear implicit finite difference approximation equation. In Sect. 3, the formulation and algorithm for family of Newton–SOR methods are shown. Then in Sect. 4, two numerical examples are included to illustrate the practical implementation of the proposed Newton–SOR methods. The accuracy performance of the numerical results obtained from the proposed methods at three different values of Reynolds number, *RE*, is compared with the exact solution. Meanwhile, the conclusion is given in the final section.

2 Half-Sweep Finite Difference Approximation Equation

Using the approach of second-order Crank–Nicolson (CN) discretization scheme, the full- and half-sweep finite difference approximation equations for problem Eq. (1) can be easily shown in a general form as follows

$$\begin{aligned} &\alpha_p U_{i,j+1}\left(U_{i+p,j+1} - U_{i-p,j+1}\right) \\ &\qquad -\beta_p U_{i-p,j+1} + \left(1+2\beta_p\right) U_{i,j+1} - \beta_p U_{i+p,j+1} = G_{i,j} \end{aligned} \tag{3}$$

where

$$\alpha_p = \frac{\Delta t}{4\,(ph)}, \quad \beta_p = \nu \frac{\Delta t}{2\,(ph)^2},$$

$$G_{i,j} = -\alpha_p U_{i,j}\left(U_{i+p,j} - U_{i-p,j}\right) + \beta_p U_{i-p,j} + \left(1+2\beta_p\right) U_{i,j} - \beta_p U_{i+p,j}$$

for $i = p, 2p, ..., m-p$ and $j = 0, 1, 2, 3, ..., M-1$. Therefore, the value of p, which equals 1 or 2, represents the full-sweep or half-sweep case. Clearly, it can seen that the half-sweep CN finite difference approximation is derived for problem (1) by using the equal grid spacing $2h$. Both full- and half-sweep CN finite difference approximation equations in (3) have been classified to be unconditionally stable schemes and second-order accurate in space and time. Due to the stability of this CN scheme, there is no restriction in choosing mesh size. Also both of these approximation equations lead nonlinear systems. Before applying the Newton method to these nonlinear systems in order to construct the corresponding linear systems, define the nonlinear function f based on Eq. (3) for each interior grid point $\left(x_i, t_{j+1}\right)$ at any time level t_{j+1} as follows:

$$\begin{aligned} f_i\left(U_{p,j+1}, U_{2pj+1}, ..., U_{m-p,j+1}\right) &= \alpha_p U_{i,j+1}\left(U_{i+p,j+1} - U_{i-p,j+1}\right) \\ &\quad -\beta_p U_{i-p,j+1} + \left(1+2\beta_p\right) U_{i,j+1} - \beta_p U_{i+p,j+1} - G_{i,j} \end{aligned} \tag{4}$$

for $i = p, 2p, ..., m-p$. Now by taking into account all interior node points in the solution domain Eq. (1), Eq. (4) leads to the form of nonlinear system as

$$f_i\left(U_{p,j+1}, U_{2pj+1}, ..., U_{m-p,j+1}\right) = 0 \tag{5}$$

where $U^{(k)}_{i,j+1}$, $i = p, 2p, ..., m-p$ is indicated as the kth estimation for corresponding exact solutions at time level. Then it is clear that for each time level t_{j+1}, we have to construct and solve a system of nonlinear equations (5). To solve the nonlinear system in Eq. (5), the Newton method was used to linearize and transform the original system into a linear system, which can be rewritten in the matrix form as

$$J\left(\bar{U}^{(k)}\right)\bar{\Delta h} = -f_j\left(\bar{U}^{(k)}\right) \tag{6}$$

where

$$J\left(\bar{U}^{(k)}\right) = \begin{bmatrix} \frac{\partial f_{1p,j+1}}{\partial u_{1p}} & \frac{\partial f_{1p,j+1}}{\partial u_{2p}} & \cdots & \frac{\partial f_{1p,j+1}}{\partial u_{m-p}} \\ \frac{\partial f_{2p,j+1}}{\partial u_{1p}} & \frac{\partial f_{2p,j+1}}{\partial u_{2p}} & \cdots & \frac{\partial f_{2p,j+1}}{\partial u_{m-p}} \\ \vdots & \vdots & \ddots & \vdots \\ \frac{\partial f_{m-p,j+1}}{\partial u_{1p}} & \frac{\partial f_{m-p,j+1}}{\partial u_{2p}} & \cdots & \frac{\partial f_{m-p,j+1}}{\partial u_{m-p}} \end{bmatrix},$$

$$\bar{\Delta h} = \left[\Delta h_{1p}, \Delta h_{2p}, \quad \ldots, \Delta h_{m-p}\right]^{T}.$$

From Eq. (6), the value of the vector $\bar{\Delta h}$ needs to be calculated by solving the linear system. Then estimate solutions of $\bar{U}^{(k)}$ can be determined by using the following expression

$$U_{i,j}^{(k+1)} = U_{i,j}^{(k)} + \Delta h_i, \quad i = 1p, 2p, \cdots, m-p. \tag{7}$$

3 Derivation of Family of SOR Methods

As mentioned in the Sect. 2, the coefficient matrix A of linear systems in Eq. (6) is sparse and large. Let the linear system in Eq. (6) be rewritten as

$$A\bar{\Delta h} = \bar{F}. \tag{8}$$

Consequently, iterative methods are proposed as being the natural options for efficient solutions for a sparse linear system. To solve the linear system (8), we propose and construct FSSOR and HSSOR iterative methods being used as linear solvers. Actually, Young [23], [24] initiated the SOR method, which is one of the most known and widely used iterative techniques to solve any linear systems. To derive the formulation for FSSOR and HSSOR iterative methods, let the coefficient matrix, A in Eq. (8) be decomposed as

$$A = D + L + V \tag{9}$$

where L, D, and V are lower triangular, diagonal, and upper triangular matrices respectively. Now using the decomposition in Eq. (9) and determining values of matrices D, L, and V, therefore, the general scheme for FSSOR and HSSOR methods can be stated as [23], [24], [54]

$$\bar{\Delta h}^{(k+1)} = (1-\omega)\bar{\Delta h}^{(k)} + \omega(D+L)^{-1}(-V\bar{\Delta h}^{(k)} + \bar{F}) \tag{10}$$

where ω and $\bar{\Delta h}^{(k)}$ represent a relaxation factor and an unknown vector at the k_{th} iteration respectively. The choice of relaxation factor depends upon the properties of the coefficient matrix A. In addition, a good choice of parameter can improve the convergence rate of iteration process. In practice, the optimal value of ω in the range $1 \leq \omega < 2$ will be obtained by implementing several computer programs, and then the best approximate value of ω is chosen in which its number of iterations is the smallest.

Due to the advantage of half-sweep iteration concept applied onto the HSSOR method for reducing its computational complexity, we examine the efficiency of FSSOR and HSSOR iterative methods in solving Eq. (1). Taking $\omega = 1$, SOR method reduces to Gauss–Seidel (GS) method. Later GS method will be used as control method. From the implementation of point iterations, the experimental values of ω were obtained within 0.001 by running the Newton–FSSOR and Newton–HSSOR methods for different values of ω and choosing the one(s) that produce the minimum number of iterations. In addition to this, numerical results by applying the standard Gauss–Seidel iterative method with Newton scheme, namely, Newton–FSGS for solving the nonlinear system Eq. (5) are also included.

4 Numerical Experiments

In order to validate the effectiveness of the FSSOR and HSSOR iterative methods together with the Newton approach, namel,y Newton–FSSOR and Newton–HSSOR, three nonlinear example problems were tested. For the sake of comparison, three criteria will be considered for a family of SOR methods, such as number of iterations, execution time, and maximum absolute error. In the implementation of the iterative methods, the convergence test considered the tolerance error $\epsilon = 10^{-10}$.

Example 1 [48] The analytic solution of the Burgers' equation (1) is given by

$$U(x,t) = \frac{(P(x,t)/10) + Q(x,t)/2) + R(x,t))}{P(x,t) + Q(x,t) + R(x,t)} \tag{11}$$

where

$$\begin{aligned} P(x,t) &= e^{(0.3-x-4.95t)/(20\nu)}, \\ Q(x,t) &= e^{(0.8-x-0.75t)/(4\nu)}, \\ R(x,t) &= e^{(0.375-x)/(2\nu)}. \end{aligned}$$

Example 2 [48] The analytic solution of the Burgers' equation (1) is given by

$$U(x,t) = \frac{(x-2t)/(t+0.1)}{1+\nu^2 A(x,t)} \tag{12}$$

where

$$A(x,t) = e^{(x-2t)^2/(4\nu(t+0.1))}$$

Following these two examples, results of numerical experiments obtained from implementation of the Newton–FSGS, Newton–FSSOR and Newton–HSSOR iterative methods have been summarized in Tables 1 and 2.

Table 1 Comparison of number of iterations K, the execution time (seconds) and maximum errors for the iterative methods using Example 1

Example 1										
m	Method	RE = 2			RE = 10			RE = 50		
		K	Time	Max. error	K	Time	Max. error	K	Time	Max. error
256	FSGS	4531	3.09	2.62e−7	1238	0.87	3.84e−6	327	0.25	2.23e−4
	FSSOR	489	0.41	1.89e−8	262	0.24	3.83e−6	136	0.14	2.23e−4
	HSSOR	265	0.15	4.57e−8	141	0.10	7.97e−6	74	0.07	4.64e−4
512	FSGS	16073	21.75	1.38e−6	4268	5.68	2.97e−6	1114	1.53	1.63e−4
	FSSOR	910	1.40	1.36e−8	485	0.75	2.81e−6	246	0.41	1.63e−4
	HSSOR	489	0.42	1.89e−8	262	0.23	3.83e−6	136	0.15	2.23e−4
1024	FSGS	57109	154.22	5.61e−6	14626	38.89	4.69e−6	3913	10.45	1.48e−4
	FSSOR	1716	5.15	2.27e−8	918	2.69	2.58e−6	455	1.38	1.48e−4
	HSSOR	910	1.42	1.36e−8	485	0.77	2.82e−6	246	0.42	1.63e−4
2048	FSGS	199805	1079.51	2.28e−5	50957	273.61	1.69e−5	13561	72.54	1.44e−4
	FSSOR	3318	19.79	6.81e−8	1742	10.13	2.55e−6	846	5.08	1.44e−4
	HSSOR	1716	5.26	2.27e−8	918	2.76	2.58e−6	455	1.46	1.48e−4
4096	FSGS	684790	34364.91	9.31e−5	179969	1937.79	6.25e−5	46294	501.70	1.47e−4
	FSSOR	6504	77.58	1.11e−7	3309	38.47	2.62e−6	1618	19.33	1.43e−4
	HSSOR	3318	20.25	6.81e−8	1742	10.38	2.55e−6	846	5.22	1.44e−4

Table 2 Comparison of number of iterations K, the execution time (seconds) and maximum errors for the iterative methods using Example 2

Example 2										
m	Method	RE = 2			RE = 10			RE = 50		
		K	Time	Max. error	K	Time	Max. error	K	Time	Max. error
256	FSGS	4454	3.90	2.39e−5	1415	1.01	2.39e−5	354	0.30	9.70e−4
	FSSOR	552	0.47	2.40e−5	283	0.28	2.39e−5	142	0.15	9.70e−4
	HSSOR	294	0.16	2.19e−5	155	0.10	2.39e−5	76	0.07	1.37e−4
512	FSGS	17423	27.89	2.38e−5	4974	6.91	2.39e−5	1224	1.80	8.73e−4
	FSSOR	1026	1.72	2.45e−5	529	0.87	2.39e−5	262	0.45	8.73e−4
	HSSOR	554	0.50	2.40e−5	283	0.28	2.39e−5	143	0.16	9.71e−4
1024	FSGS	62064	195.29	2.17e−5	17464	49.41	2.39e−5	4189	12.40	8.45e−4
	FSSOR	1920	6.28	2.46e−5	997	3.14	2.39e−5	481	1.52	8.98e−4
	HSSOR	1029	1.74	2.45e−5	529	0.90	2.39e−5	262	0.46	8.72e−4
2048	FSGS	219750	1372.67	1.39e−5	61493	351.85	2.39e−5	14459	88.59	8.42e−4
	FSSOR	3692	24.95	2.47e−5	1911	11.97	2.39e−5	882	5.50	8.42e−4
	HSSOR	1920	6.52	2.47e−5	997	3.29	2.39e−5	481	1.59	8.48e−4
4096	FSGS	791256	79648.25	2.17e−5	213046	2473.61	2.39e−5	48566	635.02	8.41e−4
	FSSOR	6956	94.49	2.46e−5	3664	46.83	2.39e−5	1670	21.00	8.47e−4
	HSSOR	3686	25.09	2.46e−5	1911	12.48	2.39e−5	882	5.77	8.42e−4

5 Conclusions

In this paper, we have examined the performance of Newton–FSGS, Newton–FSSOR, and Newton–HSSOR iterative methods together with Newton scheme for the numerical solution of nonlinear two-point boundary value problem using second-order finite difference approximation equations. To demonstrate the efficiency of the three proposed methods, two nonlinear examples are presented. Through the numerical results obtained in Tables 1 and 2 at three different values of *RE*, Newton–HSSOR method is compared with the Newton–FSGS and Newton–FSSOR methods. Obviously, it can be concluded that the Newton–HSSOR method is superior in terms of the number of iterations and the execution time for different mesh sizes. This is because the computational complexity of Newton–HSSOR method is approximately 50% compared to the full-sweep case. Overall, the approximate solutions for all three proposed methods are in good agreement. The numerical results show that the Newton–HSSOR method with the half-sweep implicit scheme is a promising approach for solving different types of this nonlinear partial differential equation problem. For future works, the method can be extended to solve multi-dimensional nonlinear problems. Apart from single-step iterative methods, two-step iterative methods (Ruggiero and Galligani [49]; Evans and Sahimi [50]; Sahimi et al. [51]; Sulaiman et al. [52], Galligani [53], [54] are also interesting to be examined for nonlinear problems.

References

1. Ciment, M., Leventhal, S.H., Weinberg, B.C.: The operator compact implicit method for parabolic equations. J. Comput. Phys. 28(2), 135–166 (1978)
2. Hirsh, R.S.: Higher order accurate difference solutions of fluid mechanics problems by a compact differencing technique. J. Comput. Phys. 19(1), 90–109 (1975)
3. Mitchell, A.R., Griffiths, D.F.: The Finite Difference Method in Partial Differential Equations. Wiley, New York (1980)
4. Jain, P.C., Raja, M.: Splitting-up technique for Burgers' equations. Indian J. Pure Appl. Math. **10**(12), 1543–1551 (1979)
5. Arminjon, P., Beauchamp, C.: A finite element method for Burgers' equation in hydrodynamics. Int. J. Methods Eng. **12**, 415–428 (1978)
6. Iskandar, L., Mohsen, A.: Some numerical experiments on the splitting of Burgers' equation. Numer. Methods Partial. Differ. Eqn. **8**(3), 267–276 (1992)
7. Christie, I., Mitchell, A.R.: Upwinding of high order Galerkin methods in conduction–convection problems. Int. J. Methods Eng. **12**, 1764–1771 (1978)
8. Herbst, B.M., Schoombie, S.W., Griffiths, D.F., Mitchell, A.R.: Generalized Petrov–Galerkin methods for the numerical solution of Burgers' equation. Int. J. Numer. Methods Eng. **20**, 1273–1289 (1984)
9. De Maerschalck, B., Gerritsma, M.I.: The use of Chebyshev polynomials in the space-time least-squares spectral element method Numer. Alg. **38**, 173–196 (2005)
10. De Maerschalck, B., Gerritsma, M.I.: Least-squares spectral element method for non-linear hyperbolic differential equations. J. Comput. Appl. Math. **215**, 357–367 (2008)
11. Heinrichs, W.: An adaptive spectral least-squares scheme for the Burgers equation. Numer. Alg. **44**(1), 1–10 (2007)

12. Ali, A.H.A., Gardner, G.A., Gardner, L.R.T.: A collocation solution for Burgers' equation using cubic B-spline finite elements. Comput. Methods Appl. Mech. Engrg. **100**(3), 325–337 (1992)
13. Lohar, B.L., Jain, P.C.: Variable mesh cubic spline technique for N-wave solution of Burgers' equation. J. Comput. Phys. **39**, 433–442 (1981)
14. Rubin, S.G., Khosla, P.K.: Higher order numerical solutions using cubic splines. AIAA J. **14**(7), 851–858 (1976)
15. Jain, P.C., Lohar, B.L.: Cubic spline technique for coupled non-linear parabolic equations. Comput. Math. Appl. **5**(3), 179–185 (1979)
16. Ramadan, M.A., El-Danaf, T.S., Abd Alaal, F.E.I.: Application of the non-polynomial spline approach to the solution of the Burgers' equation. Open Appl. Math. J. **1**, 15–20 (2007)
17. Jain, P.C., Shankar, R., Singh, T.V.: Numerical technique for solving convective-reaction-diffusion equation. Math. Comput. Modelling **22**, 113–125 (1995)
18. Dag, I., Saka, B., Boz, A.: B-spline Galerkin methods for numerical solutions of the Burgers' equation. Appl. Math. Comput. **166**, 506–522 (2005)
19. Dag, I., Irk, D., Saka, B.: A numerical solution of the Burgers' equation using cubic B-splines. Appl. Math. Comput. **163**, 199–211 (2005)
20. Abbasbandy, S., Darvishi, M.T.: A numerical solution of Burgers' equation by modified Adomian method. Appl. Math. Comput. **163**(3), 1265–1272 (2005)
21. Abbasbandy, S., Darvishi, M.T.: A numerical solution of Burgers' equation by time discretization of Adomian's decomposition method. Appl. Math. Comput. **170**, 95–102 (2005)
22. Abdou, M.A., Soliman, A.A.: Variational iteration method for solving Burger's and coupled Burger's equations. J. Comput. Appl. Math. **181**(2), 245–251 (2005)
23. Young, D.M.: Iterative methods for solving partial difference equations of elliptic type. Trans. Amer. Math. Soc. **76**, 92–111 (1954)
24. Young, D.M.: Iterative Solution of Large Linear Systems. Academic, London (1971)
25. Hackbusch, W.: Iterative Solution of Large Sparse Systems of Equations. Springer, New York (1995)
26. Saad, Y.: Iterative Method for Sparse Linear Systems. International Thomas, Boston (1996)
27. Abdullah, A.R.: The four point explicit decoupled group (EDG) method: A fast Poisson solver. Intern. J. Comput. Math. **38**, 61–70 (1991)
28. Aruchunan, E., Sulaiman, J.: Half-sweep conjugate gradient method for solving first order linear Fredholm integro-differential equations. Aust. J. Bas. Appl. Sci. **5**, 38–43 (2011)
29. Ibrahim, A., Abdullah, A.R.: Solving the two-dimensional diffusion equation by the four point explicit decoupled group (EDG) iterative method. Intern. J. Comput. Math. **58**, 253–256 (1995)
30. Sulaiman, J., Hasan, M.K., Othman, M.: Red-black EDGSOR iterative method using triangle element approximation for 2D Poisson equations. In: Hutchison, D., et al. (eds.) Lecture Notes in Computer Science, vol. 4707, pp. 298–308. Springer, Berlin (2007)
31. Sulaiman, J., Hasan, M.K., Othman, M.: Red-black half-sweep iterative method using triangle finite element approximation for 2D Poisson equations. In: Hutchison, D., et al. (eds.) Lecture Notes in Computer Science, vol. 4487, pp. 326–333. Springer, Berlin (2007)
32. Sulaiman, J., Hasan, M.K., Othman, M.: Nine point-EDGSOR iterative method for the finite element solution of 2D Poisson equations. Lecture Notes in Computer Science, vol. 5592, pp. 764–774 (2009)
33. Yousif, W.S., Evans, D.J.: Explicit de-coupled group iterative methods and their implementations. Parallel Algorithms Appl. **7**, 53–71 (1995)
34. Akhir, M.K.M., Othman, M., Sulaiman, J., Majid, Z.A., Suleiman, M.: Half-sweep iteration for solving two-dimensional Helmholtz equations. Int. J. Appl. Math. Stat. **29**(5), 101–109 (2012)
35. Akhir, M.K.M., Othman, M., Sulaiman, J., Majid, Z.A., Suleiman, M.: Half-sweep modified successive overrelaxation for solving two-dimensional Helmholtz equations. Aust. J. Bas. Appl. Sci. **5**(12), 3033–3039 (2011)
36. Fauzi, N.I., Sulaiman, J.: Half-sweep modified successive overrelaxtion method for solving second order two-point boundary value problems using cubic spline. Int. J. Contemp. Math. Sciences **7**(32), 1579–1589 (2012)

37. Sulaiman, J., Hasan, M.K., Othman, M.: The half-sweep iterative alternating decomposition explicit (HSIADE) method for diffusion equations. In: Hutchison, D., et al. (eds.) Lecture Notes on Computer Science, vol. 3314, pp. 57–63. Springer, Berlin (2004)
38. Muthuvalu, M.S., Sulaiman, J.: Half-sweep arithmetic mean method with composite trapezoidal scheme for solving linear Fredholm integral equations. Appl. Math. Comput. **217**(12), 5442–5448 (2011)
39. Muthuvalu, M.S., Sulaiman, J.: An implementation of the 2-point block arithmetic mean iterative method for first kind linear Fredholm integral equations. World J. Model. Simul. **8**(4), 293–301 (2012)
40. Muthuvalu, M.S., Sulaiman, J.: Half-sweep geometric mean iterative method for the repeated Simpson solution of second kind linear Fredholm integral equations. Proyecc. J. Math. **31**(1), 65–79 (2012)
41. Othman, M., Sulaiman, J., Abdullah, A.R.: A parallel halfsweep multigrid algorithm on the shared memory multiprocessors. Malays. J. Comput. Sci. **13**(2), 1–6 (2000)
42. Sulaiman, J., Othman, M., Hasan, M.K.: Half-sweep algebraic multigrid (HSAMG) method applied to diffusion equations. In: Bock, H.G., et al. (eds.) Modeling, Simulation and Optimization of Complex Processes, pp. 547–556. Springer, Berlin (2008)
43. Hasan, M.K., Othman, M., Abbas, Z., Sulaiman, J., Ahmad, F.: Parallel solution of high speed low order FDTD on 2d free space wave propagation. In: Hutchison, D., et al. (eds.) Lecture Notes in Computer Science, vol. 4706, pp. 13–24. Springer, Berlin (2007)
44. Saudi, A., Sulaiman, J.: Path planning for mobile robot with half-sweep successive over-relaxation (HSSOR) iterative method. In: Symposium on Progress in Information and Communication Technology (SPICT'09), pp. 57–62, 7–8. Royale Bintang. (2009)
45. Saudi, A., Sulaiman, J.: Red-black strategy for mobile robot path planning. Lecture Notes in Engineering and Computer Science. **2812**(1), 2215–2220 (2010)
46. Saudi, A., Sulaiman, J.: Indoor path planning for mobile robot using half-sweep Gauss–Seidel via nine-point Laplacian (HSGS9L). Int. J. Eng. Res. Technol. **1**(7), 17–29 (2012)
47. Dhawan, S., Kapoor, S., Kumar, S., Rawat, S.: Contemporary review of techniques for the solution of nonlinear Burgers equation. J. Comput. Sci. **3**(5), 405–419 (2012)
48. Ruggiero, V., Galligani, E.: An iterative method for large sparse systems on a vector computer. Comput. Math. Appl. **20**(1), 25–28 (1990)
49. Evans, D.J., Sahimi, M.S.: The alternating group explicit iterative method (AGE) to solve parabolic and hyperbolic partial differential equations. Ann. Rev. Num. Fluid Mech. Heat Trans. **2**, 283–389 (1988)
50. Sahimi, M.S., Ahmad, A., Bakar, A.A.: The iterative alternating decomposition explicit (IADE) method to solve the heat conduction equation. Intern. J. of Comput. Math. **47**, 219–229 (1993)
51. Sulaiman, J., Othman, M., Hasan, M.K.: A new quarter-sweep arithmetic mean (QSAM) method to solve diffusion equations. Chamchuri J. Math. **1**(2), 89–99 (2009)
52. Galligani, E.: The Newton-arithmetic mean method for the solution of systems of nonlinear equations. Appl. Math. Comput. **134**, 9–34 (2003)
53. Galligani, E.: The Arithmetic mean method for solving systems of nonlinear equations in finite differences. Appl. Math. Comput. **181**, 579–597 (2006)
54. Sulaiman, J., Hasan, M.K., Othman, M., Karim, S.A.A.: Numerical solutions of nonlinear second-order two-point boundary value problems using half-sweep SOR with Newton method. J. Concr. Appl. Math. **11**(1), 112–120 (2013)

MIX
Papier aus verantwortungsvollen Quellen
Paper from responsible sources
FSC® C105338

If you have any concerns about our products,
you can contact us on
ProductSafety@springernature.com

In case Publisher is established outside the EU,
the EU authorized representative is:
Springer Nature Customer Service Center GmbH
Europaplatz 3, 69115 Heidelberg, Germany

Printed by Libri Plureos GmbH
in Hamburg, Germany